15th International Conference on Numerical Modeling of Space Plasma Flows 2023

Journal of Physics: Conference Series Volume 2742

Pasadena, California, USA
26-30 June 2023

ISBN: 978-1-7138-9730-9
ISSN: 1742-6588

Printed from e-media with permission by:

Curran Associates, Inc.
57 Morehouse Lane
Red Hook, NY 12571

Some format issues inherent in the e-media version may also appear in this print version.

Institute of Physics
Dirac House, Temple Back
Bristol BS1 6BE UK

Phone: 44 1 17 929 7481
Fax: 44 1 17 920 0979

techtracking@iop.org

Additional copies of this publication are available from:

Curran Associates, Inc.
57 Morehouse Lane
Red Hook, NY 12571 USA
Phone: 845-758-0400
Fax: 845-758-2633
Email: curran@proceedings.com
Web: www.proceedings.com

TABLE OF CONTENTS

PREFACE

The Center for Space Plasma and Aeronomic Research (CSPAR) at the University of Alabama in Huntsville (UAH) and Maison de la Simulation, a joint laboratory at the French Alternative Energies and Atomic Energy Commission (CEA), National Center for Scientific Research (CNRS), University of Paris-Saclay, and University of Versailles-St Quentin, organized the 15th International Conference on Numerical Modeling of Space Plasma Flows (ASTRONUM-2023) on June 26—30, 2023 in Pasadena, California, USA.

The Program Committee consisted of Nikolai Pogorelov (University of Alabama in Huntsville/CSPAR, USA, chair), Edouard Audit (CEA Maison de la Simulation, Gif-sur-Yvette, France, co-chair), Wes Bethel (San Francisco State University, USA), Amitava Bhattacharjee (Princeton Plasma Physics Laboratory, USA), Phillip Colella (Lawrence Berkeley National Laboratory, USA), Tomoyuki Hanawa (Chiba University, Japan), Maria Elena Innocenti (Ruhr University Bochum, Germany), Kanya Kusano (Nagoya University, Japan), Dongwook Lee (University of California, Santa Cruz), Jon Linker (Predictive Science Inc., USA), Anthony Mezzacappa (University of Tennessee, Knoxville, USA), Ewald Müller (Max-Planck-Institute for Astrophysics, Garching, Germany), Dongsu Ryu (Ulsan National Institute of Science and Technology, Korea), James Stone (Princeton University), and Gary P. Zank (University of Alabama in Huntsville, USA).

The conference attracted over 90 scientists from 15 countries representing different branches of the plasma simulation community. The distinctive feature of this conference is a combination of diverse research topics, all of which are essential for performing high-resolution, continuum mechanics and particle, simulations of physical phenomena in space physics and astrophysics. Among such topics were software packages for modeling and analyzing plasma flows; advanced numerical methods for space and astrophysical flows; large-scale fluid-based, kinetic, and hybrid simulations; turbulence and cosmic ray transport; and magnetohydrodynamics. The discussed applications included cosmology and galaxy formation, supernova explosions, physics of the Sun-heliosphere-magnetosphere environments, space weather, the interstellar medium and star formation, stellar physics, experimental plasma physics, astrophysical accretion, numerical methods for ideal and non-ideal, relativistic and nonrelativistic MHD, etc. The proceedings volume is structured so that it covers all of these topics.

The structure of the ASTRONUM conference series is based on the idea that modelers working in seemingly distant fields should have an opportunity to share their scientific achievements with the broad community of computational scientists performing numerical experiments. As in previous ASTRONUM meetings, we were interested in physical systems that are coupled across a multiplicity of spatial and temporal that incorporate diverse physical processes.

The contributors to this volume are both young researchers and renowned experts in space physics and astrophysics, applied mathematics, and computational physics. This volume describes the application of numerical methods and the algorithms themselves, allowing us to discuss the challenges that theory imposes on numerical schemes for solving partial differential equations describing collisional and collisionless processes in space and astrophysical plasmas.

We would like to thank the participants who submitted their papers to Proceedings of ASTRONUM-2023 and especially to those who reviewed manuscripts thus ensuring the high quality of this publication. We also are grateful to Adele Corona (ICNSMeetings.com) for the excellent management of the conference.

This publication will be useful to graduate and postgraduate students majoring in space physics, astrophysics, numerical, engineering, and applied mathematics. It is also aimed at specialists in applied mathematics and various fields of physics that involve flows of partially ionized plasmas at both the collisional and collisionless levels.

Nikolai V. Pogorelov, Edouard Audit, and Gary P. Zank

March 3, 2024

Journal of Physics: Conference Series **2742** (2024) 011002 doi:10.1088/1742-6596/2742/1/011002

Peer Review Statement

All papers published in this volume have been reviewed through processes administered by the Editors. Reviews were conducted by expert referees to the professional and scientific standards expected of a proceedings journal published by IOP Publishing.

- **Type of peer review:** Single Anonymous
- **Conference submission management system:** Morressier
- **Number of submissions received:** 28
- **Number of submissions sent for review:** 28
- **Number of submissions accepted:** 28
- **Acceptance Rate (Submissions Accepted / Submissions Received × 100):** 100
- **Average number of reviews per paper:** 1.04
- **Total number of reviewers involved:** 27
- **Contact person for queries:**
  Name: Nikolai Pogorelov
  Email: np0002@uah.edu
  Affiliation: The University of Alabama in Huntsville

Journal of Physics: Conference Series **2742** (2024) 012001 doi:10.1088/1742-6596/2742/1/012001

Sensitivity of 3D Convective Urca Simulations to Changes in Urca Reactions

B Boyd[1,2], A Smith Clark[1], A C Calder[1,2], D M Townsley[3], and M Zingale[1]

[1] Department of Physics and Astronomy, Stony Brook University, Stony Brook, NY 11794-3800, USA

[2] Institute for Advanced Computational Science, Stony Brook University, Stony Brook, NY 11794-5250, USA

[3] Department of Physics and Astronomy, University of Alabama, Tuscaloosa, AL 35487-0324, USA

E-mail: boyd.brendan@stonybrook.edu

Abstract. A proposed setting for thermonuclear (Type Ia) supernovae is a white dwarf that has gained mass from a companion to the point of carbon ignition in the core. There is a simmering phase in the early stages of burning that involves the formation and growth of a core convection zone. One aspect of this phase is the convective Urca process, a linking of weak nuclear reactions to convection that may alter the composition and structure of the white dwarf. Convective Urca is not well understood and requires 3D fluid simulations to realistically model. Additionally, the convection is relatively slow (Mach number less than 0.005) so a low-Mach method is needed to make simulating computationally feasible. Using the MAESTROeX low-Mach hydrodynamics code, we investigate recent changes to how the weak reactions are modeled in the convective Urca simulations. We present results that quantify the changes to the reaction rates and their impact on the evolution of the simulation.

1. Introduction

Type Ia Supernovae (SNe Ia) are extremely bright thermonuclear explosions. SNe Ia light curves are standardizable, allowing them to be used as distance indicators, providing value to many areas of astronomy, particularly cosmology [1, 2]. It is well accepted that SNe Ia arises from the thermonuclear explosion of white dwarf stars in a binary system. However, the specific progenitor system of SNe Ia is still an open question. One proposed setting is the single degenerate case in which a degenerate white dwarf accretes material from a companion star. As the white dwarf mass approaches the Chandrasekhar mass ($\sim 1.4\ M_\odot$), the core becomes hot and dense enough to start carbon burning. Initially, this burning is relatively slow and will last about 1,000 to 10,000 years prior to the final explosive event. This phase is called the carbon simmering phase, and it may impact the composition of the progenitor, which can alter the final nucleosynthesis observed in the SN Ia.

During the simmering phase, the carbon burning releases enough energy to drive core convection in the white dwarf. This convection combines with weak nuclear reactions to produce local cooling due to neutrino emission, called the convective Urca process [3]. To investigate this poorly understood mechanism, we need to model both the convection and weak reactions accurately. Previous work [4–6] on the convective Urca process has largely made either 1D or 2D approximations to model the turbulent convection, an inherently 3D process. We use the low Mach hydrodynamic code MAESTROeX to model convective Urca in 3D, building on an earlier study [7, 8]. Additionally, we use weak reaction rates that depend on temperature and the electron density (which is important for this highly degenerate regime) [9]. In investigating the convective Urca process, slight changes have been made to how we estimate these weak reaction rates. We present the results of these changes and the sensitivity of our simulations to such changes.

In section 2, we describe the convective Urca process in greater detail. Next, in section 3, we briefly describe our simulation setup. In section 4, we show the changes to the Urca reaction rates. In section 5, we look at the impact of these changes on a test run of our simulation. Finally, in section 6 we conclude with our takeaways from this analysis.

2. The Convective Urca Process

The Urca process describes the combination of a beta decay and electron capture that link two isotopes, called an Urca pair. The reactions work as follows for a pair of nuclei with the same atomic mass number, A, and proton numbers, $Z-1$ and Z respectively:

$$\begin{aligned} (Z-1,A) &\longrightarrow (Z,A)+e^-+\bar{\nu}_e \\ (Z,A)+e^- &\longrightarrow (Z-1,A)+\nu_e \end{aligned} \tag{1}$$

The process provides some local cooling due to neutrino emission in each weak reaction. In a degenerate white dwarf, these weak reactions depend on temperature and the electron density. In most regions only one reaction will be active, i.e. electron captures at higher densities and beta decays at lower densities. The temperature dependence on the rate is relatively minor at the densities of interest, but it works to increase both the electron capture and beta decay reaction rates. The transition between these regions, where the reaction rates are equal, is called the Urca shell. In a convection zone that encompasses the Urca shell, material cyclically rises above and falls below the shell. This allows a single nucleus to undergo repeated reactions and drive the neutrino cooling. The cyclic nature is key to convective Urca, enabling small abundances of the Urca pair to have a meaningful impact on the white dwarf's evolution. In addition to neutrino cooling, the convective Urca process may impede the convective flow [4, 5], but a conclusive 3D analysis is needed to better understand this effect. We look to investigate this idea in future work.

There are many Urca pairs relevant to simmering white dwarfs, but we focus this paper on the important $A=23$ Urca pair, ^{23}Na-^{23}Ne. The $A=23$ pair has a relatively low Q-value, $Q=4.38$ MeV [9], and is relatively abundant in the white dwarf at the onset of carbon burning, $X(^{23}\text{Na}) \approx 10^{-4}$ [10–12]. The $A=23$ Urca shell is approximately located at a density $\rho_{\text{urca}} \sim 1.7\times10^9\ \text{g}\,\text{cm}^{-3}$ [9].

3. Numerical Methods and Simulation Setup

We use the MAESTROeX hydrodynamic code [13], which is part of the AMReX-Astro suite of hydrodynamic codes for reactive astrophysical flows. MAESTROeX is designed specifically to simulate stellar interiors and atmospheres in low Mach environments (i.e. when the fluid speed is slower than the sound speed). When modeling low Mach flow, a standard compressible finite-volume method is limited by the acoustic wave timescale as opposed to the advection timescale of the fluid. MAESTROeX addresses this issue by using an elliptic constraint on the velocity field, derived from a low Mach approximation (Ma $\lesssim 0.1$). From this velocity constraint, MAESTROeX uses a projection method to effectively filter out any acoustic waves while still accounting for density fluctuations due to heating, stratification, etc. This approach allows MAESTROeX to efficiently and accurately model slow moving flows like core convection. Nuclear reactions are coupled to the fluid equations via Strang-Splitting.

Our presented simulations are full 3D of a near Chandrasekhar mass white dwarf with a convective core. We use a Cartesian grid with 3 levels of refinement (20, 10, and 5 km resolution). The levels of refinement are defined at specified density cutoffs and we ensure the entire convective core is resolved to the finest level. The outer regions of the white dwarf are only resolved to the point that hydrostatic equilibrium is maintained. Further resolution outside the convective core would only increase the computational expense without providing further insight or accuracy to the mechanisms in the core region.

The initial conditions are based on 1D stellar evolution models [10] and were implemented in MAESTROeX by a previous study [7, 8]. We use a reaction network generated by the pynucastro python package [14]. This network incorporates a simple carbon burning scheme and the $A = 23$ Urca reactions, see Figure 1. Rates related to carbon burning come from JINA REACLIB [15]. We include the weak Urca reaction rates by interpolating a table of calculated values [9].

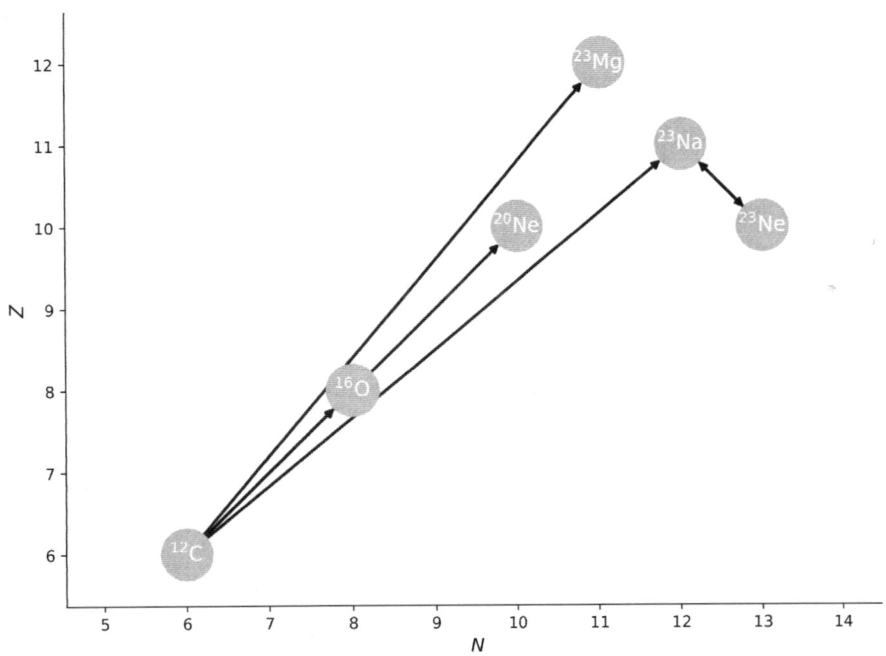

Figure 1. Graphical description of the reaction network. The x-axis displays the neutron number and the y-axis displays the proton number. Isotopes included in the network are labeled. The arrows indicate the direction of the reactions (double-sided arrows indicate both directions are included).

In our analysis, we use a simulation that has developed a fully convective core in a quasi-steady state. The convection zone has a typical Mach number of $\sim 10^{-3}$, which indicates the need for a low Mach number method like MAESTROeX. The simulation has a central temperature of 5.5×10^8 K and a central density of 4.5×10^9 g cm^{-3}. The temperature-density profile can be seen in Figure 2. This illustrates the isentropic core and the convectively stable isothermal envelope. Note the dip in the profile is due to the convection zone, and thus isentropic region, expanding outward from our initial state. The white dwarf is primarily made of 40% Carbon and 60% Oxygen, with a trace amount of $A = 23$ Urca pair $(X(^{23}\text{Na}) + X(^{23}\text{Ne}) \approx 10^{-4})$.

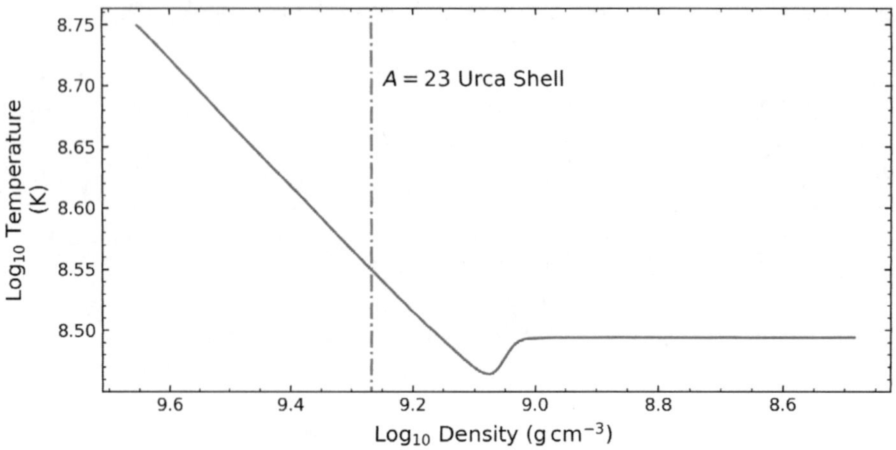

Figure 2. The profile of density vs temperature for the white dwarf star. The grey vertical line marks the location of the Urca shell.

4. Changes to Urca Reactions

The Urca reaction rates we use are tabulated over a range of temperatures, T, and electron densities, ρY_e. We calculate the reaction rate for a given temperature and electron density by interpolating over this table. Prior to version 2.1 of pynucastro this interpolation had been done strictly bi-linearly, i.e. interpolating between values of T and ρY_e. However, the tables were constructed in log space and thus it's more accurate to interpolate in log space as well, i.e. interpolating between values of $\log_{10} \mathrm{T}$ and $\log_{10} \rho Y_e$. The shift from linear-space to log-space interpolation decreases the estimated reaction rate for values far from the table's spacing, $\Delta \log_{10} \mathrm{T} = 0.05$ and $\Delta \log_{10} \rho Y_e = 0.02$. The significance of this shift depends on the temperature, density and composition.

To quantify the change in the Urca rates for the conditions in our simulation, we generate a radial profile of the white dwarf and calculate the average reaction rate at a given radius for each interpolation scheme. These Urca reactions primarily depend on the density, so we plot the difference in rates vs density (see Figure 3 and 4). The two figures suggest the difference in rates can be large, up to 80%, but only in the regions where the rates are relatively small and the reaction is negligible (see solid grey lines in Figure 3 and 4). Looking at the zoomed in plots in Figure 3 and 4, we see that the difference between interpolation scheme is upwards of 5% in these regions where the reaction is active. The vertical, dashed lines indicate the density spacing of the tabulated calculations which is why the difference approaches zero for these densities.

5. Results

We investigated the sensitivity of our simulation to the change in the Urca rates by comparing two runs with different interpolation schemes. The initial state, described in Section 3, is from a simulation that had settled to a quasi-steady state, using the linear-space interpolation scheme. This allows us to start with a realistic velocity field and distribution of the Urca pair. We ran each simulation for 300 seconds of simulation time (about 4000 timesteps), which accounts for several convective turnovers.

In Figure 5, we plot the time evolution of the nuclear energy generation rate and the kinetic energy of the white dwarf. We calculate the specific energy generation rate by integrating the reaction network (see Figure 1) using an implementation of the VODE integrator [16]. The total nuclear energy generation rate for the star is then the mass integrated sum of the specific

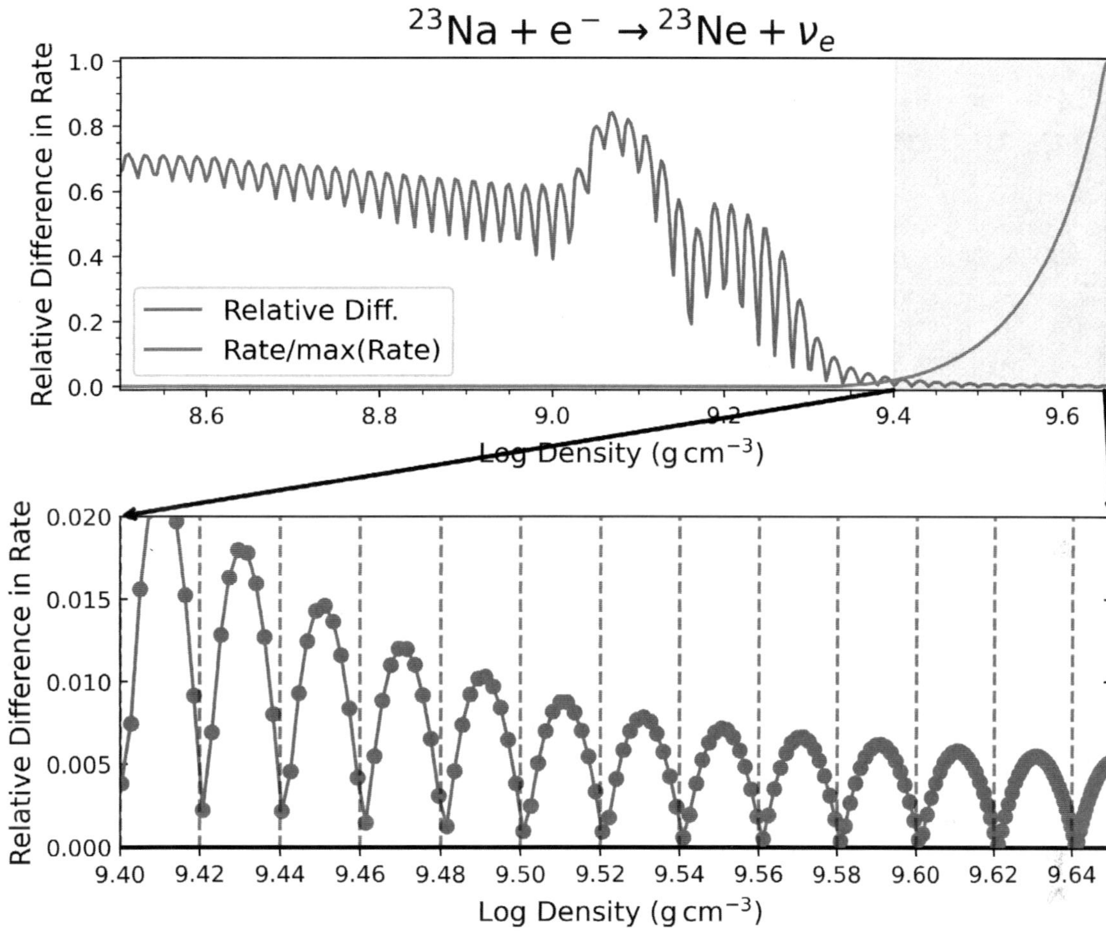

Figure 3. A Radially averaged profile of the electron capture reaction rates. The bottom plot is a zoomed in version of the shaded region in the top plot. In each plot, the blue curve is the relative difference between the Linear-space and Log-space interpolation schemes for the initial state of the simulation (see Figure 2). In the top plot, the grey curve is a normalized profile of the reaction rate. In the bottom plot, the grey vertical dashed lines mark the spacing of the tabulated values.

energy generation rate. The energy is primarily generated by the core carbon burning, but changes to the Urca reactions can easily be seen as well. For the first 70 seconds, the log-space interpolation has slightly higher energy generation due to fewer Urca reactions (i.e. less neutrino cooling). After about 70 seconds, the two simulations diverge in the energy being generated. The total kinetic energy is calculated by summing the kinetic energy of each cell in the grid. The evolution of this energy follows a similar pattern to the nuclear energy generation rate (i.e. the simulations diverge after $\sim$70 seconds). These differences indicates the changes to the Urca reactions have had an impact on the evolution of the simulation.

To further quantify the impact on the velocity field, we calculate the primary direction of the flow, i.e. the dipole moment. This is done by taking the density weighted average of the radial velocity in the x, y, and z directions as shown:

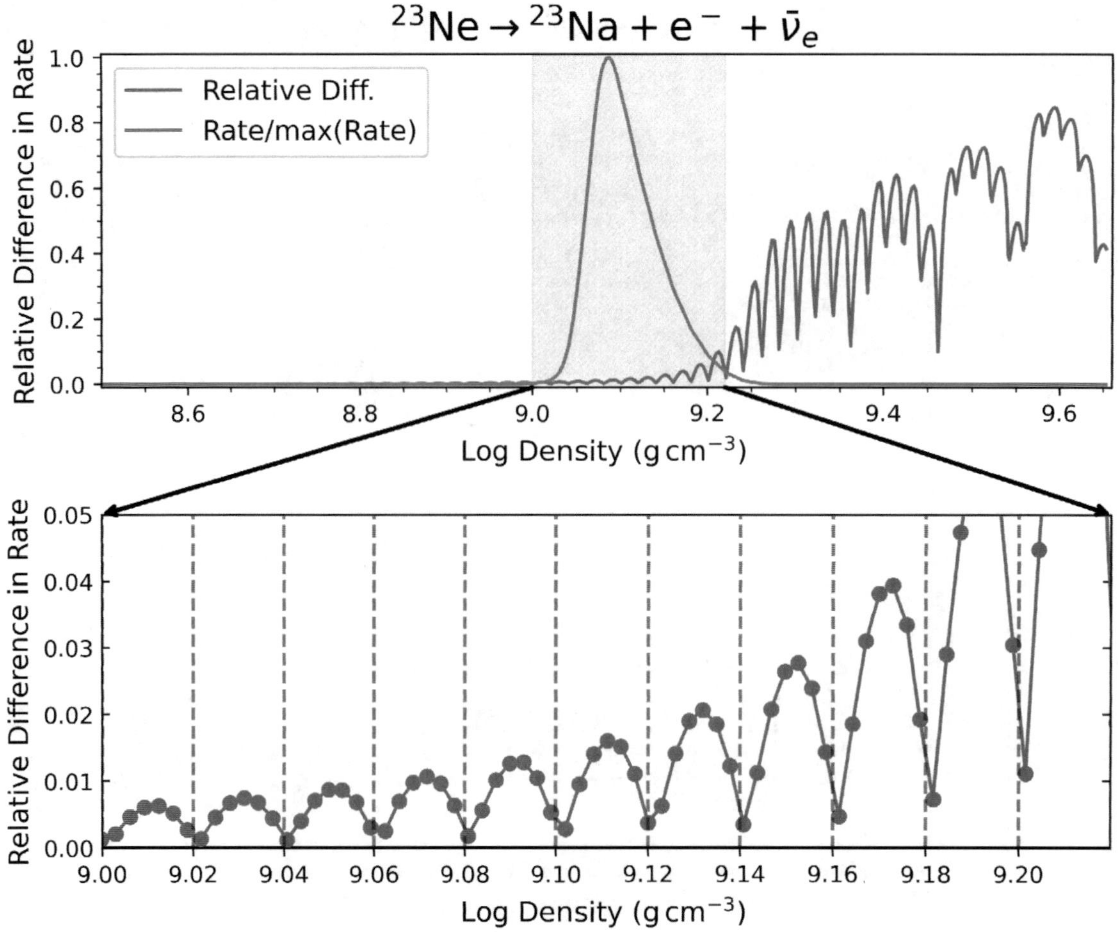

Figure 4. A Radially averaged profile of the beta decay reaction rates. The bottom plot is a zoomed in version of the shaded region in the top plot. In each plot, the blue curve is the relative difference between the Linear-space and Log-space interpolation schemes for the initial state of the simulation (see Figure 2). In the top plot, the grey curve is a normalized profile of the reaction rate. In the bottom plot, the grey vertical dashed lines mark the spacing of the tabulated values.

$$\langle v_r \rangle_x = \left(\sum \rho v_r \frac{x - x_c}{r} \right) / \sum \rho \tag{2}$$

Here we are summing over each cell in the simulation. x_c is the x coordinate for the center of the star and r is the radial coordinate from the center. $\langle v_r \rangle_y$ and $\langle v_r \rangle_z$ are calculated analogously. Using these values, we calculate the angle θ from the z-axis and the azimuthal angle ϕ (in the x-y plane) as follows:

$$\theta = \arctan \left(\frac{\sqrt{\langle v_r \rangle_x^{\,2} + \langle v_r \rangle_y^{\,2}}}{\langle v_r \rangle_z} \right) \tag{3}$$

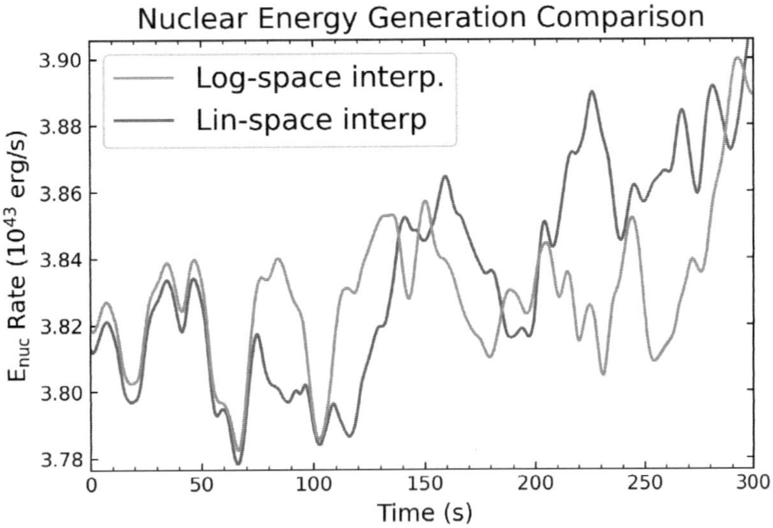

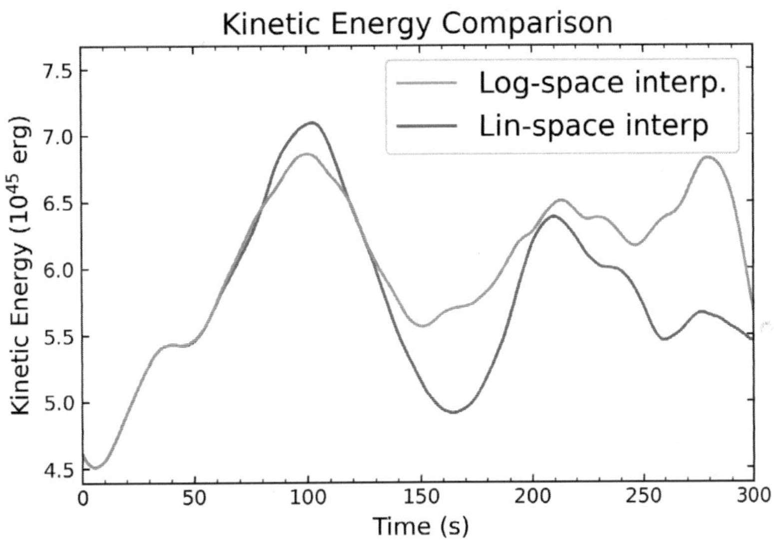

Figure 5. The top plot displays the time evolution of the nuclear energy generation rate integrated over the whole star. The bottom plot displays the time evolution of the total kinetic energy of the whole star. The blue curves represent the simulation using the linear-space interpolation scheme. The orange curves represent the simulation using the log-space interpolation scheme.

$$\phi = \arctan\left(\frac{\langle v_r \rangle_y}{\langle v_r \rangle_x}\right) \tag{4}$$

The results are shown in Table 1. Although the general direction is similar (downward in the negative z direction), there is clear deviations in the net flow between the two simulations.

From the changes in energy generation rate, total kinetic energy and the directional dependence of the flow, it is clear that the small changes in the Urca reaction rates can alter a simulation. However, these simulations are very chaotic (i.e. sensitive to initial conditions)

	θ	ϕ
Initial State	167°	348°
Log-space Interp.	168°	7°
Lin-space Interp.	174°	49°

Table 1. Angular direction of the velocity field's dipole moment. θ is the polar angle (from the z-axis) ranging from 0 to 180 degrees. ϕ is the azimuthal angle (in the x-y plane) ranging from 0 to 360 degrees. The initial state is included for reference.

due to the highly non-linear system of equations we are solving. This chaotic nature means the new interpolation changes may alter the simulation without having a significant impact on the conclusions we draw from averaged values over many convective turnovers. We compare the changes due to interpolation relative to the natural evolution of the simulation, see Figure 6 and 7. In our regions of interest, the reaction rates tend to vary more significantly in time (orange curves) than by the change in interpolation scheme (blue curves). This indicates the dynamic nature of the simulation leads to comparable or larger shifts in the rates over just a few convective turnovers, reducing the relative impact the new interpolation scheme has on the reaction rates over a long periods of time. The exception to this is the region from $\log_{10}\rho$ 9.15 to 9.20 in Figure 7. Here, the difference in interpolation scheme (blue curves) is larger and the reaction rate is still significant (10-20% the peak value).

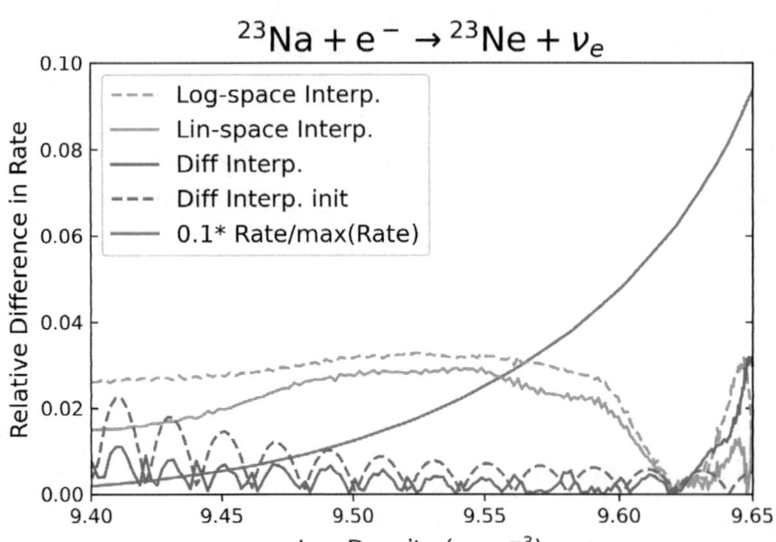

Figure 6. The radially averaged profile of the electron capture reaction rate. The blue curves are the relative difference between the Log-space and Linear-space interpolations at the initial state (dashed) and after 300 seconds of evolution (solid). The orange curves are the relative differences between the initial state and the end state (after 300 seconds) for a given interpolation scheme (Log-space dashed, Lin-space solid). The grey curve is the normalized rate profile for reference.

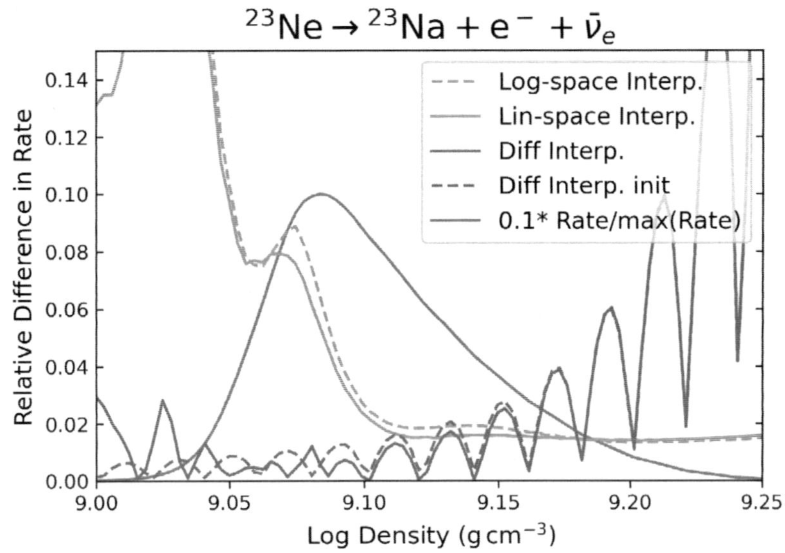

Figure 7. The radially averaged profile of the beta decay reaction rate. The blue curves are the relative difference between the Log-space and Linear-space interpolations at the initial state (dashed) and after 300 seconds of evolution (solid). The orange curves are the relative differences between the initial state and the end state (after 300 seconds) for a given interpolation scheme (Log-space dashed, Lin-space solid). The grey curve is the normalized rate profile for reference.

6. Conclusions

The convective Urca process is an interesting mechanism to provide cooling in a simmering white dwarf star. The reaction rates for these Urca reactions need to be tabulated and recent changes to pynucastro (v2.1) has altered how the tables are interpolated. In the regions where the rates are most active, we found differences of up to 5%. We investigate the sensitivity of our 3D simulations to these changes by comparing two simulations that had run for 300 s, a few convective turnovers. We found that these two simulations deviated in their energy generation rate, total kinetic energy and in the general direction of the flow. However, we account much of this to the highly non-linear nature of our equations. This makes the simulation particularly sensitive to even slight changes to initial conditions. We found that the reaction rates largely varied in time more so than they varied in interpolation scheme for our regions of interest. We interpret this finding to mean that although the new interpolation scheme is more accurate, it does not significantly alter the quasi-steady state of the simulation and the time averaged values that we are interested in. Further time evolving these simulations, through more convective turnovers, may uncover more subtle effects which alter the simulation on longer timescales. With particular interest in the region above the Urca shell where differences in interpolation of the beta-decay reaction are most significant. In future work, we look to further investigating the sensitivity of the convective Urca process to changes in the reaction network, including adding a more accurate prescription of neutron decay [17], using bi-cubic interpolation, adding additional Urca pairs, and using a larger carbon burning network.

Acknowledgments

MAESTROeX is freely available on GitHub (`https://github.com/AMReX-Astro/`), and all problem setup files for the calculations presented here are in the code repository. This research was supported in part by the US Department of Energy (DOE) under grant DE-

FG02-87ER40317 The reaction networks were generated using the pynucastro library [14]. This research used resources of the National Energy Research Scientific Computing Center, which is supported by the Office of Science of the U.S. Department of Energy under Contract No. DE-AC02-05CH11231. Visualizations and part of this analysis made use of yt [18].

References

[1] Phillips M M 1993 *ApJ* **413** L105

[2] Perlmutter S, Aldering G, Goldhaber G, Knop R A, Nugent P, Castro P G, Deustua S, Fabbro S, Goobar A, Groom D E, Hook I M, Kim A G, Kim M Y, Lee J C, Nunes N J, Pain R, Pennypacker C R, Quimby R, Lidman C, Ellis R S, Irwin M, McMahon R G, Ruiz-Lapuente P, Walton N, Schaefer B, Boyle B J, Filippenko A V, Matheson T, Fruchter A S, Panagia N, Newberg H J M, Couch W J and The Supernova Cosmology Project 1999 *ApJ* **517** 565 (*Preprint* astro-ph/9812133)

[3] Paczyński B 1972 *Astrophys. Lett.* **11** 53

[4] Stein J and Wheeler J C 2006 *ApJ* **643** 1190–1197 (*Preprint* astro-ph/0512580)

[5] Lesaffre P, Podsiadlowski P and Tout C A 2005 *MNRAS* **356** 131–144 (*Preprint* astro-ph/0411016)

[6] Piersanti L, Bravo E, Straniero O, Cristallo S and Domínguez I 2022 *ApJ* **926** 103 (*Preprint* 2112.02943)

[7] Willcox D E 2018 *The Single Degenerate Progenitor Scenario for Type Ia Supernovae and the Convective Urca Process* Ph.D. thesis SUNY Stony Brook, New York

[8] Calder A C, Willcox D E, DeGrendele C J, Shangase D, Zingale M and Townsley D M 2019 *Journal of Physics Conference Series* (*Journal of Physics Conference Series* vol 1225) p 012002 (*Preprint* 1904.02238)

[9] Suzuki T, Toki H and Nomoto K 2016 *ApJ* **817** 163 (*Preprint* 1512.00132)

[10] Martínez-Rodríguez H, Piro A L, Schwab J and Badenes C 2016 *ApJ* **825** 57 (*Preprint* 1602.00673)

[11] Schwab J, Martínez-Rodríguez H, Piro A L and Badenes C 2017 *ApJ* **851** 105 (*Preprint* 1711.04780)

[12] Piersanti L, Bravo E, Cristallo S, Domínguez I, Straniero O, Tornambé A and Martínez-Pinedo G 2017 *ApJ* **836** L9 (*Preprint* 1701.06453)

[13] Fan D, Nonaka A, Almgren A S, Harpole A and Zingale M 2019 *ApJ* **887** 212 (*Preprint* 1908.03634)

[14] Smith A I, Johnson E T, Chen Z, Eiden K, Willcox D E, Boyd B, Cao L, DeGrendele C J and Zingale M 2023 *The Astrophysical Journal* **947** 65 URL https://dx.doi.org/10.3847/1538-4357/acbaff

[15] Cyburt R H, Amthor A M, Ferguson R, Meisel Z, Smith K, Warren S, Heger A, Hoffman R D, Rauscher T, Sakharuk A, Schatz H, Thielemann F K and Wiescher M 2010 *ApJS* **189** 240–252

[16] Brown P N, Byrne G D and Hindmarsh A C 1989 *SIAM journal on scientific and statistical computing* **10** 1038–1051

[17] Langanke K and Wiescher M 2001 *Reports on Progress in Physics* **64** 1657–1701

[18] Turk M J, Smith B D, Oishi J S, Skory S, Skillman S W, Abel T and Norman M L 2011 *The Astrophysical Journal Supplement Series* **192** 9 (*Preprint* 1011.3514)

Journal of Physics: Conference Series **2742** (2024) 012002 doi:10.1088/1742-6596/2742/1/012002

Amplification and Dissipation of Magnetic Fields in Accreting Compact Objects

M. Bugli[1,2,3], J. Guilet[2], L. Del Zanna[4,5], A. Mignone[1,3], G. Mattia[5], V. Berta[1], R. Raynaud[6] and A. Reboul-Salze[7]

[1]Dipartimento di Fisica, Università di Torino, I-10125 Torino, Italy
[2]Université Paris-Saclay, Université Paris Cité, CEA, CNRS, AIM, F-91191 Gif-sur-Yvette, France
[3]INFN, Sezione di Torino, I-10125 Torino, Italy
[4]Dipartimento di Fisica e Astronomia, Università di Firenze, I-50019 Sesto Fiorentino (FI), Italy
[5]INFN , Sezione di Firenze, I-50019 Sesto Fiorentino (FI), Italy
[6]Université Paris Cité, Université Paris-Saclay, CEA, CNRS, AIM, F-91191 Gif-sur-Yvette, France
[7]Max Planck Institute for Gravitational Physics (Albert Einstein Institute), D-14476 Potsdam, Germany

E-mail: `matteo.bugli@unito.it`

Abstract. Magnetic fields play a crucial role in shaping the dynamics of accreting compact objects. Whether we consider the formation of a proto-neutron star during the gravitational collapse of a massive star or the accretion disk around a black hole after a compact binary merger, a key process that remains challenging to include in large-scale simulations is the amplification and dissipation of magnetic fields driven by turbulent fluid motions. Despite the enormous increase in computational power currently available, the large separation between all the relevant spatial and temporal scales still poses severe limits to what can be achieved with ideal fluid simulations. One way to tackle such issue is to rely on sub-grid models, which however need to be appropriately tuned in light of models probing the small-scale dynamics. In this work we present the current state-of-the-art of dynamo models in proto-neutron stars, which aim at describing the amplification of magnetar-like magnetic fields during the gravitational collapse of a massive star. We also review some of the works from the past few years that included turbulent dynamos in accretion disks around a black holes, relying on a mean-field formalism. Finally, we will present a recent study on polar jets with explicit turbulent resistivity which showcases the importance of employing highly accurate numerical schemes.

1. Introduction

Accretion onto compact objects is one of the most efficient mechanisms in the Universe to convert gravitational energy into other forms of energy (either kinetic, magnetic or thermal), energetic particles, and gravitational waves. The efficiency for converting accretion power to luminous output can be as high as $\sim 42\%$ in the case of a maximally rotating black hole (as opposed to the $\sim 0.7\%$ efficiency of mass to energy conversion during hydrogen fusion in stellar cores) which motivates the central role of accreting black holes and neutron stars in the modeling of astrophysical high-energy sources such as X-ray binaries, active galactic nuclei, and gamma-ray bursts. Observational evidences for non-thermal spectral components and significant degrees of polarization are found in virtually all high-energy sources, and are usually interpreted as smoking guns for particle acceleration processes and large-scale magnetic fields, respectively. The role of magnetic fields is particularly crucial for accreting compact objects. They enable the accretion of matter via magnetohydrodynamic (MHD) turbulence produced by the onset of the Magneto-Rotational Instability (MRI, [1]) and when distributed over large-scales they can launch relativistic outflows. Moreover, magnetic fields can directly energize the outflow through reconnection events, where particles are accelerated at the expenses of magnetic energy due to the change in magnetic field topology [2].

The numerical modeling of accretion and ejection flows around compact objects remains a challenging endeavour to this day, given the large scale separation (both in space and time) that divides the local plasma dynamics, the central engine, and the emission sites of these sources A self-consistent description of the non-collisional nature of astrophysical plasmas (such as in PIC simulations) allows to capture fundamental kinetic and dissipative effects. However, the elevated computational costs that come with it limits this approach to relatively short temporal and spatial scales compared to those of typical astrophysical compact objects. At the same time, fluid approximations such as MHD can probe the accretion flow dynamics and the launch of relativistic jets, but they generally ignore the microphysics behind the dissipative and kinetic properties of the plasma.

There are several approaches that can be adopted to face these challenges and try to bridge the gaps that separate all the relevant scales of the problem. One might attempt to focus on different parts of the accreting systems, resolving smaller and larger scales with dedicated numerical simulations. The connection between the different scales can then be established either with suitable boundary or initial conditions (such as in the case of jet launching models not including the central engine ([3], [4]) or with a well-calibrated sub-grid model (e.g. to include unresolved turbulent dynamics of the plasma [5]–[7]). Hybrid MHD-PIC methods are another useful tool to describe the acceleration of non-thermal particles embedded in a thermal fluid (such as in the case of cosmic rays [8] or electrons accelerated at collisionless shocks [9]). Finally, the advent of GPU-accelerated astrophysical codes [10]–[12] has paved the way to numerical studies with unprecedented resolutions [13]–[15] that probed dynamical regimes so far unattainable.

Whether we consider a still forming proto-neutron star (PNS) during the onset of a core-collapse supernova (CCSN) or the accretion disk around a supermassive black hole, the fundamental ingredients describing the amplification and dissipation of magnetic fields and their dynamical impact on the accretion/ejection dynamics are remarkably similar. In this work we review some of the most recent developments in the numerical modeling of magnetized accreting compact objects, which involve some of the aforementioned strategies to tackle the scale-separation problem. In §2 we will present results from studies focusing on PNS dynamo studies and CCSN large-scale simulations, i.e. the early stages of the formation of stellar compact objects. We will then show in §3 current state-of-the-art simulations of accreting black holes and relativistic jets that include non-ideal effects to model dynamo and diffusive processes in the plasma. Finally, in §4 we will present a new 4th-order accurate finite volume method for classical and relativistic MHD that can be implemented to boost the effective resolution of numerical models and increase their efficiency.

2. Core-collapse supernovae

CCSN are the result of the gravitational collapse of massive stars at the end of their life, and represent one the main channels for the formation of stellar compact objects. Once nuclear densities are reached at the core of the star, the equation of state (EoS) suddenly stiffens and a shock wave starts propagating from a central newly formed PNS. Due to the ram pressure of the infalling stellar material and the loss of energy via photo-dissociation of Fe-group elements [16], the shock stalls and cannot further propagate unless some other process

Journal of Physics: Conference Series **2742** (2024) 012002 doi:10.1088/1742-6596/2742/1/012002

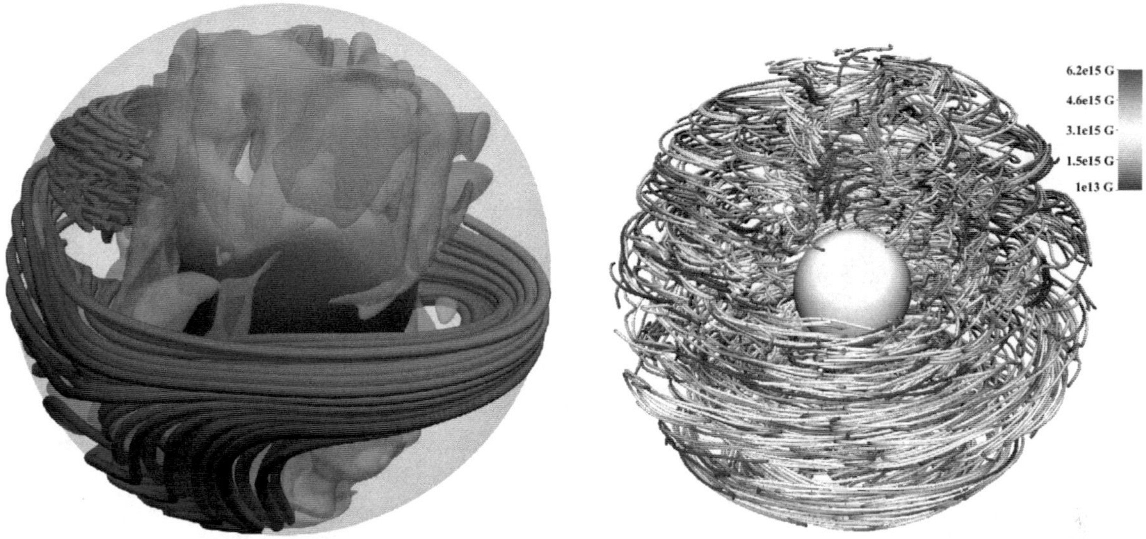

Figure 1. Left panel: 3D rendering of the magnetic field produced by the convective dynamo in the strong regime [22]. Blue (red) isosurfaces represent the radial velocity, i.e. downflows (outflows) of the order of 10^8 cm/s. Right panel: magnetic field lines within the MRI unstable zone of the PNS from a numerical simulation in the incompressible MHD regime [23]. Blue and red lines indicate, respectively, low and high intensity magnetic fields.

provides a suitable amount of energy. In the vast majority of CCSN events the shock is revived by the so-called *neutrino-heating* mechanism: a small fraction of the flux of neutrinos emitted from the PNS surface (which overall amount to $\sim 99\%$ of the total gravitational binding energy lost by the explosion) can be absorbed behind the shock, stirring convection and aiding the outward propagation of the shock wave [16]. However, outstanding stellar explosions such as superluminous supernovae [17] and Type Ic-BL [18] cannot be explained by neutrinos alone, and need to consider an additional source of energy. A common suggestion is that stellar progenitor with fast rotation can produce, in presence of strong large-scale magnetic fields, a *magneto-rotational explosion*: the rotational energy within the stellar core is extracted either via magnetic braking (for a proto-neutron star) or a Blandord-Znajek mechanism (for a black hole with and accretion disk), launching energetic bipolar outflows [19]–[21].

2.1. PNS dynamos

The exact origin of the magnetic fields required to obtain magneto-rotational CCSN remains unclear, as they could be produced within the stellar progenitor in the pre-collapse phase ([24], [25]), amplified during stellar mergers [26] or directly in-situ within the PNS during the explosion via dynamo processes [27]–[29]. In a series of recent studies we explored the latter scenario using the pseudo-spectral code `MagIC` ([30], [31]) to focus on specific layers of the PNS where magnetic field amplification can occur through the action of mainly two physical processes. The first one is the convection triggered in the inner regions of the PNS ($10 \lesssim r \lesssim 25$ km) predominantly by negative lepton gradients due to persistent losses of electron neutrinos ν_e [32] about ~ 50 ms after shock formation. When rotation in the PNS is sufficiently fast, the Coriolis and Lorentz forces balance each other, thus producing a magnetostrophic balance that leads to the saturation of a dynamo magnetic field close to magnetar-like values (up to $\sim 10^{16}$ G) [22]. The resulting field is dominated by the axisymmetric toroidal component (left panel of Fig. 1) and non-axisymmetric structures, and can lead to a significant emission of GW over the spans of several seconds [33].

Journal of Physics: Conference Series **2742** (2024) 012002 doi:10.1088/1742-6596/2742/1/012002

Dynamo action can also take place in the stably stratified layers of the PNS that surround its convective region ($25 \lesssim r \lesssim 50$ km) in presence of differential rotation, as the MRI can develop and further amplify the magnetic fields [34]. Starting from small-scale magnetic seeds (but with intensity sufficiently high so as to properly resolve the MRI) the onset of the instability leads to the amplification of large-scale magnetic fields [35], which are also dominated by the toroidal component (right panel of Fig. 1). Moreover, the low multipolar (i.e. large-scale) components of the field are generally misaligned with respect to the PNS rotational axis. When the density stratification of the PNS is taken into account, the amplification process assumes characteristics typical of an $\alpha\Omega$-dynamo, with magnetic field reversals and periodic patterns that give rise to butterfly-diagrams [23].

2.2. Magneto-rotational explosions and multi-messenger signals

Regardless of the specific process at the core of PNS dynamos, the large-scale magnetic fields that they amplify tend to be dominated by non-axisymmetric structures and multipolar components beyond a simple magnetic dipole. To self-consistently evaluate the impact of such magnetic fields on the explosion mechanism and the ejecta launch would require to cover a range of spatial and temporal scales spanning at least 5 orders of magnitude (from a few tens of meters to thousands of km). A more approximated (but also affordable) approach is to perform CCSN simulations with initial magnetic fields having a topology more complex than a standard aligned dipolar field (as usually assumed in magneto-rotational explosion models) to qualitatively capture the possible effects of PNS dynamos on the explosion mechanism [37]. We considered the stellar progenitor `350C` [38] with its original fast rotation rate and superimposed different magnetic field configurations: aligned dipole (model L1-0), aligned quadrupole (L2-0A and L2-0B), equatorial dipole (L1-90), and without any magnetic field (H). The strength of the magnetic field in the iron core is set to $\sim 10^{12}$ G, which is a typical value employed in magneto-rotational explosion models [21], [39]. We performed our 3D relativistic MHD simulations with the `AENUS-ALCAR` code [40], which includes nuclear EoS and a multi-dimensional M1-neutrino transport scheme to include all the relevant interactions between matter and neutrinos. The two-moment closure approximation we use to describe the neutrino dynamics is adopted by several other research groups [41]–[43], as it allows to capture the multidimensional effects of neutrino-matter interactions at a significantly lower computational cost w.r.t. the full-Boltzmann approach. However, it should be noted that this approximation can become inaccurate when transitioning between the optically thick and thin regimes if the structure of the flow is spatially complex (as in the PSN and the post-shock region; [44], [45]) and lead to quantitatively different results [46], [47]. We include gravity via a simple Newtonian treatment with general relativistic corrections to the gravitational potential [48] (which captures the main impact that general relativity has on the compression of the PNS) and compute GW strains by means of the quadrupole formula described in [49]. As shown in Fig. 2, configurations with higher multipoles (L2-0) or inclined magnetic axis (L1-90) produce less energetic explosions and less collimated outflows. Moreover, magnetic fields play a crucial role in quenching the onset of the low-$T/|W|$ instability, i.e. a hydrodynamic co-rotational instability that can affect a PNS with very fast rotation [50]. While in the hydrodynamic case we see around 200 ms post-bounce the emergence of large-scale spiral structures that protrude from the inner regions of the PNS through its surface (left panel of Fig. 3), in none of the magnetized models we observe such dynamics (right panel). This is a direct consequence of the magnetic transport of angular momentum, which is quite efficient at flattening the rotation profile within the PNS, thus stabilizing the PNS against the onset of the low-$T/|W|$ instability.

This qualitative deviation between magnetized and unmagnetized models reflects also on the properties of the GW and neutrino emission during the explosion [51]. The onset of the low-$T/|W|$ instability corresponds to a burst of GW that lasts for several hundreds of seconds, which has a distinctive spectral signature in the frequency domain that appears around ~ 200 ms p.b.

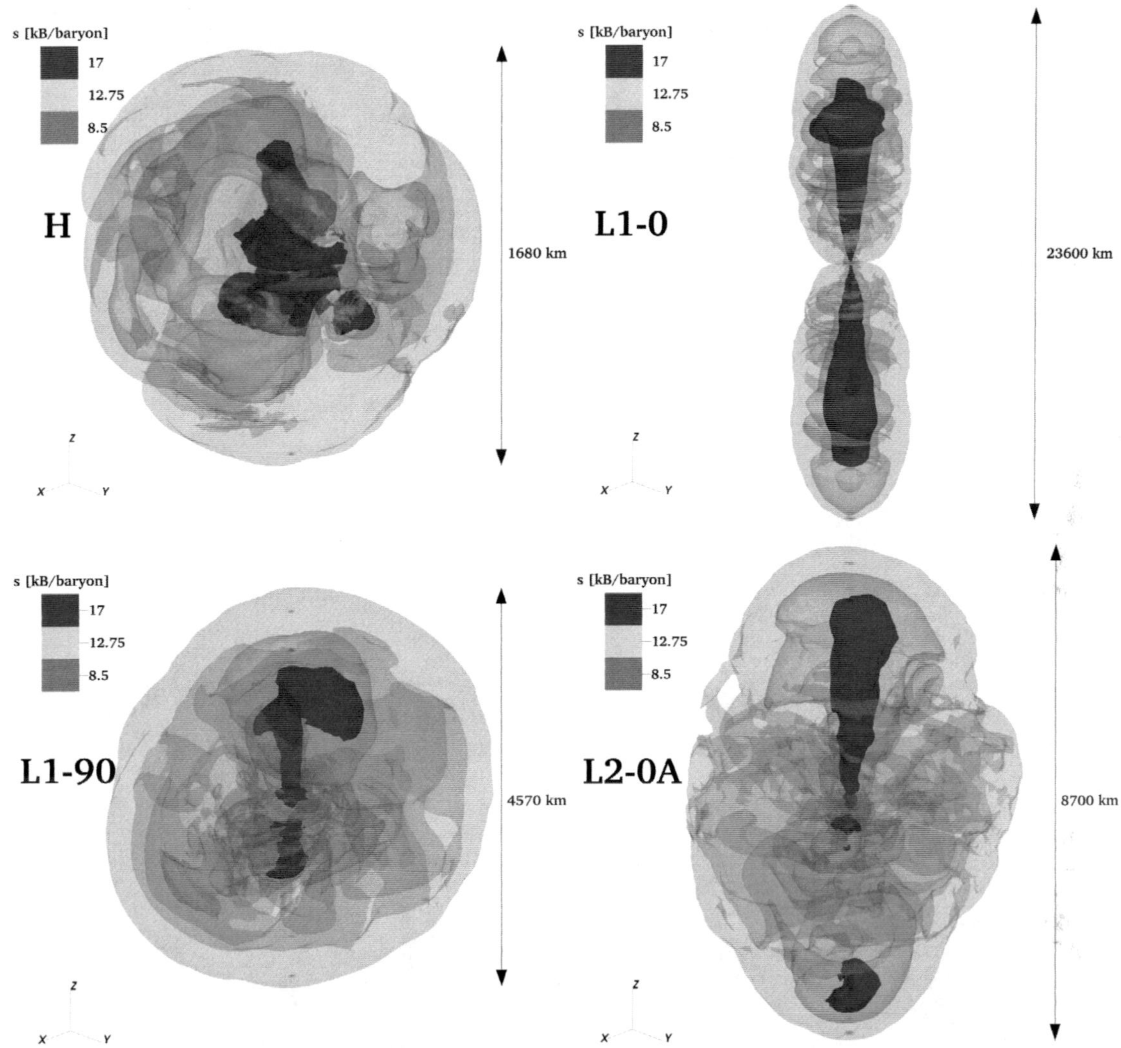

Figure 2. Volume rendering of the specific entropy for CCSN models with different initial magnetic field configurations: no magnetic field (top left panel), aligned dipole (top right), equatorial dipole (lower left panel), and aligned quadrupole (lower right panel) at $t = 410$ ms p.b. [36]

and increases frequency with time due to the compression of the PNS (left column of Fig. 4). On the other hand, no such strong emission is observed in the magnetized models, which produce a GW signal with a rather broad spectrum that goes beyond the rotational frequencies within the PNS (white curves), as they have been narrowed by the outer transport of angular momentum. The neutrino emission is also deeply affected by the action of magnetic fields. All magnetized models have neutrino lightcurves along the equatorial plane which are systematically lower than the hydrodynamic case. This is due to the more oblate shape of the PNS, which is induced once again by the Maxwell stresses acting in the PNS and leads to the emission along the equatorial plane of neutrinos with a lower mean energy (since they decouple from the matter at larger distances). Moreover, the time-frequency diagrams of such lightcurves show features that, in

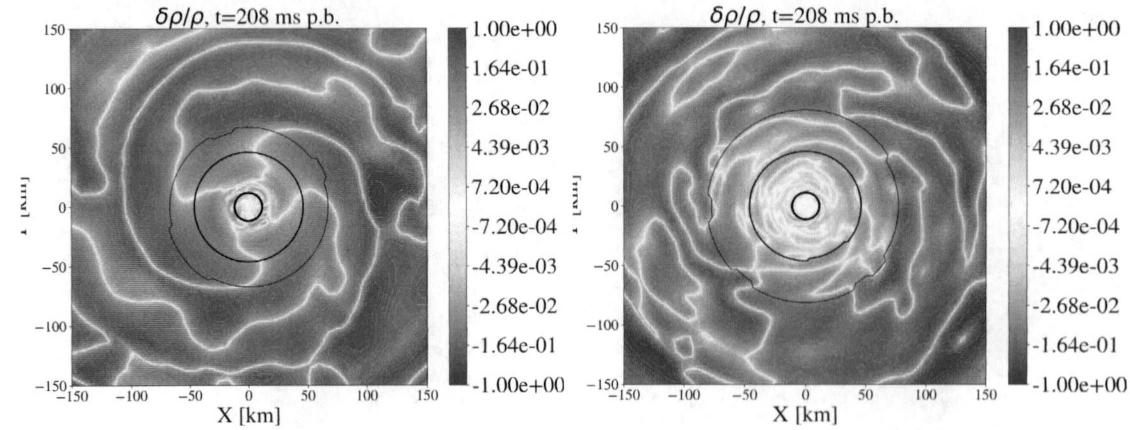

Figure 3. Equatorial non-axisymmetric density distributions from a hydrodynamic model (left) and a magnetized (right) CCSN model with fast rotation [51]. The black curves are isocontours of density 10^{11}, 10^{12}, and 10^{14} g/cm^3.

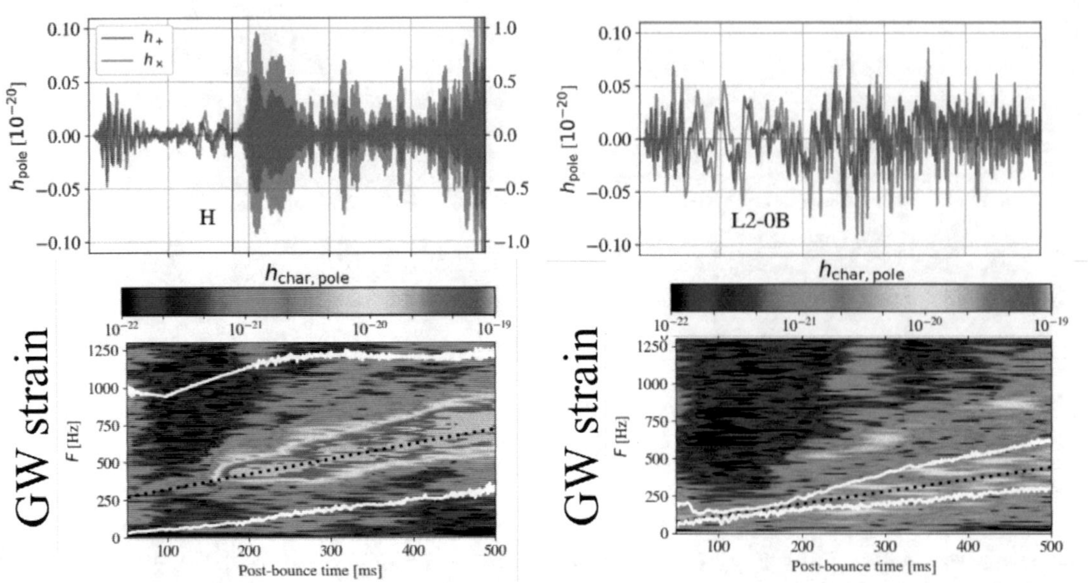

Figure 4. Top row: Gravitational wave strains (+ and × polarizations) for the hydrodynamic CCSN model (left) and the one with an aligned quadrupolar magnetic field (right) [51]. Bottom row: time-frequency spectrograms of the characteristic strain $h_{\rm char}$ relative to the the strains above. The white lines track twice the rotational frequency of the PNS surface and center [51].

the hydrodynamic case, correlate quite well with the onset of the low-$T/|W|$ instability and the its associated GW emission (see Fig. 12 of [51]), showcasing the important synergy between GW and neutrino detectors.

3. Accretion disks and jets

We now move to a later stage in the activity of compact objects, namely the case of a black hole surrounded by an accretion disk. Since at least two decades general relativistic

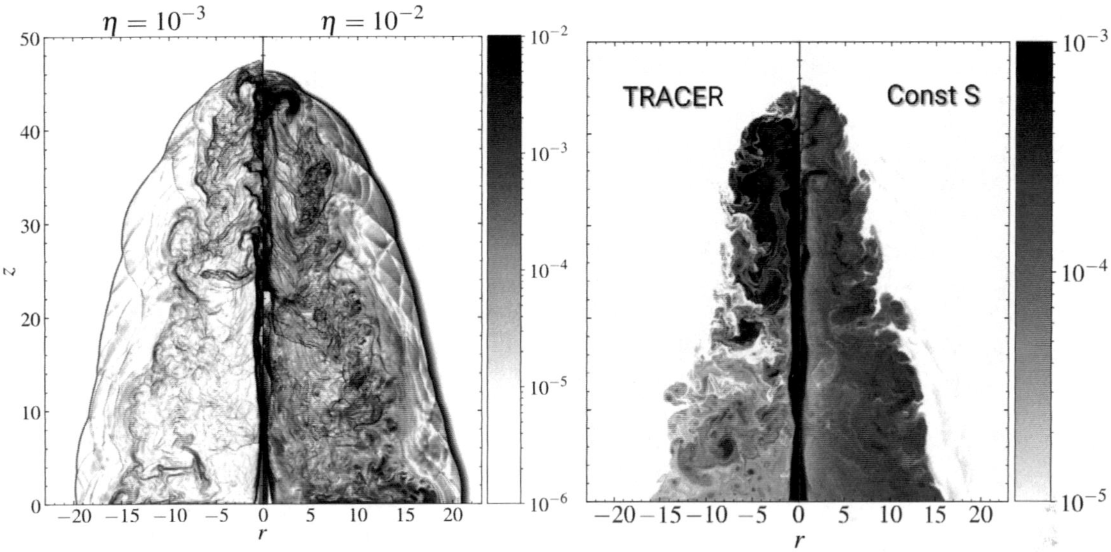

Figure 5. Left panel: dissipated electromagnetic power in a relativistic jet with constant resistivity $\eta = 10^{-3}$ (left side) and $\eta = 10^{-2}$ (right). Right panel: non-constant resistivity profiles in relativistic jets set by jet-tracing particles (left side) or by imposing a constant Lundquist number $S = 10^3$ [63].

magnetohydrodynamics (GRMHD) provide the fundamental framework for the study of such systems. The scientific community has independently developed several ideal GRMHD codes that have proved to be able to consistently produce models in quantitative agreement which each other [52]. Nonetheless, in recent years there have been several efforts to go beyond the ideal GRMHD formulation by introducing kinetic dissipative effects [53], [54], electron thermodynamics [55], viscosity/anisotropic pressure [56], collisionless moment closures [57], magnetic resistivity [58]–[60], and mean-field dynamo processes [61], [62]. In the following we focus on the last two effects, which are examples of sub-grid models aim at including physical processes that occur at unresolved spatial scales.

3.1. Resistive (G)RMHD

The introduction of an explicit resistivity η allows to model the physical dissipation of magnetic fields and avoid to be dominated by the numerical diffusivity (which is always present in any grid-based code). This is particularly relevant for highly magnetized environments such as the corona of an accretion disk or a relativistic jet close to its central engine, since the formation of current sheets and reconnection events play a crucial role in particle acceleration and magnetic dissipation [2], [64]. However, the relativistic regime poses an additional challenge with respect to the classical case, since the displacement current cannot, in principle, be neglected in Ampère's law. Moreover, the electric field $\mathbf{E}$ cannot simply be expressed in terms of the magnetic field $\mathbf{B}$ and the fluid velocity $\mathbf{v}$ as in the ideal GRMHD case (where $\mathbf{E} = -\mathbf{v} \times \mathbf{B}$), but it is instead coupled to the current density $\mathbf{J}$ via the resistive Ohm's law [58]

$$\mathbf{E} + \mathbf{v} \times \mathbf{B} = \frac{\eta}{\Gamma}(\mathbf{J} - q\mathbf{v}) + (\mathbf{E} \cdot \mathbf{v})\mathbf{v}, \quad (1)$$

where Γ is the fluid's Lorentz factor and q is the electric charge density The combination of these two aspects promotes Ampère's law to a new evolution equation for the electric field (which effectively becomes a new independent variable) [58]:

$$\gamma^{-1/2}\partial_t\left(\gamma^{1/2}\mathbf{E}\right)+\nabla\times(-\alpha\mathbf{B}+\boldsymbol{\beta}\times\mathbf{E})=-(\alpha\mathbf{J}-q\boldsymbol{\beta}),\tag{2}$$

where γ is the determinant of the 3D spatial metric tensor, α is the lapse function and $\boldsymbol{\beta}$ is the shift vector. Since Eq. 2 can become stiff for low values of η, a common strategy to ensure stability to the integration of the resistive (G)RMHD equations is to employ IMEX schemes [65], which are designed to treat implicitly the stiff equations and explicitly the rest of the system.

We recently used this framework (specialized to a flat space-time) to perform, for the first time, axisymmetric simulations of relativistic magnetized jets with a finite physical resistivity propagating through a magnetized ambient medium [63] with the resistive RMHD module of the `PLUTO` code. The resistivity we considered in this study does not represent the Ohmic dissipation within the plasma (which is generally extremely small and occurring at microscopic scales) but rather a *turbulent* resistivity, which tracks the dissipation of magnetic fields due to unresolved turbulent motions. While such effects are generally dependent on the dynamic and thermodynamic state of the plasma, for simplicity we first treated η as a completely free parameter. We tested the impact of a constant $\eta = 10^{-6}, 10^{-3}, 10^{-2}$ across the domain (in code units, i.e. in units of $4\pi Lc$ with $L = 10^{-6}$ pc) and verified that for a higher resistivity the jet injected less electromagnetic energy and the magnetic field was distributed over larger spatial scales. In all cases, the bulk of the dissipation occurred in the jet's spine, close to the symmetry axis (left panel of Fig. 5). We also tested two different profiles for a non-constant resistivity, by either: 1) modulating its value between 10^{-6} and 10^{-3} using a passive tracer for the jet; 2) imposing a constant Lundquist number $S = v_A L/\eta = 10^3$ with v_A the Alfvèn speed and L the characteristic length scale of the system which we identified with the injection radius $r_j = 1$. The resultant spatial distributions for η (right panel of Fig. 5) differ significantly from one another, with the tracer prescription leading to a slightly stronger mean value for η over the course of the simulation.

3.2. *Mean-field dynamos in accretion disks*

A further generalization of Eq. 1 can include the effects of a mean-field dynamo action, which models unresolved small-scale fluctuations in velocity and magnetic field as an additional large-scale electromotive force in the fluid comoving frame that is proportional to the magnetic field [61]. Such an assumption leads to a slightly modified Ohm's law [61]

$$\mathbf{E}+\mathbf{v}\times\mathbf{B}=\frac{\eta}{\Gamma}(\mathbf{J}-q\mathbf{v})+(\mathbf{E}\cdot\mathbf{v})\mathbf{v}+\xi[\mathbf{B}-\mathbf{v}\times\mathbf{E}-(\mathbf{B}\cdot\mathbf{v})\mathbf{v})],\tag{3}$$

where the parameter ξ is the dynamo parameter defined as $-\alpha_{\rm dyn}$, i.e. the opposite of the parameter after which the α-effect of $\alpha\Omega$-dynamos is named. On the other hand, all the rest of the relevant equations coincide with the purely resistive case. With tested this prescription by studying both kinematic [67] and fully dynamic [5] phases of an $\alpha\Omega$-dynamo in thick accretion disks around black holes. An arbitrarily weak magnetic field seed can be efficiently amplified by such mechanism (even in axisymmetry) and the process self-regulates itself through the modulation of the mass accretion rate onto the black hole, without any explicit quenching mechanism. The resulting magnetic and mass fluxes and the black hole event horizon are consistent with those retrieved by 3D ideal GRMHD models with higher resolution, validating its implementation [68]. Moreover, when applied to 3D models the relativistic $\alpha\Omega$-dynamo induces large-scale non-axisymmetric structures [66] that can sometimes be found in ideal 3D models (Fig. 6). The emergence of $\alpha\Omega$-dynamos in relativistic disks has been recently obtained by very high-resolution 3D models [69], which furthermore demonstrates the potential of a relativistic dynamo closure to study more efficiently accretion flows around black holes.

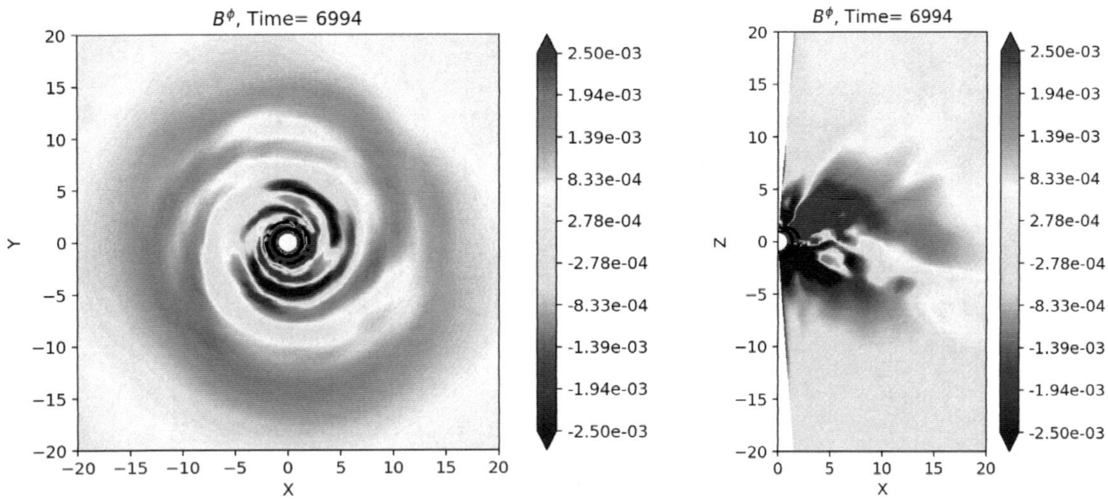

Figure 6. Toroidal magnetic field distribution in the equatorial (left panel) and meridional (right) plane for a GRMHD model of accretion disk around a Kerr black hole with spin parameter $a = 0.9375$ and mean-field dynamo action [66]. The $xy-$axes and the magnetic field are express in code units, i.e. in units of the black hole gravitational radius $R_G = GM_{BH}/c^2$ and $\sqrt{4\pi/G^3}c^4/M_{BH}$, respectively.

4. A new 4th-order FV scheme

We finally present another recent development that aims at increasing the efficiency of numerical simulations of astrophysical plasmas and therefore mitigating the difficulties of studying systems with a large scale-separation. We developed a new genuinely 4th-order accurate FV method based on pointwise reconstructions that can significantly increase the effective resolution of a typical astrophysical simulation and thus reduce its computational cost [70].

Consider the following FV discretization of a generic conservation law over a uniform Cartesian grid with spacing $(\Delta x, \Delta y, \Delta z)$

$$\frac{d\langle U\rangle_{\mathbf{c}}}{dt} = -\left(\frac{\hat{F}_{\mathbf{x}_f} - \hat{F}_{\mathbf{x}_f-\hat{\mathbf{e}}_x}}{\Delta x} + \frac{\hat{F}_{\mathbf{y}_f} - \hat{F}_{\mathbf{y}_f-\hat{\mathbf{e}}_y}}{\Delta y} + \frac{\hat{F}_{\mathbf{z}_f} - \hat{F}_{\mathbf{z}_f-\hat{\mathbf{e}}_z}}{\Delta z}\right), \tag{4}$$

where $\langle U\rangle_{\mathbf{c}}$ is the cell volume average of the conservative quantity U, $\hat{F}_{\mathbf{x}_f}$ is the surface average of the x-component of the flux over a face perpendicular to the x-direction, $\hat{F}_{\mathbf{x}_f-\hat{\mathbf{e}}_x}$ is the surface average of the same component over the opposite side of the cell, and similarly for the other flux terms along the y- and z-directions. For schemes that are less than 4th-order accurate, one can identify the volume and surface averages in Eq. 4 with the point values of U and F evaluated at the cell center and at its sides centers, respectively, which effectively makes low-order FV and finite differences (FD) formulations equivalent to each other. However, this is no longer true for schemes that aim at achieving higher orders of accuracy, which then need to calculate suitable high-order approximations of both sides of Eq. 4. Starting from a high-order cell volume average of the conserved quantity U, we adopt the following general strategy (see [70] for a complete description):

- We convert the volume average $\langle U\rangle_{\mathbf{c}}$ into its point value at the cell center U_c using the

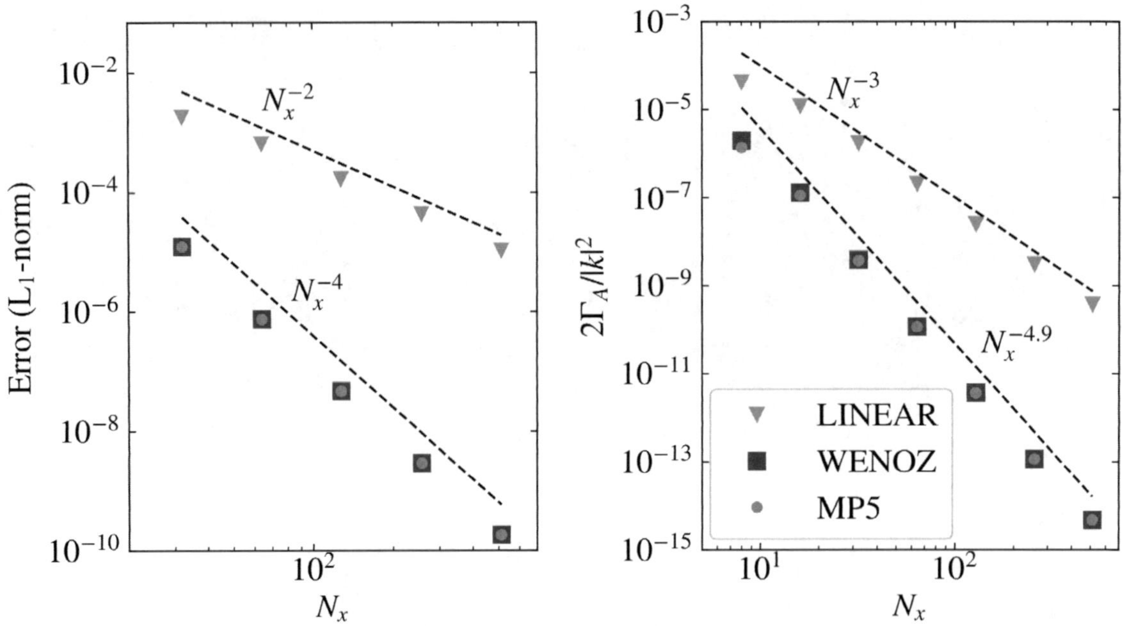

Figure 7. Left panel: convergence scaling computed for 3D Alfvén waves propagating along a diagonal for a 2$^{\rm nd}$-order (triangles) and 4$^{\rm th}$-order (squares and circles) FV schemes [70]. Right panel: scaling with resolution of the estimated numerical dissipation from the aforementioned test problem.

expression

$$U_{\bf c} = \langle U \rangle_{\bf c} - \frac{\Delta \langle U \rangle_{\bf c}}{24}, \tag{5}$$

where $\Delta()$ is the Laplacian operator introduced by [71].

- From the ensemble of cell centered conservative variables $\{U_{\bf c}\}$ we retrieve the primitive variables $\{V_{\bf c}\}$.
- We apply a pointwise reconstruction to obtain the primitive variables at the cell's interfaces and compute the upwind fluxes.
- We convert the point-valued fluxes into high-order surface averages using

$$\hat{F}_{{\bf x}_f} = F_{{\bf x}_f} + \frac{\Delta_\perp^x F_{{\bf x}_f}}{24}, \tag{6}$$

where $\Delta_\perp^x$ is a transverse Laplacian operator along the x-direction (and similarly for the other components).

- We integrate Eq. 4 in time using a 4$^{\rm th}$-order accurate Runge-Kutta scheme.

As shown in Fig. 7, this scheme exhibits a clear 4$^{\rm th}$-order scaling and greatly improves the solution accuracy even at modest resolutions. Its numerical dissipation rate (right panel) is more than 10 times lower with respect to a 2$^{\rm nd}$-order scheme even at very low resolutions (~ 8 points), which demonstrate its potential at allowing a more realistic description of magnetic diffusion when applied to a resistive framework.

5. Conclusion

We presented a series of recent results from numerical simulations aimed at studying the dynamics of magnetic fields in different kinds of accreting compact objects, such as proto-neutron stars in CCSN and black holes surrounded by an accretion disk. Despite the different scientific context, microphysics, and numerical tools required, all these studies demonstrate how an accurate description of the amplification and dissipation of magnetic fields can significantly impact the accretion process, the launch of polar outflows, and the acceleration of particles. Several strategies can be implemented to overcome the challenges posed by the large scale-separation between all the relevant physical processes involved, from multi-scale studies to sub-grid models, from the use of GPU-accelerated HPC code to the adoption of more accurate and higher order numerical schemes.

References

[1] S. A. Balbus and J. F. Hawley, *ApJ*, vol. 376, pp. 214–233, Jul. 1991.

[2] L. Sironi and A. Spitkovsky, *ApJ*, vol. 783, p. L21, Mar. 2014.

[3] R. Bollig, N. Yadav, D. Kresse, H.-T. Janka, B. Müller, and A. Heger, *ApJ*, vol. 915, no. 1, 28, p. 28, Jul. 2021.

[4] O. Just *et al.*, *ApJl*, vol. 951, no. 1, L12, p. L12, Jul. 2023.

[5] N. Tomei, L. Del Zanna, M. Bugli, and N. Bucciantini, *MNRAS*, vol. 491, pp. 2346–2359, Jan. 2020.

[6] C. Palenzuela, R. Aguilera-Miret, F. Carrasco, R. Ciolfi, J. V. Kalinani, W. Kastaun, B. Miñano, and D. Viganò,, vol. 106, no. 2, 023013, p. 023 013, Jul. 2022.

[7] G. Mattia and C. Fendt, *ApJ*, vol. 935, p. 22, Aug. 2022.

[8] A. Mignone, G. Bodo, B. Vaidya, and G. Mattia, *ApJ*, vol. 859, p. 13, May 2018.

[9] V. Nedora, T. Dietrich, M. Shibata, M. Pohl, and L. Crosato Menegazzi, *MNRAS*, vol. 520, pp. 2727–2746, Apr. 2023.

[10] M. T. P. Liska *et al.*, *ApJ Supplement Series*, vol. 263, p. 26, Dec. 2022.

[11] G. R. J. Lesur, S. Baghdadi, G. Wafflard-Fernandez, J. Mauxion, C. M. T. Robert, and M. Van den Bossche, *A&A*, vol. 677, A9, Sep. 2023.

[12] D. Bégué, A. Pe'er, G.-Q. Zhang, B.-B. Zhang, and B. Pevzner, *ApJ Supplement Series*, vol. 264, no. 2, p. 32, Jan. 2023.

[13] B. Ripperda, M. Liska, K. Chatterjee, G. Musoke, A. A. Philippov, S. B. Markoff, A. Tchekhovskoy, and Z. Younsi, *ApJ*, vol. 924, p. L32, Jan. 2022.

[14] O. Gottlieb, M. Liska, A. Tchekhovskoy, O. Bromberg, A. Lalakos, D. Giannios, and P. Mösta, *arXiv:2204.12501 [astro-ph]*, Apr. 2022.

[15] G. Wafflard-Fernandez and G. Lesur, *A&A*, vol. 677, A70, Sep. 2023.

[16] H.-T. Janka, *Annual Review of Nuclear and Particle Science*, vol. 62, pp. 407–451, Nov. 2012.

[17] C. Inserra, *Nature Astronomy*, vol. 3, pp. 697–705, Aug. 2019.

[18] J. Hjorth and J. S. Bloom. Nov. 2012, pp. 169–190.

[19] S. Akiyama, J. C. Wheeler, D. L. Meier, and I. Lichtenstadt, en, *ApJ*, vol. 584, no. 2, pp. 954–970, Feb. 2003.

[20] A. Burrows, L. Dessart, E. Livne, C. D. Ott, and J. Murphy, en, *ApJ*, vol. 664, no. 1, p. 416, Jul. 2007.

[21] M. Obergaulinger and M. Á. Aloy, *MNRAS*, vol. 503, pp. 4942–4963, May 2021.

[22] R. Raynaud, J. Guilet, H.-T. Janka, and T. Gastine, *Science Advances*, vol. 6, eaay2732, Mar. 2020.

[23] A. Reboul-Salze, J. Guilet, R. Raynaud, and M. Bugli, *A&A*, vol. 667, A94, A94, Nov. 2022.

[24] N. Langer, *Magnetic Fields throughout Stellar Evolution*, vol. 302, p. 1, Aug. 2014.

[25] D. R. Aguilera-Dena, N. Langer, J. Antoniadis, and B. Müller,, vol. 901, no. 2, 114, p. 114, Oct. 2020.

[26] F. R. N. Schneider, S. T. Ohlmann, P. Podsiadlowski, F. K. Röpke, S. A. Balbus, R. Pakmor, and V. Springel, en, *Nature*, vol. 574, no. 7777, p. 211, Oct. 2019.

[27] C. Thompson and R. C. Duncan, en, *ApJ*, vol. 408, p. 194, May 1993.

[28] P. Barrère, J. Guilet, A. Reboul-Salze, R. Raynaud, and H.-T. Janka, *A&A*, vol. 668, A79, A79, Dec. 2022.

[29] P. Barrère, J. Guilet, R. Raynaud, and A. Reboul-Salze, *MNRAS*, vol. 526, no. 1, pp. L88–L93, Nov. 2023.

[30] T. Gastine and J. Wicht, *Icarus*, vol. 219, no. 1, pp. 428–442, May 2012.

[31] N. Schaeffer, *Geochemistry, Geophysics, Geosystems*, vol. 14, no. 3, pp. 751–758, Mar. 2013.

[32] H. Nagakura, A. Burrows, D. Radice, and D. Vartanyan, *MNRAS*, staa261, Jan. 2020.

[33] R. Raynaud, P. Cerdá-Durán, and J. Guilet, *MNRAS*, vol. 509, pp. 3410–3426, Jan. 2022.

[34] J. Guilet, A. Reboul-Salze, R. Raynaud, M. Bugli, and B. Gallet, *MNRAS*, Sep. 2022.

[35] A. Reboul-Salze, J. Guilet, R. Raynaud, and M. Bugli, *A&A*, vol. 645, A109, Jan. 2021.

[36] M. Bugli, J. Guilet, and M. Obergaulinger, *MNRAS*, vol. 507, pp. 443–454, Oct. 2021.

[37] M. Bugli, J. Guilet, M. Obergaulinger, P. Cerdá-Durán, and M. A. Aloy, *MNRAS*, vol. 492, no. 1, pp. 58–71, Feb. 2020.

[38] S. E. Woosley and A. Heger, en, *ApJ*, vol. 637, no. 2, p. 914, Feb. 2006.

[39] T. Kuroda, A. Arcones, T. Takiwaki, and K. Kotake,, vol. 896, no. 2, 102, p. 102, Jun. 2020.

[40] O. Just, M. Obergaulinger, and H.-T. Janka, *MNRAS*, vol. 453, pp. 3386–3413, Nov. 2015.

[41] E. O'Connor, *The Astrophysical Journal Supplement Series*, vol. 219, no. 2, p. 24, Aug. 2015.

[42] T. Kuroda, K. Kotake, and T. Takiwaki, *ApJ*, vol. 829, p. L14, Sep. 2016.

[43] D. Vartanyan, A. Burrows, D. Radice, M. A. Skinner, and J. Dolence, *Monthly Notices of the Royal Astronomical Society*, vol. 477, no. 3, p. 3091, Jul. 2018.

[44] W. Iwakami, H. Okawa, H. Nagakura, A. Harada, S. Furusawa, K. Sumiyoshi, H. Matsufuru, and S. Yamada, *The Astrophysical Journal*, vol. 903, no. 2, p. 82, Nov. 2020.

[45] W. Iwakami *et al.*, *The Astrophysical Journal*, vol. 933, p. 91, Jul. 2022.

[46] R. M. Cabezón, K.-C. Pan, M. Liebendörfer, T. Kuroda, K. Ebinger, O. Heinimann, A. Perego, and F.-K. Thielemann,, vol. 619, A118, A118, Nov. 2018.

[47] E. P. O'Connor and S. M. Couch, en, *ApJ*, vol. 854, p. 63, Feb. 2018.

[48] A. Marek, H. Dimmelmeier, H.-T. Janka, E. Müller, and R. Buras, en, *Astronomy & Astrophysics*, vol. 445, no. 1, pp. 273–289, Jan. 2006.

[49] S. Scheidegger, T. Fischer, S. C. Whitehouse, and M. Liebendörfer, *Astronomy and Astrophysics*, vol. 490, pp. 231–241, Oct. 2008.

[50] T. Takiwaki, K. Kotake, and T. Foglizzo,, vol. 508, no. 1, pp. 966–985, Nov. 2021.

[51] M. Bugli, J. Guilet, T. Foglizzo, and M. Obergaulinger, *MNRAS*, vol. 520, pp. 5622–5634, Apr. 2023.

[52] O. Porth *et al.*, en, *ApJ Supplement Series*, vol. 243, no. 2, p. 26, Aug. 2019.

[53] B. Ripperda, O. Porth, L. Sironi, and R. Keppens, *MNRAS*, vol. 485, no. 1, pp. 299–314, May 2019.

[54] S. Selvi, O. Porth, B. Ripperda, F. Bacchini, L. Sironi, and R. Keppens, *The Astrophysical Journal*, vol. 950, p. 169, Jun. 2023.

[55] S. M. Ressler, A. Tchekhovskoy, E. Quataert, M. Chandra, and C. F. Gammie, *MNRAS*, vol. 454, pp. 1848–1870, Dec. 2015.

[56] M. Chandra, C. F. Gammie, F. Foucart, and E. Quataert, en, *ApJ*, vol. 810, no. 2, p. 162, Sep. 2015.

[57] E. R. Most, J. Noronha, and A. A. Philippov, *MNRAS*, vol. 514, pp. 4989–5003, Aug. 2022.

[58] C. Palenzuela, L. Lehner, O. Reula, and L. Rezzolla, *MNRAS*, vol. 394, pp. 1727–1740, Apr. 2009.

[59] Q. Qian, C. Fendt, S. Noble, and M. Bugli, *ApJ*, vol. 834, p. 29, Jan. 2017.

[60] B. Ripperda, F. Bacchini, and A. A. Philippov, *ApJ*, vol. 900, p. 100, Sep. 2020.

[61] N. Bucciantini and L. Del Zanna, en, *MNRAS*, vol. 428, no. 1, pp. 71–85, Jan. 2013.

[62] L. Del Zanna and N. Bucciantini, *MNRAS*, vol. 479, no. 1, pp. 657–666, Sep. 2018.

[63] G. Mattia, L. Del Zanna, M. Bugli, A. Pavan, R. Ciolfi, G. Bodo, and A. Mignone, *A&A*, vol. 679, 2023.

[64] L. Del Zanna, E. Papini, S. Landi, M. Bugli, and N. Bucciantini, *MNRAS*, vol. 460, pp. 3753–3765, Aug. 2016.

[65] L. Pareschi and G. Russo, *JSC*, vol. 25, no. 1-2, p. 129, 2005.

[66] L. Del Zanna, N. Tomei, K. Franceschetti, M. Bugli, and N. Bucciantini, *Fluids*, vol. 7, p. 87, Feb. 2022.

[67] M. Bugli, L. Del Zanna, and N. Bucciantini, *MNRAS*, vol. 440, pp. L41–L45, May 2014.

[68] N. Tomei, L. Del Zanna, M. Bugli, and N. Bucciantini, *Universe*, vol. 7, p. 259, Jul. 2021.

[69] M. Liska, A. Tchekhovskoy, and E. Quataert, *MNRAS*, vol. 494, pp. 3656–3662, May 2020.

[70] V. Berta, A. Mignone, M. Bugli, and G. Mattia, *Journal of Computational Physics*, vol. 499, 112701, p. 112 701, Feb. 2024.

[71] P. Colella, M. Dorr, J. Hittinger, D. F. Martin, and P. McCorquodale, vol. 180, p. 012 010, Jul. 2009.

Journal of Physics: Conference Series **2742** (2024) 012003 doi:10.1088/1742-6596/2742/1/012003

pynucastro 2.1: an update on the development of a python library for nuclear astrophysics

Alexander Smith Clark[1], Eric T. Johnson[1], Zhi Chen[1], Kiran Eiden[3], Michael Zingale[1], Brendan Boyd[1,2], Parker T. Johnson[4], Luis Rangel DaCosta[5]

[1] Department of Physics and Astronomy, Stony Brook University, Stony Brook, NY 11794-3800, USA
[2] Institute for Advanced Computational Science, Stony Brook University, Stony Brook, NY 11794-5250, USA
[3] Department of Astronomy, University of California, Berkeley, Berkeley, CA 94720-3411, USA
[4] Department of Physics and Astrophysics, University of North Dakota, Grand Forks, ND 58202, USA
[5] Department of Materials Science and Engineering, University of California, Berkeley, Berkeley, CA 94720-1760, USA

E-mail: alexander.smithclark@stonybrook.edu

Abstract. pynucastro[1] is an open-source python library that provides visualization and analyze techniques to classify, construct, and evaluate nuclear reaction rates and networks. It provides tools that allow users to determine the importance of each rate in the network, based on a specified list of thermodynamic properties. Additionally, pynucastro can output a network in C++ or python for use in simulation codes, include the AMReX-Astrophysics simulation suite. We describe the changes in pynucastro since the last major release, including new capabilities that allow users to generate reduced networks and thermodynamic tables for conditions in nuclear statistical equilibrium.

1. Introduction

In stars, nuclear reactions are the primary source of energy, and their composition drives their evolution [1, 2]. Understanding stars means understanding how nuclear reactions work, and their impact on the whole network under different environments and conditions. Simulations of stars model their steady evolution and explosive deaths, requiring accurate and efficient nuclear reaction networks. In the current exascale era, GPU-enabled reaction networks are also required. pynucastro was developed to meet this need—it provides an open-source framework for exploring reaction rates and networks interactively in python and for exporting a reaction network to python or (GPU-enabled) C++ code for use in a simulation. pynucastro complements other reaction libraries in the field, for example, `SkyNet` [3] and `WinNet` [4], but its goals differ slightly: the emphasis in pynucastro is on interactive exploration of a network, to allow the user to create a network tailored to a particular science problem, and on the ability to export the network in a format that can be used in a multidimensional simulation code. This also allows

[1] https://github.com/pynucastro/pynucastro

us to easily explore the effect of the network size on a simulation, as done, for example, in a recent study of flames in X-ray bursts [5].

Recently pynucastro hit the 2.0 milestone [6], with the addition of weak rates, a nuclear statistical equilibrium solver, the ability to derive reverse rates through detailed balance, support for approximate rates, C++ output for the AMReX-Astrophysics suite of simulation codes, and some basic methods to help understand the importance of different rates in a network. The development has since continued, with a lot of new features leading to the 2.1 release, which we summarize here.

2. Overview of New Features

2.1. Expanded weak rates

The sensitivity of weak rates in core-collapse supernovae, novae, X-ray bursts, massive stars, and high-temperature environments plays a significant role in their nucleosynthesis and energy generation [7, 8]. In pynucastro 2.0, we included the weak-rate tabular data [9, 10] for nuclei $A = 17$ to 28. This was driven by our desire to simulate convective Urca [11]. From these tables, we extracted the β^--decay rate (β^-), electron-capture rate (e^-), gamma-energy rate (Γ_e), and the neutrino-energy rate loss (Γ_ν), reformatted in cgs units. We know that both weak reactions, β^--decay and electron-capture, produce the following changes in a nucleus (Z, A):

$$\begin{aligned} (Z, A) &\rightarrow (Z+1, A) + e^- + \nu_e, \\ (Z, A) + e^- &\rightarrow (Z-1, A) + \bar{\nu}_e, \end{aligned} \tag{1}$$

However, the positron-capture and the β^+-decay reactions produce the same effects on the nuclei (Z, A):

$$\begin{aligned} (Z, A) + e^+ &\rightarrow (Z+1, A) + \bar{\nu}_e, \\ (Z, A) &\rightarrow (Z-1, A) + e^+ + \nu_e, \end{aligned} \tag{2}$$

respectively. These additional channels become important at hotter and denser conditions than seen in the Urca simulations, and some rate sources tabulate them in addition to their electron counterparts. In pynucastro 2.1, we combine the bare β^--decay and the positron-capture rates in a single rate, and similarly, the bare electron-capture and the β^+-decay rates. This means that each sequence $(Z, A) \rightarrow (Z+1, A)$ and $(Z, A) \rightarrow (Z-1, A)$ are represented by a single effective link in a network.

In pynucastro 2.1, we have improved our weak-rate tabular data availability by including the $A = 45$ to 65 rates in [12–14] to support simulations of massive stars. We also reformatted our tables to consider the original logged tabular entries instead, in cgs units, and in our effective rate convention: $\log(\beta^-)$, $\log(e^-)$, $\log(\Gamma_e)$, and $\log(\Gamma_\nu)$ to each tabulated pair $(\log \rho Y_e, \log T)$. A comparison between pynucastro 2.0 and pynucastro 2.1 weak rate nuclei availability is depicted in Fig 1. Our addition of rates is driven by our science goals, and more rates will be added as needed.

2.2. A new bilinear interpolation scheme for the tabular rates implementation

In `pynucastro` 2.0, the tabular rate evaluation in a `PythonNetwork` simply returned the closest tabular entry for the $(\rho Y_e, T)$ input pair, while the `AmrexAstroCxxNetwork` C++ network class performed a bi-linear interpolation over the tabular input pairs. Thus, in `pynucastro` 2.1, we implemented a bi-linear interpolation scheme in `PythonNetwork`, making it consistent with the same functionality in `AmrexAstroCxxNetwork`.

In our new release, we reformatted our tabular entries in terms of $\log_{10}(\rho Y_e)$ and $\log_{10}(T)$ and implemented a bi-linear interpolation scheme for both network classes. For the `PythonNetwork`

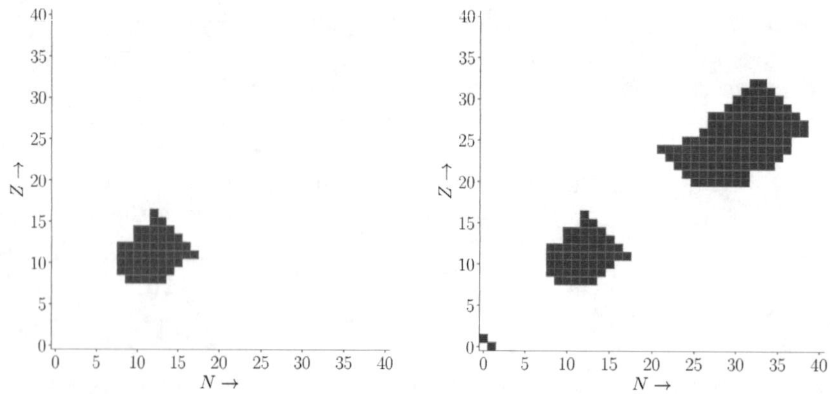

(a) pynucastro 2.0 weak-rate nuclei availability

(b) pynucastro 2.1 weak-rate nuclei availability

Figure 1: (a) *Left*: the original weak rate group of nuclei supported pynucastro 2.0. (b) *Right*: the original weak rate group of nuclei plus the contribution from the [12–14] rates.

class, we implemented a bi-linear interpolation scheme stored in the `TableInterpolator` class that can be used both for interactive evaluation of the rates and when the network is written out to a python module for use elsewhere. An example of the new interpolation interface is shown in the following script:

```
def na23__ne23(rate_eval, T, rhoY):
    # na23 --> ne23
    na23__ne23_interpolator = TableInterpolator(*na23__ne23_info)
    r = na23__ne23_interpolator.interpolate(np.log10(rhoY), np.log10(T),
                                            TableIndex.RATE.value)
    rate_eval.na23__ne23 = 10.0**r
```

Note, that we construct first the `TableInterpolator` object for the `na23__ne23` rate, and then we evaluate the rate. A comparison that illustrates how sensitive the inclusion of weak-rates may affect the behavior of a simulation can be found in [15].

2.3. *A Simpler C++ Network*

The original C++ network support in pynucastro is through `AmrexAstroCxxNetwork`, and utilizes the data structures in the AMReX library [16] to provide the ability to offload the reaction network onto GPUs. This is compatible with any simulation code that uses AMReX, like Castro [17] or MAESTROeX [18].

pynucastro now supports a pure C++ network, through the `SimpleCxxNetwork` class, that can output the RHS and Jacobian of the network. This was developed to allow other simulation codes to use pynucastro networks. At the moment, electron screening [19–22] is not included in this implementation (but it is easy to add a hook for a user-supplied screening function) and tabulated rates are not supported. For codes written in C, the C++ interfaces can be accessed via an `extern "C" {}` interface. This network output will be expanded as community demand dictates and we hope this new interface will help spur the adoption of pynucastro-generated networks by other simulation codes.

2.4. *Support for custom rates*

The base `Rate` class in pynucastro was improved to make it easier to add a custom rate. This can allow a user to explore variations or approximations to the standard rates in the pynucastro library. For example, suppose we want to add a simple powerlaw rate:

$$r = r_0 \left(\frac{T}{T_0}\right)^{\nu} \tag{3}$$

where r_0 is the rate evaluated at the temperature T_0, on which, the exponent ν is evaluated. This can be accomplished by creating a new class that inherits from `Rate`, stores the parameters r_0 and T_0, and calls the super-class initialization, e.g.

```
class MyRate(pyna.Rate):
    def __init__(self, reactants=None, products=None,
                 r0=1.0, T0=1.0, nu=0):

        # we set the chapter to custom so the network knows how to deal with it
        self.chapter = "custom"

        # call the Rate init to do the remaining initialization
        super().__init__(reactants=reactants, products=products)

        self.r0 = r0
        self.T0 = T0
        self.nu = nu
```

This sets up all of the logic needed to understand how to use our new rate, `MyRate`, in a network. Then we only need to add a few additional methods. If we only want to use the rate in python, then we just need to write an `eval()` method that computes the rate given temperature, such as:

```
    def eval(self, T, rhoY=None):
        return self.r0 * (T / self.T0)**self.nu
```

If we want to export the network to python or C++ code, then we need to add one additional method that writes the function string. This custom rate can then be used in any function in pynucastro that takes a `Rate` object.

2.5. *The DRGEP modified reduction method implementation*

In pynucastro 2.0, we provided a simple `find_unimportant_rates` method to help reduce the size of a network. In the latest release, a more extensive reduction method framework is included.

The idea of selecting the most important nodes of a graph is also a problem with important applications in combustion and chemical kinetics [23–25]. The general idea behind these methods starts with a large-sized directed graph $\mathcal{G}$ representing the network and a user-defined selection of fixed target nodes T. For each pair of nodes (A, B) of the graph $\mathcal{G}$, one can define the following matrix coefficients $r_{A,B}$:

$$r_{A,B} = \frac{\left|\sum_{i=1}^{N_{\max}} \nu_{A,i}\omega_i\delta_{i,B}\right|}{\max(P_A, C_A)}. \tag{4}$$

Here, P_A are all the rates on which A is a product, similarly C_A are all the rates on which A is a reactant, $\nu_{A,i}$ is the stoichiometric coefficient of A in the reaction i, ω_i is the i-th rate evaluation, and $\delta_{i,B}$ is the Kronecker delta object. From each target nucleus T, a weight R_{TB} is defined for each connected nuclei B to the target T as follows:

$$R_{TB} = \max_{\text{all path p}} \left(\prod_{i=1}^{n-1} r_{i,i+1,p}\right) \tag{5}$$

where the index $1 \le i \le n$ represents all the nuclei between T and B for a given path p, including both. The weights R_{TB} can be calculated using a modified Dijkstra's algorithm [25–27]. From

(5), all of the paths from a nucleus B to a given target T will have weights less than R_{TB}. For each target T, we can flag whether a nucleus B is considered unimportant by comparing to a user-specified tolerance choice ϵ_T:

$$R_{TB} < \epsilon_T. \tag{6}$$

We can then remove nuclei with R_{TB} less than ϵ_T for all possible targets. The reduced network will be a sub-graph $\mathcal{G}' \subset \mathcal{G}$ that will contain at least T nodes:

$$\mathcal{G} \to (\mathcal{G}', T) \tag{7}$$

Following the particular case of the DRGEP modified method described in [25–27], we have implemented the DRGEP modified reduction method in pynucastro. A direct application of this method is to identify all of the connections that are important for a particular subset of nuclei from a known reaction network set of nuclei. By setting each nucleus from the known network as a target, we can explore which rates are significant from all the currently supported rate libraries, up to an endpoint nucleus given by the user. For example, let us explore how several nuclei may be connected from the known CNO cycle to all the species up to ^{56}Ni. The importance of ^{56}Ni endpoint species is explained, for example in [28]. We start by creating a pool of rates with all the currently supported REACLIB rates up to ^{56}Ni with a fixed solar composition. This two-step process is achieved by the `load_network()` method with a fixed network `endpoint` in ^{56}Ni. This will create a `PythonNetwork` containing all the nuclei with A and Z less than the endpoint and all the rates connecting them. Then, by selecting 25 representative nuclei of the original hydrogen-CNO burning reaction network, we construct the `targets` list. Finally, each link that is ultimately connected to each target, will be evaluated by a weight-constraint function. The `tols` list contains a tolerance that will determine whether each link should be severed or not based on the previous evaluation. From the original pool network composed of a total of 6758 rates with 584 nuclei, we can drastically reduce it to 607 rates with 63 nuclei. Thus, from all the rates REACLIB rates with nuclei up to ^{56}Ni, we were able to extract the most important rates connected to the original network nuclei targets.

```
from pynucastro import Nucleus
from pynucastro.reduction import drgep
from pynucastro.reduction.generate_data import dataset
from pynucastro.reduction.load_network import load_network

net = load_network(endpoint=Nucleus('ni56'),
                   library_name='reaclib_default2_20220329')

b_rho = (1e2, 1e6)  # density (g/cm^3)
b_T = (1.0e7, 3.0e8)  # temperature (K)

data = list(dataset(net, n=10, permute=False, b_rho=b_rho, b_T=b_T, b_Z=None))

targets = ["p", "d", "he3", "he4", "li7", "be7", "be8", "b8", "c12",
           "n13", "n14", "n15", "o14", "o15", "o16", "o17", "o18",
           "f17", "f18", "f19", "f20", "ne18", "ne19", "ne20", "ne21"]

targets = [Nucleus(n) for n in targets]

# The tols, for each target, indicate if the nuclei connected to it is
# important or not

tols = [0.3, 0.3, 0.3, 0.3, 0.3, 0.3, 0.3, 0.3, 0.3, 0.3, 0.3, 0.3, 0.3,
        0.3, 0.3, 0.3, 0.3, 0.3, 0.3, 0.3, 0.3, 0.5, 0.5, 0.5, 0.5]

reduced_net = drgep(net=net, conds=data, targets=targets, tols=tols,
                    returnobj='net', use_mpi=False, use_numpy=True)
```

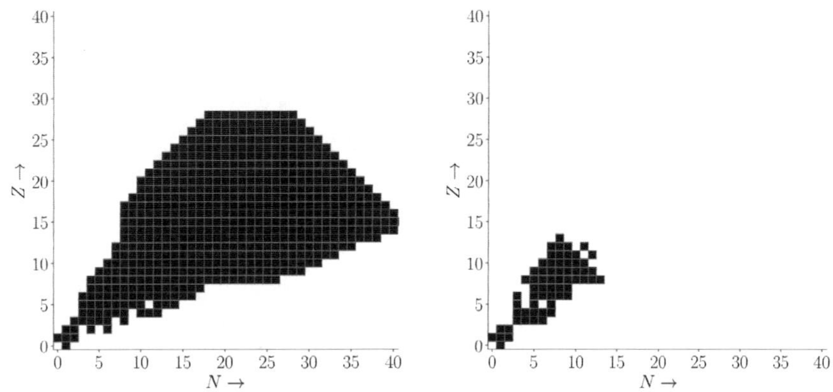

(a) The available pool of rate nuclei (b) The reduced network nuclei

Figure 2: (a) *Left*: the available rate nuclei up to ^{56}Ni, extracted from REACLIB. (b) *Right*: the reduced network availability, by selecting a network up to ^{21}Ne.

An exploration of alternative reduction method implementations, energy-based weight-constraints functions, and the analysis of their implications in large nuclear reaction networks should be studied further. The exploration of reduction network algorithms in chemistry has been highlighted in [24, 29], and now with its implementation in the context of nuclear reaction networks, new applicability studies are feasible, opening the possibility of further explorations. These algorithms may prove to be useful in highlighting the most important species and rates from an extensive nuclear reaction network, improving our understanding of the group of reactions that dominate our problem and the efficiency of our calculations.

2.6. Interfacing with Julia

The SciPy library [30] provides an ODE integration function `solve_ivp()` that provides many different ODE solvers, including BDF solvers, tailored for stiff systems. This is the main way we suggest how a `PythonNetwork` network should be integrated. However, many ODE methods are not implemented in SciPy, some of which can be more efficient for our problems.

The library `diffeqpy` makes possible the communication to `DifferentialEquations.jl` in Julia runtime. The Julia library `DifferentialEquations.jl` [31] expands the available ODE methods that we may consider to integrate a particular reaction network. Now, in pynucastro 2.1, we can integrate a `PythonNetwork` using the Julia solvers. The performance relies heavily on Numba acceleration [32], opening the door to further exploration of different ODE solvers. This new interface makes it straightforward to test a variety of solvers on a particular network of interest. Below is an example using a simple CNO nuclear reaction network with an initial composition of a standard solar mix.

```
import numba
import numpy as np
from diffeqpy import de
import cno_test_integrate as cno # importing network created

rho = 150. #g/cm^3
T = 1.5e7  #K
screen_func = None

Y0 = np.zeros((cno.nnuc), dtype=np.float64)
Y0[cno.jp] = 0.7 # Solar mix of mass fractions
```

```
Y0[cno.jhe4] = 0.28
Y0[cno.jc12] = 0.02

Y0[:] = Y0[:]/cno.A[:] # Convert to molar fractions

tspan = (0.0, 1.e20)
p = np.array([rho, T]) # initial parameters
alg = de.FBDF() # Choosing Julia ODE solver

def f(u, p, t):
    """u = array of unknowns, p = array of parameters, t = time"""
    rho, T = p
    Y = u
    return cno.rhs_eq(t, Y, rho, T)

def f_jac(u, p, t):
    rho, T = p
    Y = u
    return cno.jacobian_eq(t, Y, rho, T)

# create Numba compiled versions of the functions
numba_f = numba.njit(f)
numba_jac = numba.njit(f_jac)
ff = de.ODEFunction(numba_f, jac=numba_jac)

prob = de.ODEProblem(ff, Y0, tspan, p)
sol = de.solve(prob, alg, abstol=1.e-9, reltol=1.e-9)
```

This code integrates the system over the specified time using the `FBDF` solver in the Julia package. Note that the Julia interface requires that the righthand side and Jacobian functions should be combined into a single interface via `ODEFunction`. After integration, `sol.stats` can provide a variety of different statistics including the number of function evaluations iterations, and linear algebra operations it performs. These can be useful in determining the performance of the solver.

2.7. NSE table construction

In pynucastro 2.0 we introduced a robust nuclear statistical equilibrium (NSE) solver [33]. Now, in version 2.1, the NSE solver has been split off into its own network type, `NSENetwork`, allowing the addition of new features that support creating NSE tables for use in simulation codes. To support this capability, a new method `Composition.bin_as()` has been added to bin a large `Composition`, for example, representing the NSE state from 100 nuclei, down to a smaller `Composition` that will be what is carried on the grid in a simulation code. We've also added the ability to supply an optional rate filter function when evaluating the RHS of a network. For the NSE evolution, this can be used to only compute the evolution due to weak interactions, which cause an evolution in the electron molar fraction Y_e, further evolving the NSE state. Likewise, the neutrino loss from weak rates can be returned separately when evaluating the energy production in a network. Together these new methods allow us to easily create an NSE table like that in [34].

3. Summary and Future Work

This new release of pynucastro added many features: an expanded weak rate library with a new interpolation scheme over the tabular rates, a new template to cover bare C++ support, the ability to construct a custom rate, network reduction capabilities, a new Julia integration interface, and the capability of creating an NSE table from the network. All of these newly introduced features are driven by science applications that the developers are working on, including studies of convective burning in massive stars leading up to core-collapse.

The next major developments will likely be the addition of the remaining pieces needed to support a self-heating burn completely in python. This includes the addition of an equation of state (to get the temperature evolution) and neutrino cooling sources. Pull requests with initial implementations of each are under review, with the equation of state based on the Helmholtz equation of state [35, 36]. Once the previous step is completed, we may explore the addition of other equations of state, for example, [37].

Finally, all of the major machine learning libraries have python interfaces and much AI research relies on these libraries. By providing a clean, extensible python interface to nuclear reaction rates and networks, we enable the use of pynucastro for machine-learning applications. In particular, since reaction network integration can be one of the most time-consuming parts of a simulation, using neural networks to augment or replace the ODE integration is a promising area [38].

Acknowledgments
The work at Stony Brook was supported by DOE/Office of Nuclear Physics grant DE-FG02-87ER40317. LRD and KE were supported by the U.S. Department of Energy, Office of Science, Office of Advanced Scientific Computing Research, Department of Energy Computational Science Graduate Fellowship under Award Number DE-SC0021110. This research has made use of NASA's Astrophysics Data System Bibliographic Services.

References

[1] Eddington A S 1920 *The Observatory* **43** 341–358

[2] Bethe H A 1939 *Phys. Rev.* **55**(5) 434–456 URL https://link.aps.org/doi/10.1103/PhysRev.55.434

[3] Lippuner J and Roberts L F 2017 *ApJS* **233** 18 (*Preprint* 1706.06198)

[4] Reichert M, Winteler C, Korobkin O, Arcones A, Bliss J, Eichler M, Frischknecht U, Fröhlich C, Hirschi R, Jacobi M, Kuske J, Martínez-Pinedo G, Martin D, Mocelj D, Rauscher T and Thielemann F K 2023 *arXiv e-prints* arXiv:2305.07048 (*Preprint* 2305.07048)

[5] Chen Z, Zingale M and Eiden K 2023 *The Astrophysical Journal* **955** 128 URL https://dx.doi.org/10.3847/1538-4357/acec72

[6] Smith A I, Johnson E T, Chen Z, Eiden K, Willcox D E, Boyd B, Cao L, DeGrendele C J and Zingale M 2023 *The Astrophysical Journal* **947** 65 URL https://dx.doi.org/10.3847/1538-4357/acbaff

[7] Sullivan C, O'Connor E, Zegers R G T, Grubb T and Austin S M 2015 *The Astrophysical Journal* **816** 44 URL https://dx.doi.org/10.3847/0004-637X/816/1/44

[8] Fujimoto M Y, Hanawa T and Miyaji S 1981 *ApJ* **247** 267 ISSN 0004-637X, 1538-4357 URL https://doi.org/10.1086/159034

[9] Toki H, Suzuki T, Nomoto K, Jones S and Hirschi R 2013 *Phys. Rev. C* **88**(1) 015806 URL https://link.aps.org/doi/10.1103/PhysRevC.88.015806

[10] Suzuki T, Toki H and Nomoto K 2016 *The Astrophysical Journal* **817** 163 URL https://dx.doi.org/10.3847/0004-637X/817/2/163

[11] Calder A C, Willcox D E, DeGrendele C J, Shangase D, Zingale M and Townsley D M 2019 *Journal of Physics: Conference Series* **1225** 012002 URL https://dx.doi.org/10.1088/1742-6596/1225/1/012002

[12] Caurier E, Langanke K, Martínez-Pinedo G and Nowacki F 1999 *Nuclear Physics A* **653** 439–452 ISSN 0375-9474 URL https://www.sciencedirect.com/science/article/pii/S0375947499002407

[13] Langanke K and Martínez-Pinedo G 2000 *Nuclear Physics A* **673** 481–508 ISSN 0375-9474 URL https://www.sciencedirect.com/science/article/pii/S0375947400001317

[14] Langanke K and Martínez'-Pinedo G 2001 *Atomic Data and Nuclear Data Tables* **79** 1–46 ISSN 0092-640X URL https://www.sciencedirect.com/science/article/pii/S0092640X01908654

[15] Chen Z, Zingale M and Eiden K 2023 *ApJ* **955** 128 (*Preprint* 2306.16320)

[16] Zhang W, Almgren A, Beckner V, Bell J, Blaschke J, Chan C, Day M, Friesen B, Gott K, Graves D, Katz M, Myers A, Nguyen T, Nonaka A, Rosso M, Williams S and Zingale M 2019 *JOSS* **4** 1370 ISSN 2475-9066 URL https://doi.org/10.21105/joss.01370

[17] Almgren A, Sazo M B, Bell J, Harpole A, Katz M, Sexton J, Willcox D, Zhang W and Zingale M 2020 *Journal of Open Source Software* **5** 2513 URL https://doi.org/10.21105/joss.02513

[18] Fan D, Nonaka A, Almgren A, Willcox D, Harpole A and Zingale M 2019 *Journal of Open Source Software* **4** 1757 URL https://doi.org/10.21105/joss.01757

[19] Salpeter E E 1954 *Australian Journal of Physics* **7** 373

[20] Dewitt H E, Graboske H C and Cooper M S 1973 *ApJ* **181** 439–456

[21] Chugunov A I 2021 *Journal of Physics: Conference Series* **1787** 012047 URL https://dx.doi.org/10.1088/1742-6596/1787/1/012047

[22] Kushnir D, Waxman E and Chugunov A I 2019 *MNRAS* **486** 449–452 (*Preprint* 1805.08788)

[23] Lu T and Law C K 2005 *Proceedings of the Combustion Institute* **30** 1333–1341 ISSN 1540-7489 URL https://www.sciencedirect.com/science/article/pii/S0082078404001973

[24] Lu T and Law C K 2006 *Combustion and Flame* **144** 24–36 ISSN 0010-2180 URL https://www.sciencedirect.com/science/article/pii/S0010218005001720

[25] Pepiot-Desjardins P and Pitsch H 2008 *Combustion and Flame* **154** 67–81 ISSN 0010-2180 URL https://www.sciencedirect.com/science/article/pii/S0010218007003264

[26] Niemeyer K E and Sung C J 2011 *Combustion and Flame* **158** 1439–1443 ISSN 0010-2180 URL https://www.sciencedirect.com/science/article/pii/S0010218010003640

[27] Dijkstra E W 1959 *Numerische Mathematik* **1** 269–271 ISSN 0945-3245 URL https://doi.org/10.1007/BF01386390

[28] Alexakis A, Calder A C, Heger A, Brown E F, Dursi L J, Truran J W, Rosner R, Lamb D Q, Timmes F X, Fryxell B, Zingale M, Ricker P M and Olson K 2004 *The Astrophysical Journal* **602** 931 URL https://dx.doi.org/10.1086/381086

[29] Lu T and Law C K 2006 *Combustion and Flame* **146** 472–483 ISSN 0010-2180 URL https://www.sciencedirect.com/science/article/pii/S0010218006001398

[30] Virtanen P, Gommers R, Oliphant T E, Haberland M, Reddy T, Cournapeau D, Burovski E, Peterson P, Weckesser W, Bright J, van der Walt S J, Brett M, Wilson J, Millman K J, Mayorov N, Nelson A R J, Jones E, Kern R, Larson E, Carey C J, Polat İ, Feng Y, Moore E W, VanderPlas J, Laxalde D, Perktold J, Cimrman R, Henriksen I, Quintero E A, Harris C R, Archibald A M, Ribeiro A H, Pedregosa F, van Mulbregt P and SciPy 10 Contributors 2020 *Nature Methods* **17** 261–272

[31] Rackauckas C and Nie Q 2017 *Journal of Open Research Software* **5** 15

[32] Lam S K, Pitrou A and Seibert S 2015 *Proceedings of the Second Workshop on the LLVM Compiler Infrastructure in HPC* pp 1–6

[33] Smith A I, Johnson E T, Chen Z, Eiden K, Willcox D E, Boyd B, Cao L, DeGrendele C J and Zingale M 2023 *The Astrophysical Journal* **947** 65 URL https://dx.doi.org/10.3847/1538-4357/acbaff

[34] Seitenzahl I R, Townsley D M, Peng F and Truran J W 2009 *Atomic Data and Nuclear Data Tables* **95** 96–114

[35] Timmes F X and Arnett D 1999 *ApJS* **125** 277–294

[36] Timmes F X and Swesty F D 2000 *ASTROPHYS J SUPPL S* **126** 501–516 ISSN 0067-0049, 1538-4365 source code obtained from http://cococubed.asu.edu/codes/eos.shtml/helmholtz.tbz URL https://doi.org/10.1086/313304

[37] Jermyn A S, Schwab J, Bauer E, Timmes F X and Potekhin A Y 2021 *ApJ* **913** 72 (*Preprint* 2104.00691)

[38] Fan D, Willcox D E, DeGrendele C, Zingale M and Nonaka A 2022 *The Astrophysical Journal* **940** 134 URL https://dx.doi.org/10.3847/1538-4357/ac9a4b

Journal of Physics: Conference Series **2742** (2024) 012004 doi:10.1088/1742-6596/2742/1/012004

Exponential amplification of the magnetic field in the primordial star-forming cloud

Shingo Hirano[1,2], Masahiro N. Machida[3,4] and Shantanu Basu[4]

Department of Astronomy, School of Science, University of Tokyo, Tokyo 113-0033, Japan
Department of Applied Physics, Faculty of Engineering, Kanagawa University, Kanagawa 221-0802, Japan
Department of Earth and Planetary Sciences, Faculty of Science, Kyushu University, Fukuoka 819-0395, Japan
Department of Physics and Astronomy, University of Western Ontario, London, ON N6A 3K7, Canada

E-mail: shingo-hirano@kanagawa-u.ac.jp

Abstract. In the study of the initial mass function associated with the first generation of stars, known as Population III (Pop III) stars, a fundamental yet unresolved question pertains to the ultimate destiny of the secondary protostars emerging within the accretion disk – specifically, their likelihood of either merging or persisting as distinct entities. Our research concentrates on the magnetic influences affecting the genesis of these first stars under the conditions set by the cosmological initial magnetic field strength. We employ ideal magnetohydrodynamic simulations, utilizing a stiff equation-of-state (EOS) model, to accurately depict the magnetic field structure interconnecting these protostars. We observe that the magnetic field experiences rapid intensification due to the gas near the protostar completing multiple tens of orbital rotations in the initial decade following the formation of the protostar. Concurrently, as mass accretion continues, the region influenced by the significant magnetic field expands outward. This process of magnetic braking effectively curtails the disk fragmentation that would typically occur without a magnetic field. The resulting exponential augmentation of the magnetic field is posited to facilitate the formation of supermassive first stars.

1. Introduction

In contemporary cosmological research, a paramount and unresolved issue pertains to elucidating the formation mechanisms of the initial stellar generation, designated as Population III (Pop III) stars. To date, empirical observation of these primordial stars still needs to be explored. However, advancements in theoretical astrophysics, most notably the employment of progressively sophisticated numerical simulations, have substantially contributed to our understanding of the intrinsic characteristics of these earliest stars [1].

In the theoretical framework, a crucial yet unresolved question centers on the influence of magnetic phenomena, such as magnetic braking, on this process. Prior investigations, premised on the hypothesis of substantial magnetic fields within the primordial star-forming gas clouds, have highlighted several consequential outcomes: the retardation of gas contraction within the dark matter (DM) minihalo, thereby impacting the timing of initial star formation [2]; the inhibition of disk fragmentation owing to effective angular momentum transmission via magnetic fields [3, 4]; and the diminution of protostellar rotational velocity, which in turn may

Journal of Physics: Conference Series **2742** (2024) 012004 doi:10.1088/1742-6596/2742/1/012004

dictate the ultimate fate of Population III (Pop III) stars [5]. Nevertheless, it is recognized that the universe's primordial magnetic field, measured at a mere 10^{-18} G [6], is significantly weaker compared to the magnetic fields observed in contemporary star-forming regions, which are approximately 10^{-6} G. Consequently, the amplification of the magnetic field through flux freezing during the cloud collapse phase, described by $B \propto n^{2/3}$, is deemed inadequate to substantially impact the formation of the first stars.

In the context of first star formation, preceding investigations have primarily focused on the intensification of the magnetic field during the collapse phase [7, 8]. However, the potential augmentation of this field throughout the subsequent accretion phase and its consequential impact on the growth of protostars still need to be explored.

To address this gap, we conducted three-dimensional ideal magnetohydrodynamic (MHD) simulations centered on the early stages of primordial star formation and employed the stiff-equation-of-state (EOS) methodology. Following the inception of the first Population III (Pop III) protostar, these simulations revealed that a period of exponential magnetic field amplification occurs in the immediate vicinity of the protostar, notably within the initial three years. This amplified field region effectively inhibits disk fragmentation, thereby facilitating the formation of massive stars. Additionally, our study extends to exploring magnetic field amplification in metal-free and metal-enriched atomic-cooling halos (ACHs), environments conducive to forming intermediate-mass black holes (IMBHs). Our findings validate a consistent mechanism of magnetic field amplification during the accretion phase associated with the formation of direct-collapse black holes (DCBHs) in these ACHs. We deduce that exponential magnetic field amplification is a viable process in metal-enriched ACHs, which moderates the criteria necessary for DCBH formation.

2. Numerical methodology

In our research, we undertake three-dimensional (3D) ideal magnetohydrodynamic (MHD) simulations to probe the star formation process in the early universe. To accurately depict the thermal progression within the star-forming nebula, we incorporate a barotropic equation-of-state (EOS) table, meticulously derived from chemical reaction computations [9]. As an alternative to the conventional sink particle technique, our approach utilizes a stiff EOS framework to authentically represent the magnetic field structure linked to regions of dense gas. Figure 1 delineates the derived EOS tables, showcasing models of H2-cooling with varying threshold densities at $n\mathrm{th} = 10^{16}\,\mathrm{cm}^{-3}$ and $10^{19}\,\mathrm{cm}^{-3}$ (left), alongside H-cooling models differing in metallicity levels, with $Z/Z_{\odot} = 0$, 10^{-5}, and 10^{-4} (right).

This paper elucidates selected model outcomes, initially detailed in the foundational paper. For an exhaustive discourse on the calculation methodology and a comprehensive comparison of the results across all models, please refer to the original publications ([11] for H_2-cooling models and [12, 13] for H-cooling models).

3. Results

3.1. First star formation in the H_2-cooling gas cloud [11]

We impose a uniform magnetic field B_0 in alignment with the initial cloud's rotational axis within our computational domain. Our analysis focuses on how varying magnitudes of $B_0 = 0$, 10^{-20}, 10^{-15}, and 10^{-10} G (labelled as B00, B20, B15, and B10) influence the formation of the first stars. The B20 model, with $B_0 = 10^{-20}$ G, is chosen as the fiducial model in this study, as it is below the cosmological magnetic field estimation of approximately 10^{-18} G. Demonstrating the impact of magnetic fields on forming the first stars in this model would substantiate that magnetic fields invariably influence these primordial stars. We conduct simulations across four models with different threshold densities, continuing until $t_{\mathrm{ps}} = 100$ yr for $n_{\mathrm{th}} = 10^{19}\,\mathrm{cm}^{-3}$ and 1000 yr for $n_{\mathrm{th}} = 10^{16}\,\mathrm{cm}^{-3}$.

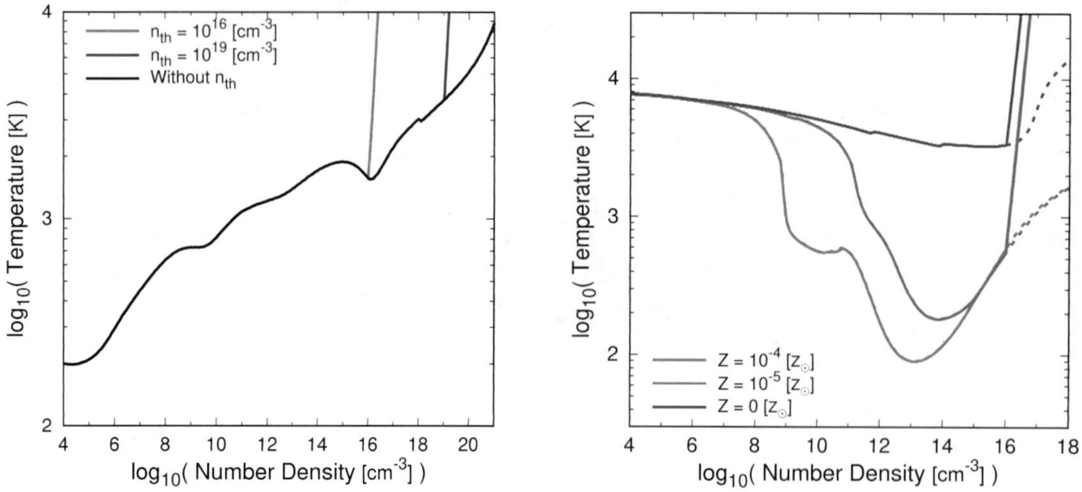

Figure 1. Thermal evolution models of star-forming clouds as a function of the gas number density. (Left panel) H_2-cooling models with zero-metallicity. In the left panel, we explore H2-cooling models under the condition of zero-metallicity. The fundamental theoretical model, represented by the black line, is derived from comprehensive chemical reaction calculations [9, 10]. The colored lines in this figure, integral to our study, represent adaptations of this base model. These adaptations are achieved through the application of the stiff equation of state (EOS) technique, with specific threshold densities of $n_{\rm th} = 10^{16}\,{\rm cm}^{-3}$ (in red) and $10^{19}\,{\rm cm}^{-3}$ (in blue), respectively. In the right panel, we examine H-cooling models that incorporate varying levels of metallicity, specifically $Z/Z_\odot = 0$, 10^{-5}, and 10^{-4}. The base models for these scenarios, illustrated by dashed lines, are theoretically formulated based on chemical reaction calculations, taking into account the influence of strong Lyman-Werner radiation at $J_{21} = 10^3$ [9, 10]. The solid lines, representing the models adopted in our study, are variations of these dashed line models, utilizing the stiff-EOS approach. These variations are characterized by a set threshold density of $n_{\rm th} = 10^{16}\,{\rm cm}^{-3}$.

Figure 2 contrasts the simulation results of the fiducial magnetized model (B20) with those of the non-magnetized model (B00) throughout the first 1000 years of the protostar accretion phase. Initially, at the inception of the first protostar (depicted in the left panels), the density structure surrounding the protostar is identical in both models. This similarity arises because the magnetic field strength near the protostar is too weak (at most pico-gauss, or 10^{-12} G) to influence the collapsing gas cloud. However, a notable divergence appears after a decade, as shown in the middle column of Figure 2. At this point, the magnetic field strength on the surface of the primary protostar (around $30\,R_\odot$) escalates to kilo-gauss levels. This strong "seed" magnetic field causes the magnetic field in a region with a 10 au radius surrounding the protostar to amplify significantly. Within this region of intense magnetic field strength, the density and velocity structures of the accreting gas are markedly influenced by the magnetic field. After 1000 years, as indicated in the right column of Figure 2, the area affected by the amplified magnetic field extends to a radius of approximately 500 au. In this expanded region, multiple protostars in the fiducial model have disappeared in the unmagnetized model. The emergence of a global spiral structure within the gas, inside the area of the amplified magnetic field, is attributed to the angular momentum transport by magnetic braking. This process allows for more efficient accretion.

Journal of Physics: Conference Series **2742** (2024) 012004 doi:10.1088/1742-6596/2742/1/012004

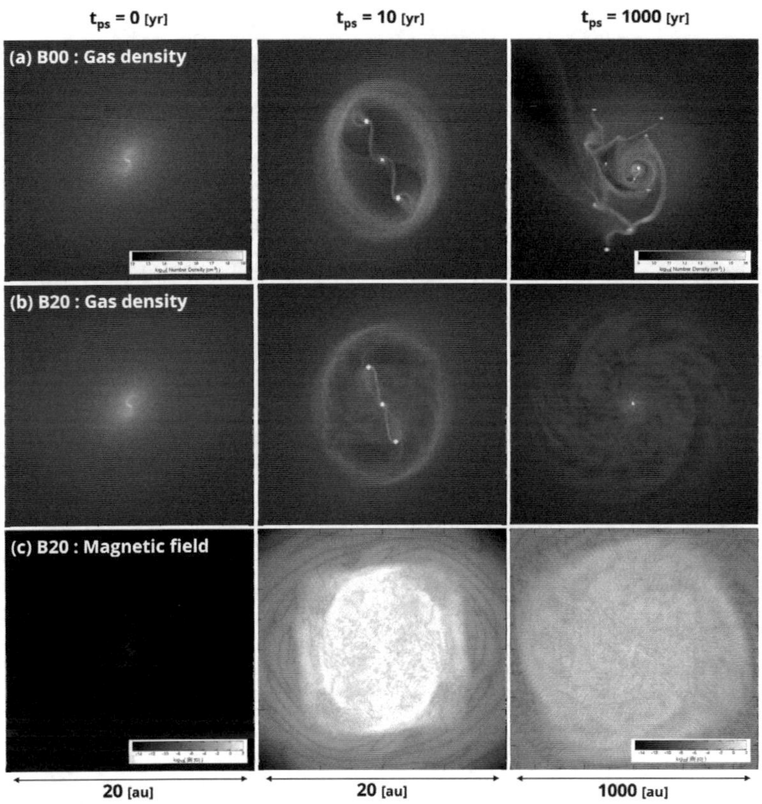

Figure 2. Cross-sectional view on the $z = 0$ plane around the most massive protostar at $t_{\rm ps} = 0$, 10, and 1000 yr after the first protostar formation from left to right. The uppermost panel delineates the gas number density in the B00 model, the central panel similarly portrays the gas number density in the B20 model, and the lowermost panel illustrates the absolute magnetic field strength in the B20 model. The spatial dimensions of the depicted regions are 20 astronomical units (au) for the left and middle panels, in contrast to a significantly broader scope of 1000 au for the panels on the right.

The remarkable escalation of the magnetic field strength from 10^{-12} G to 10^3 G near protostars originates from their rapid rotational motion, which effectively winds up magnetic fields. Within the first three years following protostar formation, the magnetic field strength near the protostar undergoes significant amplification, as depicted in Figure 3(a). In the area less than or equal to 10 au, the number of orbital rotations surpasses one at $t_{\rm ps} = 0$ yr and escalates to several dozen by $t_{\rm ps} = 3$ yr (Figure 3b). Consequently, the region affected by magnetic field amplification progressively expands, with its boundary aligning with the radius where the orbital rotation rate exceeds one ($N_{\rm rot} = 1$). This observation indicates that gravitational energy is effectively transformed into magnetic energy via kinetic (or rotational) energy following the formation of the first protostar. By the end of the accretion phase of the first stars (around 10^5 yr, as indicated by the dotted line), the region influenced by the amplified magnetic field could extend to approximately 10^4 au ~ 0.05 pc, encompassing a total gas mass of about $500\,M_{\odot}$ in this model.

Regarding the influence of this exponentially amplifying magnetic field on the formation process of the first stars, the amplified magnetic field prevents fragmentation of the gravitationally unstable accretion disk, forming a single protostar at the cloud's center. However, the stellar masses remain consistent across both magnetized and unmagnetized models. Invariably, any fragments that form near the protostar will likely merge promptly. The rotation velocities, a critical parameter in stellar evolution theory, maintain a nearly constant rate at approximately 0.05 times the Keplerian velocity, irrespective of the model or evolutionary timeline. Given the low degree of rotation, employing a non-rotational model for stellar evolution seems plausible.

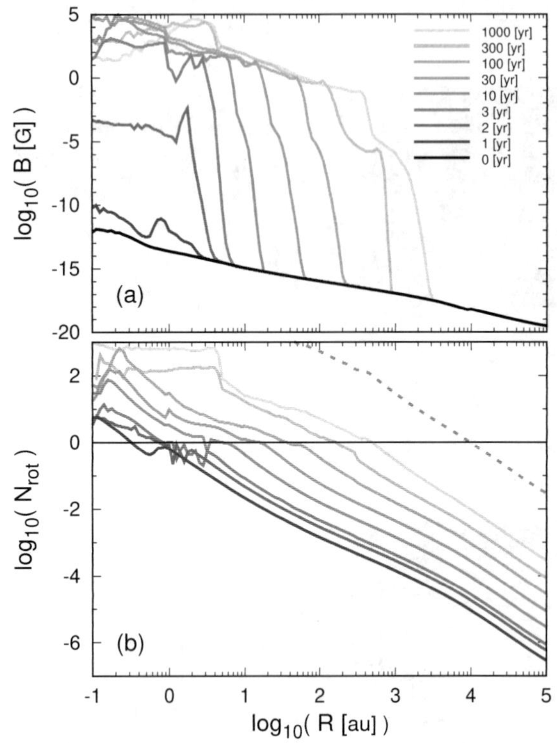

Figure 3. Radial profiles for model B20 are shown at different times after the first protostar formation: $t_{\rm ps}$ = 0, 1, 2, 3, 10, 30, 100, 300, and 1000 yr. The panels include (a) the magnetic field strength and (b) the number of orbital rotations at each time point $t_{\rm ps}$, expressed as $N_{\rm rot} = (v_{\rm rot}t_{\rm ps})/(2\pi R)$. Notably, the line for $t_{\rm ps} = 0\,{\rm yr}$ is not included in panel (b) since $N_{\rm rot} = 0$ equals zero for every R at $t_{\rm ps} = 0\,{\rm yr}$. Panel (b) also features a dotted line that estimates the expected $N_{\rm rot}$ at $t_{\rm ps} = 10^5\,{\rm yr}$, when the first star ends its accretion phase. This estimate uses the radial rotational velocity profile at $t_{\rm ps} = 1000\,{\rm yr}$. The horizontal lines in panel (b) indicate where $N_{\rm rot} = 1$.

3.2. Supermassive star formation in the H-cooling gas cloud [12, 13]

We consider two critical parameters for gas clouds: (1) metallicity and (2) initial magnetic field strength. We examine three distinct metallicities: $Z/Z_\odot = 0$, 10^{-5}, and 10^{-4}, which encompass the range for "supercompetitive accretion" ($Z/Z_\odot \leq 10^{-4}$). Additionally, five different initial magnetic field strengths are explored: $B_0/{\rm G} = 0$, 10^{-12}, 10^{-10}, 10^{-8}, and 10^{-6}. The minimum magnetic field strength of $B_0/{\rm G} = 10^{-12}$ at $n_{\rm H} = 10^4\,{\rm cm}^{-3}$ is extrapolated from the cosmological seed value (10^{-15} G at $n_{\rm H} = 1\,{\rm cm}^{-3}$ [14]) through flux freezing during cloud compression, which follows a power law $B \propto n_{\rm H}^{2/3}$. A uniform magnetic field, aligned with the initial cloud's rotation axis, is imposed in the computational domain.

Our analysis first focuses on how metallicity affects magnetic influences. As metallicity rises, gas temperature decreases, leading to a lower mass accretion rate in clouds with higher metallicity. This trend is evident in unmagnetized models, where the mass growth rate of the primary protostar diminishes with increasing metallicity. At the end of simulations at $t_{\rm ps} = 500\,{\rm yr}$, the masses of primary protostars in these models reach a few $10^3\,M_\odot$, aligning with previous unmagnetized simulations.

Incorporating magnetic effects, we find that the masses of primary protostars at $t_{\rm ps} = 500\,{\rm yr}$ in magnetized models are about 10 times larger than those in unmagnetized models across all metallicity levels examined. While the mass infall rate in distant regions is similar in magnetized and unmagnetized models, the infall rate near the center is higher in magnetized models, indicating that magnetic fields influence the mass accretion rate in these areas. In the early accretion phase ($t_{\rm ps} < 20\,{\rm yr}$), the magnetic field strength amplifies rapidly with decreasing metallicity. This early difference is mainly due to the "seed" magnetic field, amplified by protostellar spin, as lower metallicity leads to a higher accretion rate and fragmentation of the central high-density gas into multiple protostars. Eventually, all magnetized models experience significant magnetic field amplification, with angular momentum transport by the amplified field

Journal of Physics: Conference Series **2742** (2024) 012004 doi:10.1088/1742-6596/2742/1/012004

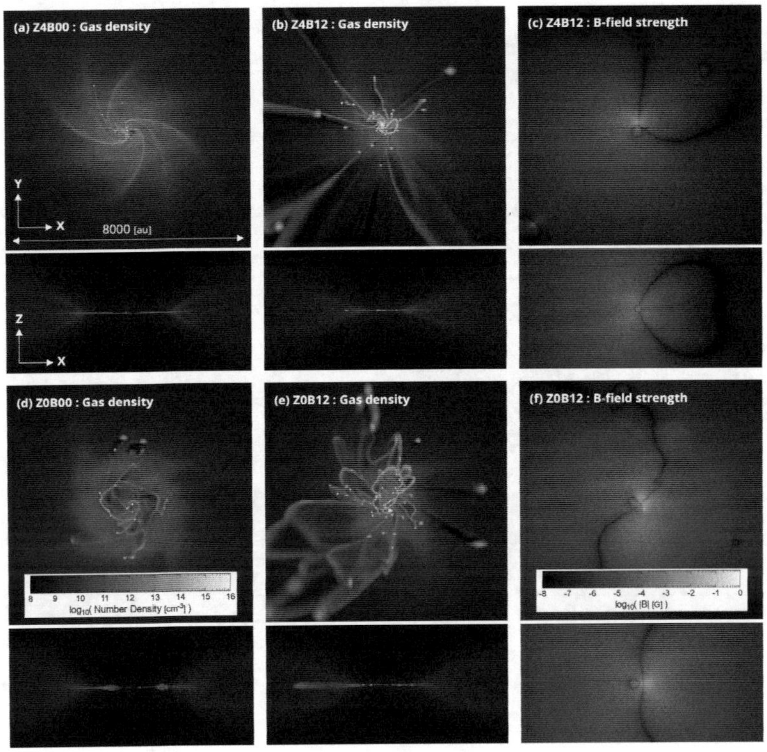

Figure 4. Cross-sectional views on the $z = 0$ and $y = 0$ planes surrounding the most massive protostar at $t_{\rm ps} = 500\,{\rm yr}$ following the formation of the first protostar. The panels include: (a, b, d, and e) showcasing the gas number density in models Z4B00, Z4B12, Z0B00, and Z0B12, and (c and f) displaying the absolute magnetic field strength in models Z4B12 and Z0B12.

enhancing gas accretion and protostellar mergers. After just 500 years, the primary protostars in these models have a mass with $\sim 10^4\,M_\odot$.

Figure 4 presents two-dimensional maps of density and magnetic field strength around the primary protostar at the end of the simulation ($t_{\rm ps} = 500\,{\rm yr}$). In the magnetized models, numerous protostars form during a fragmentation burst at $t_{\rm ps} = 400\,{\rm yr}$. These protostars tend to escape farther from the cloud center in magnetized models than unmagnetized ones due to more efficient N-body interactions, as magnetic effects lead to a more significant influx of protostars into the central region. We conclude that in magnetized models, the rapid accretion flow displaces protostars from the cloud's center more than in unmagnetized models.

Next, we focus on the impact of initial magnetic field strength, labeled as B_0. Figure 5 summarizes all models' B-n (magnetic field density) diagrams at a uniform time frame, $t_{\rm ps} = 100\,{\rm yr}$. This overview clearly shows the amplification of the magnetic field across all models. We observe that this amplification primarily occurs in two different density regions: a high-density area in the circumstellar disk ($n \geq 10^{13}\,{\rm cm}^{-3}$) and a lower-density region above the disk ($n \sim 10^{11}\,{\rm cm}^{-3}$).

4. Discussion

The magnetic field amplification post-protostar formation follows a nuanced progression, as illustrated in Figure 6:

1. Initially, the spinning motion of protostars boosts the weak magnetic field in their proximity.
2. Next, the amplification pace of the magnetic field surrounding the protostar heightens. This is in line with the induction equation, which suggests that the interplay between the velocity and the magnetic field itself dictates the magnetic field's temporal alteration.
3. Subsequently, the area influenced predominantly by rotational dynamics widens, thus escalating the magnetic field strength by the initial amplification process.

Journal of Physics: Conference Series **2742** (2024) 012004 doi:10.1088/1742-6596/2742/1/012004

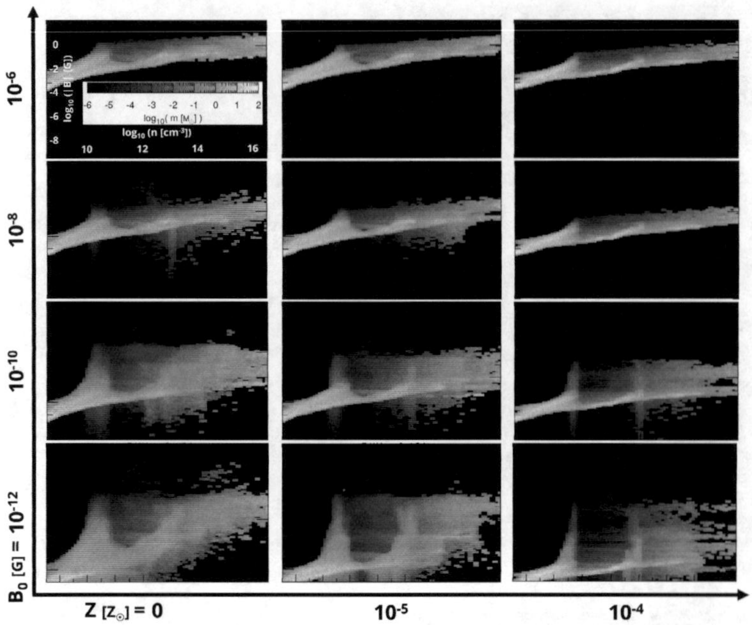

Figure 5. Phase diagrams of the absolute magnetic field strength for all models at t_{ps} = 100 yr after the first protostar formation.

However, MHD simulations employing the sink particle methodology fall short in simulating the primary amplification of the "seed" magnetic field near the protostar and its outward propagation. This shortfall is linked to the lack of interaction between the rotation within high-density regions and the magnetic field in these specific simulations.

In atomic hydrogen (H) cooling halos, the elevated accretion rate results in several fragments, boosting the magnetic field through their rotational activity. Our observations in molecular hydrogen (H_2) cooling halos indicate initial disk fragmentation, which rapidly subsides as these fragments integrate into the primary protostar. The orbital rotation around the protostar fosters magnetic field amplification without triggering further fragmentation. This phenomenon, exclusive to early Universe star formation, is absent in nearby star-forming areas, where magnetic fields peak before the protostar's accretion phase, underscoring the importance of considering strong magnetic field environments.

This magnetic field amplification mechanism plays a crucial role in preventing the fragmentation of the accretion disk. When a star-forming gas cloud exhibits substantial rotational motion, it forms a gravitationally unstable accretion disk that initially fragments. However, this fragmentation is swiftly curtailed by the augmented magnetic field. Conversely, in gas clouds with minimal rotation, magnetic field amplification is less pronounced, inhibiting the formation of a gravitationally unstable accretion disk and subsequent fragmentation. In both scenarios, the result is the emergence of a singular first star.

5. Conclusion

We propose a mechanism that increases magnetic field strength exponentially during the accretion stage of initial stellar genesis. This process remarkably enhances a minuscule magnetic seed, even under the mere influence of a cosmological magnetic field, elevating it to kilo-gauss magnitudes in under a decade from the birth of the protostar. During the accretion phase, the area of the intensified magnetic field progressively broadens, extending to approximately 10,000 astronomical units and covering a mass of roughly 500 solar masses by about 100,000 years. This is the juncture when the protostar evolves into a zero-age main-sequence star. The formidable magnetic field plays a pivotal role in preventing the fragmentation of the accretion disk, leading

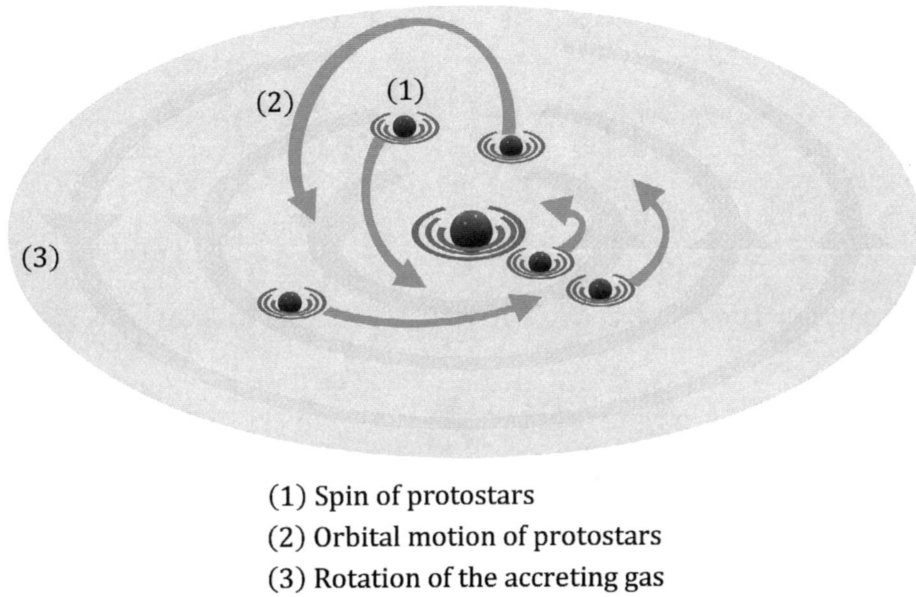

Figure 6. A schematic overview of the rotation-driven magnetic field amplification process following protostar formation includes the following stages: (1) The spinning motion of each protostar leads to the amplification of magnetic fields in their vicinity. (2) The enhanced "seed" magnetic field, a result of the initial amplification, further strengthens the magnetic fields around it due to the orbital motion of the protesters. (3) As gas continues to accrete, its rotation causes the region of the amplified magnetic field to expand outward gradually. It is important to note that this mechanism of magnetic field amplification is not typically observed in contemporary star formation processes, as the magnetic field tends to dissipate significantly within the disk in these scenarios.

to the exclusive formation of a singular protostar within each disk. We infer that magnetic fields significantly impact the formation of the earliest stars, regardless of their initial magnetic field strength being as diminutive as about 10^{-18} G.

Furthermore, our analysis corroborates the rotation-induced magnetic field amplification during the accretion phase in direct-collapse black hole (DCBH) formation within metal-rich atomic-cooling halos (ACHs). Although heightened metallicity dampens the mass accretion rate, the central region undergoing collapse is gravitationally unstable, culminating in fragmentation and the emergence of numerous protostars. These protostars play a crucial role in magnifying the magnetic field, facilitating additional gas accretion and the merging of stars. Notably, this magnetic enhancement is effective regardless of the initial intensity of the studied regime of dynamically weak magnetic fields. This amplification process remains operational even in metal-rich ACHs with an initial magnetic field strength of $B_0/\,\mathrm{G} = 10^{-12}$ at $n_\mathrm{H} = 10^4\,\mathrm{cm}^{-3}$. The rate of mass growth peaks in extremely metal-poor ACHs with $Z/Z_\odot = 10^{-5}$, marking a distinct deviation from trends observed in unmagnetized simulations. We conclude that exponential magnetic field amplification is feasible in metal-rich ACHs, thereby moderating the conditions required for DCBH formation.

Looking ahead, we aim to undertake a comprehensive parameter survey of MHD simulations, focusing on the primordial star-forming gas clouds ascertained from cosmological simulations. Although the current study indicates the complete elimination of disk fragmentation, future inquiries will assess whether gas clouds with diverse physical attributes, such as varying accretion rates and degrees of rotation, lead to comparable results. Moreover, our ongoing simulations terminate at $t_{ps} = 1000\,\mathrm{yr}$, necessitating additional research to determine the ultimate stellar mass at $t_{ps} \sim 10^5\,\mathrm{yr}$, coinciding with the first star's arrival at the zero main sequence stage. Our goal is to thoroughly revise the theory of first star formation, incorporating MHD influences, and to ascertain the formation rates of observable counterparts, including low-mass surviving stars and massive black hole binaries.

Acknowledgments

This work used the computational resources of the HPCI system provided by the supercomputer system SX-Aurora TSUBASA at Tohoku University Cyber Sciencecenter and Osaka University Cybermedia Center through the HPCI System Research Project (Project ID: hp190035, hp210004 and hp220003) and Earth Simulator at JAMSTEC provided by 2020-2022 Koubo Kadai. Numerical analyses were carried out on analysis servers at Center for Computational Astrophysics, National Astronomical Observatory of Japan. S.H. was supported by JSPS KAKENHI Grant Numbers JP18H05222, JP21H01123, and JP21K13960 and Qdai-jump Research Program 02217. M.N.M. was supported by JSPS KAKENHI Grant Numbers JP17H02869, JP17H06360, JP17K05387, JP17KK0096, JP21K03617, and JP21H00046 and University Research Support Grant 2019 from the National Astronomical Observatory of Japan (NAOJ). S.B. was supported by a Discovery Grant from NSERC.

References

[1] Klessen R S and Glover S C O 2023 *Annual Review of Astronomy and Astrophysics* **61** 65
[2] Koh D, Abel T and Jedamzik K 2021 *The Astrophysical Journal: Letters* **909** L21
[3] Machida M N and Doi K 2013 *Monthly Notices of the Royal Astronomical Society* **435** 3283
[4] Sadanari K E, Omukai K, Sugimura K, Matsumoto T and Tomida K 2021 *Monthly Notices of the Royal Astronomical Society* **505** 4197
[5] Hirano S and Bromm V 2018 *Monthly Notices of the Royal Astronomical Society* **476** 3964
[6] Ichiki K, Takahashi K, Ohno H, Hanayama H and Sugiyama N 2006 *Science* **311** 827
[7] McKee C F, Stacy A and Li P S 2020 *Monthly Notices of the Royal Astronomical Society* **496** 5528
[8] Stacy A, and McKee C F, Lee A T, Klein R I and Li P S 2022 *Monthly Notices of the Royal Astronomical Society* **511** 5042
[9] Omukai K, Schneider R and Haiman Z 2008 *The Astrophysical Journal* **686** 801
[10] Machida M N and Nakamura T 2015 *Monthly Notices of the Royal Astronomical Society* **448** 1405
[11] Hirano S and Machida M N 2022 *The Astrophysical Journal: Letters* **935** L16
[12] Hirano S, Machida M N and Basu S 2021 *The Astrophysical Journal* **917** 34
[13] Hirano S, Machida M N and Basu S 2023 *The Astrophysical Journal* **952** 56
[14] Xu H, O'Shea B W, Collins D C, Norman M L, Li H and Li S 2008 *The Astrophysical Journal: Letters* **688** L57

Journal of Physics: Conference Series **2742** (2024) 012005 doi:10.1088/1742-6596/2742/1/012005

Simulating Lateral H/He Flame Propagation in Type I X-ray Bursts

Eric T. Johnson[1], **Michael Zingale**[1]

[1]Department of Physics and Astronomy, Stony Brook University, Stony Brook, NY 11794-3800 USA

E-mail: `eric.t.johnson@stonybrook.edu`

Abstract. X-ray bursts are the thermonuclear runaway of a mixed H/He layer on the surface of a neutron star. Observations suggest that the burning begins locally and spreads across the surface of the star as a flame. Recent multidimensional work has looked in detail at pure He flames spreading across a neutron star. Here we report on progress in multidimensional modeling of mixed H/He flames and discuss the challenges.

1. Introduction

Type I X-ray bursts (XRBs) occur in accreting binary systems between a neutron star and a low-mass companion. As a layer of fuel builds up on the surface of the neutron star, gravitational compression heats it up until the conditions at the base lead to a thermonuclear runaway. A challenging aspect of modeling these systems is capturing how the flame spreads from the initial ignition point across the surface of the star. This is important for understanding oscillations observed during the initial rise of some bursts, which are believed to arise from the bright hotspot coming in and out of view as the neutron star rotates [1]. Unfortunately, one-dimensional simulations cannot capture these dynamics, as they are inherently multi-dimensional.

Most burst sources accrete a mix of ^{1}H and ^{4}He from a main sequence companion, while some ultra-compact systems accrete nearly pure ^{4}He [2, 3]. Depending on the accretion rate, the hydrogen can burn stably into helium which will then ignite a pure helium burst, or the helium could ignite before all the hydrogen is consumed, leading to a mixed H/He burst. Previous 2D and 3D simulations by our group and others have been limited to pure helium flames [4–10] as the mixed H/He reaction networks are usually larger and more difficult to integrate, and the timescales are slower due to the weak rate waiting points [11]. We recently showed [10] that the nucleosynthesis and evolution of a spreading flame in 3D matches 2D simulations very closely, so we will focus on 2D simulations here.

2. Simulations

We use the Castro [12] compressible hydrodynamics code to perform our simulations, following the simulation methodology developed in [7]. Castro uses the AMReX framework [13] for adaptive mesh refinement and parallelization across CPUs and GPUs. We use the stellar equation of state based on [14] and conductivities from [15] for the thermal diffusion operator. Castro can use an arbitrary nuclear reaction network, and we discuss some of the options below.

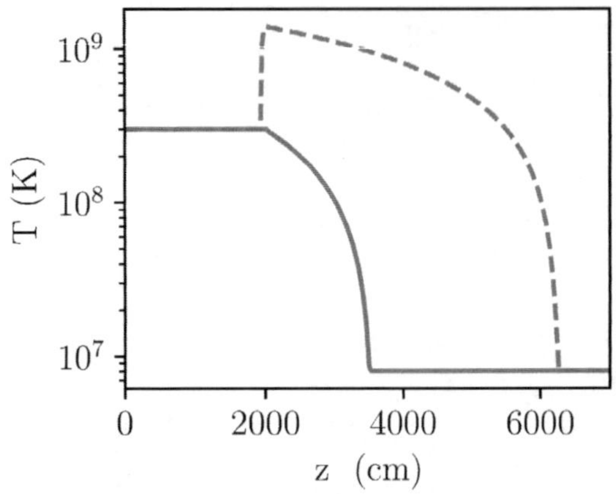

(a) `rprox` run with 74% H, 25% He

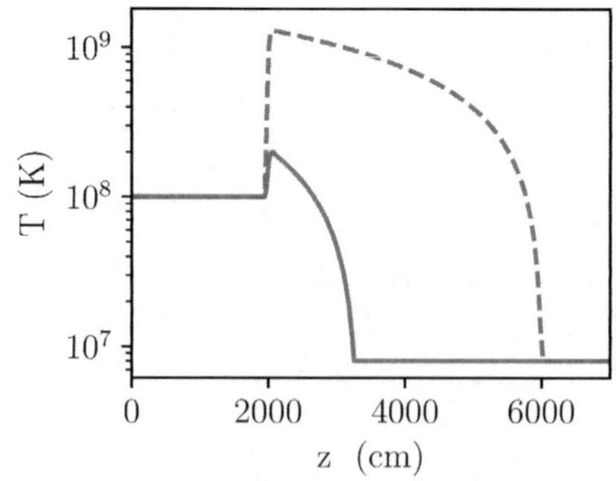

(b) `aprox19` run with 73% H, 25% He

Figure 1: Initial vertical temperature profiles for the cool (solid) and hot (dashed) regions.

All simulations use a two-dimensional axisymmetric simulation domain with a physical size of 2.4576×10^5 cm $\times$ 3.072×10^4 cm and a base grid of 1024×128 zones. The grid is subdivided using two levels of refinement, refining first by a factor of 4, then by a factor of 2. This gives a resolution of 30 cm at the finest level. We note that this is a little coarser than the He simulations we have done previously. However, since the addition of H to the atmosphere increases the scale height and H burning is less strongly peaked than He burning, we believe we can relax the resolution requirements.

Our initial runs used `rprox`, a 10 isotope nuclear reaction network introduced in [16] that models hot CNO burning, triple-α, and rp-process breakout. We vary the proportions of ^{1}H and ^{4}He in the fuel layer, and also include 1% of ^{14}O and ^{15}O to help start CNO burning. The oxygen abundance is split between the two isotopes in proportion to their respective β^+-decay lifetimes. We assume that the composition is completely mixed by a period of convection preceding the ignition of the flame, as explored in [17].

The initial model is set up as described in section 4 of [7]. In summary, it consists of a hot region behind the flame at $T = 1.4 \times 10^9$ K and a cool region ahead of it at $T = 3 \times 10^8$ K, with a density of 2×10^6 g cm^{-3} at the base of the atmosphere ($z = 2000$ cm). These two regions are constructed individually in 1D to be in hydrostatic equilibrium vertically, then smoothly blended at the interface, giving us an equilibrium starting condition that will lead to the ignition of a burning front. Figure 1a shows the temperature profiles for the two regions for one of the runs. The other profiles with different compositions are similar, but those with more helium are squashed due to the greater mass of the atmosphere above the surface. The hot region extends to $r_{\rm pert} = 6.144 \times 10^4$ cm (25% of the domain), and the transition interface between the two regions has a width of $\delta_{\rm blend} = 2.048 \times 10^3$ cm. Paired with the axisymmetric geometry, this effectively models a hotspot at the pole of the star spreading outwards towards the equator. The rotation rate of the neutron star is set to an artificially high 1000 Hz, which helps to confine the flame laterally and allows us to use a smaller simulation domain. All of our simulations were run on the OLCF Summit machine, with the entire calculation offloaded to the NVIDIA V100 GPUs there.

Figure 2 shows snapshots of the mass fraction of ash (everything heavier than ^{16}O) after 80 ms for each of the initial compositions we used. We have found that this diagnostic is much easier to compare between the different compositions than the mean molecular weight, $\bar{A}$, used in our previous work. The runs in the first two panels never produced well-defined flames. The

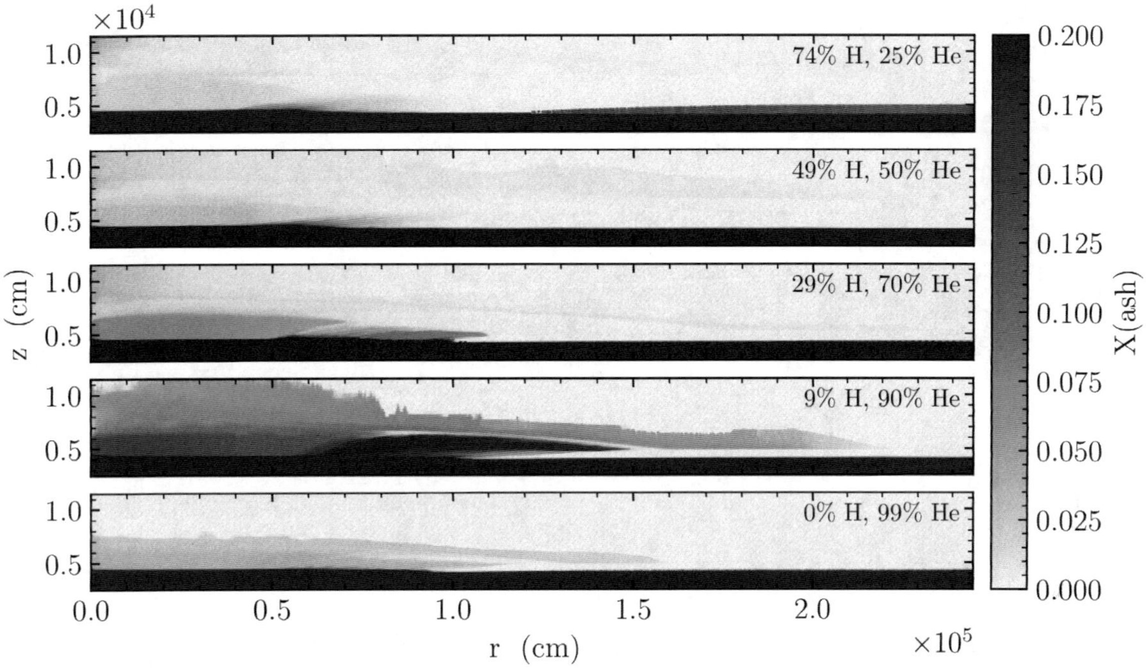

Figure 2: Total mass fraction of all species heavier than ^{16}O for `rprox` runs at $t = 80$ ms.

third panel has more burning than the previous two, but the flame front never moves beyond 10^5 cm. The most interesting is the fourth panel, with 9% H and 90% He, which shows a flame front with burning up to ^{56}Ni. The final panel models a pure He burst using the same network for comparison.

Many of our runs exhibited an unexpected high-temperature region at the base of the atmosphere across the entire simulation domain, starting from the outer radial edge of the domain and spreading toward the pole. This often ignited and consumed all the fuel before the flame could set up and start propagating. The magenta layer on the right of the first panel in figure 2, for example, is a result of this excess burning. We attributed this to our selection of the base temperature in the cool region, and used a lower temperature in later runs to try to avoid this. This is one of the main differences between a pure He burst—a pure He network will not burn much at those temperatures. However, we found that the simulation behavior was very sensitive to any changes in the initial conditions, and it was difficult to reproduce a steady burning front. To explore this behavior in more detail, we experimented with 1D flame simulations using different reaction networks, and found that `aprox19` produced a much more robust propagating flame front than `rprox`. Due to the different composition in that network, we replaced the oxygen isotopes with ^{12}C.

For our next set of runs with `aprox19`, we increased the resolution to 20 cm (using a base grid of 1536 × 192) and used a new lower boundary condition with a well-balanced pressure reconstruction, as done in [10]. We reduced the cool region temperature to 2×10^8 K and the material below the atmosphere (i.e. the star) to 1×10^8 K, as mentioned above. The new initial conditions are shown in figure 1b. Interestingly, increasing the resolution this way actually scaled slightly better than linear in the number of compute nodes, as the new grid size distributed more evenly onto Summit's 6 GPUs per node.

Figure 3 shows the same diagnostics as figure 2 for the `aprox19` runs. We can see that the run with 25% He again doesn't produce a flame, although it does have more ash present in

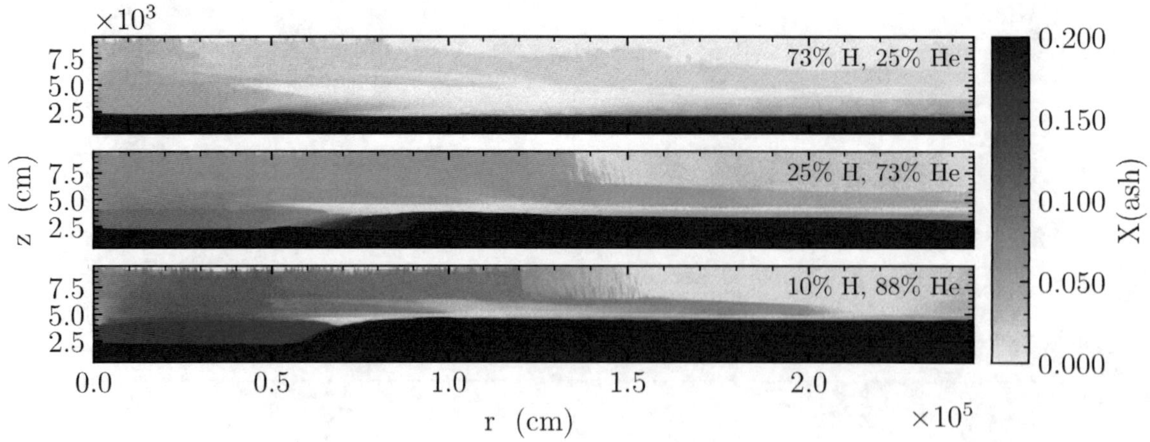

Figure 3: Total ash mass fraction at $t = 80\,\text{ms}$ for `aprox19` runs.

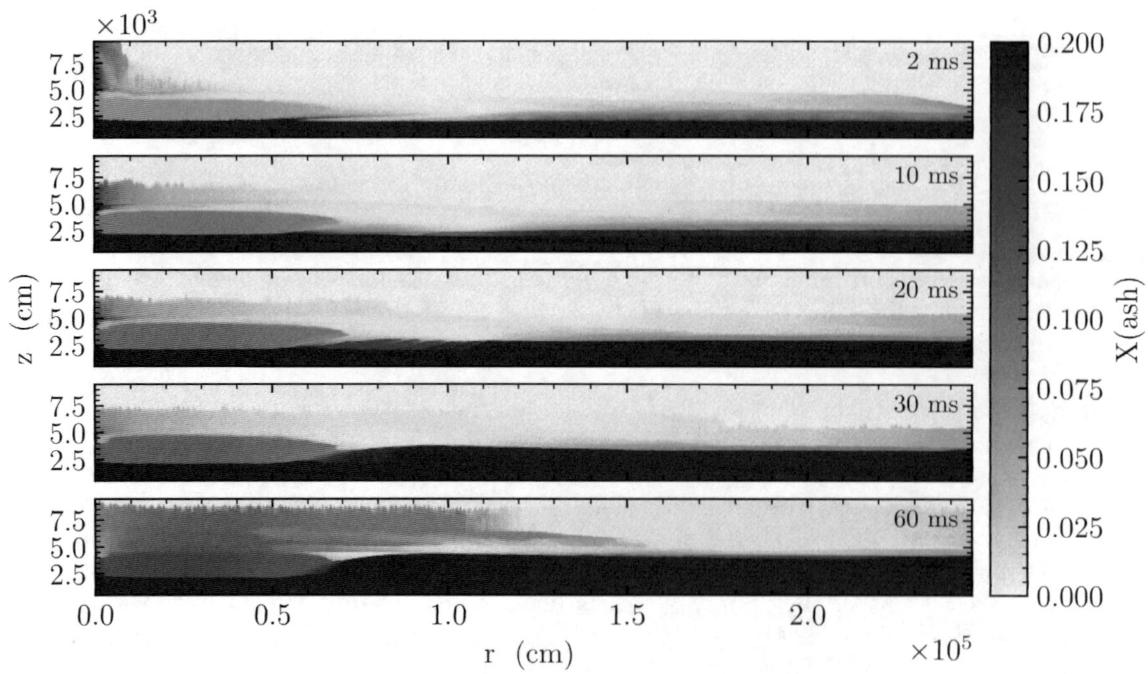

Figure 4: Ash mass fraction evolution for the `aprox19` run with 10% H and 88% He.

the atmosphere than the corresponding `rprox` run. This is due to vigorous carbon burning in the first few milliseconds of the simulation, which consumes the majority of the initial ^{12}C in the lower atmosphere. In other two runs, the right side of the atmosphere heats up and starts burning before the flame front. An example of how this progresses for the run with 10% H and 88% He is shown in figure 4.

Finally, figure 5 shows the evolution of each species in the network over time, for the same run. We can see that most of the initial ^{12}C (solid red line) is consumed in under 1 millisecond. An interesting detail in this plot is the step-like features at early times, most visible in the profiles for ^{28}Si and ^{32}S. These are fairly evenly spaced in time with a separation of $\sim 5 \times 10^{-4}$ s, which matches the width of the domain divided by the average sound speed near the base of the

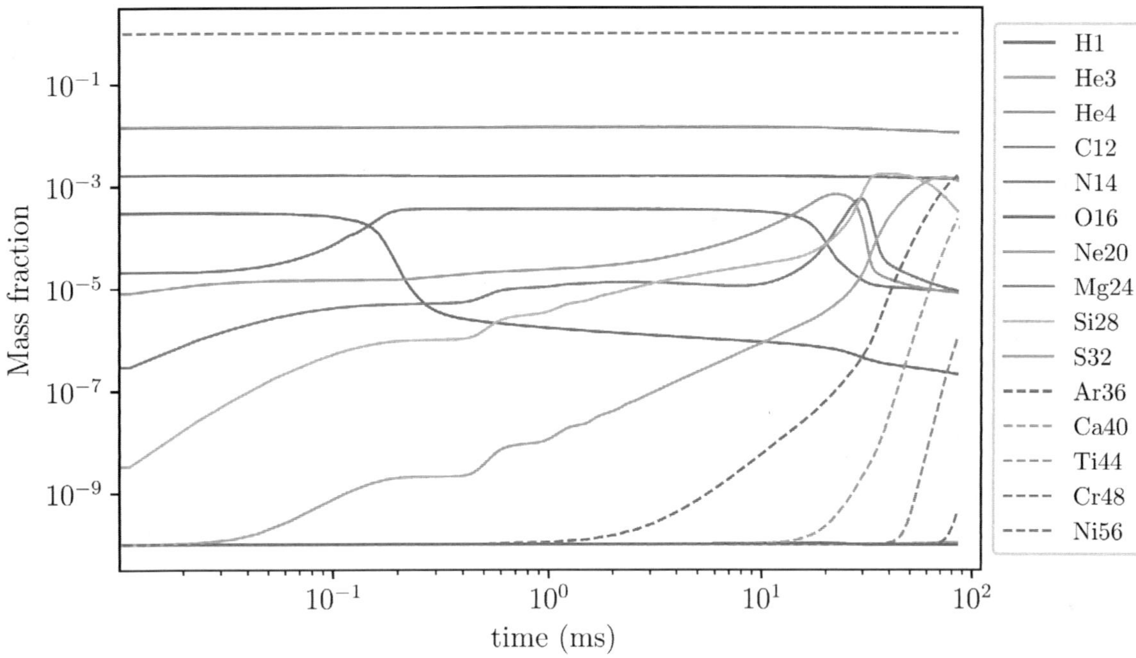

Figure 5: Total mass fractions for each species in the aprox19 run with 10% H, excluding those which never increased above 10^{-10}.

atmosphere: $L_x/c_s \approx 2.4576 \times 10^5\,\mathrm{cm}/5.2 \times 10^8\,\mathrm{cm\,s^{-1}} \approx 4.7 \times 10^{-4}\,\mathrm{s}$. This suggests they may be related to acoustic oscillations travelling across the simulation domain and reflecting off of the boundaries.

3. Summary

We ran several two-dimensional simulations of mixed H/He flames in XRBs, varying the proportions of H and He in the atmosphere. We found that the flames are very sensitive to the composition and initial temperatures, which makes it difficult to set up a flame in the first place. Using the rprox network, we were only able to produce a stable, propagating flame with an initial composition of 9% H and 90% He. The aprox19 network gave much better results in 1D simulations, but we have not yet been able to reproduce a clean H/He flame in 2D.

An unexpected hot region would often develop at the surface of the star, away from the flame, which would start burning and disrupt the flame. The source of this heating is unclear, and needs to be investigated further. We are currently testing out different reaction networks with pynucastro [18, 19], which allows us to construct and inspect arbitrary reaction networks, and generate C++ code for use in our Castro simulations. It also allows us to disable specific rates at runtime, which can help us understand which rates are relevant for H/He burning under these conditions. Finally, we may need to consider different initial models, where the pre-runaway convective mixing is confined only to the He layer at high column depth, as explored in [20].

Acknowledgments

Castro is freely available at `http://github.com/AMReX-Astro/Castro`. All of the code and problem setups used here are available in the git repo. The work at Stony Brook was supported by DOE/Office of Nuclear Physics grant DE-FG02-87ER40317. This research used resources of the Oak Ridge Leadership Computing Facility at the Oak Ridge National Laboratory, which is supported by the Office of Science of the U.S. Department of Energy under Contract No. DE-AC05-00OR22725, awarded through the DOE INCITE program. We thank NVIDIA Corporation for the donation of a Titan X and Titan V GPU through their academic grant program. This research has made use of NASA's Astrophysics Data System Bibliographic Services.

Software: AMReX [13], Castro [12, 21], GCC (https://gcc.gnu.org/), GNU Parallel [22], helmeos [14], linux (https://www.kernel.org/), matplotlib [23], NumPy [24, 25], python (https://www.python.org/), valgrind [26], VODE [27], yt [28].

References

[1] Chakraborty M and Bhattacharyya S 2014 *ApJ* **792** 4 ISSN 1538-4357 (*Preprint* 1407.0845) URL https://doi.org/10.1088/0004-637x/792/1/4

[2] Galloway D K and Keek L 2021 *Timing Neutron Stars: Pulsations, Oscillations and Explosions* (*Astrophysics and Space Science Library* vol 461) ed Belloni T M, Méndez M and Zhang C pp 209–262 (*Preprint* 1712.06227)

[3] in't Zand J J M, Jonker P G and Markwardt C B 2007 *Astronomy and Astrophysics* **465** 953–963 (*Preprint* astro-ph/0701810)

[4] Cavecchi Y, Watts A L, Braithwaite J and Levin Y 2013 *Mon. Not. R. Astron Soc.* **434** 3526–3541 ISSN 0035-8711, 1365-2966 (*Preprint* 1212.2872) URL https://doi.org/10.1093/mnras/stt1273

[5] Cavecchi Y, Watts A L, Levin Y and Braithwaite J 2015 *Mon. Not. R. Astron Soc.* **448** 445–455 ISSN 1365-2966, 0035-8711 (*Preprint* 1411.2284) URL https://doi.org/10.1093/mnras/stu2764

[6] Cavecchi Y and Spitkovsky A 2019 *ApJ* **882** 142 ISSN 0004-637X, 1538-4357 URL https://doi.org/10.3847/1538-4357/ab3650

[7] Eiden K, Zingale M, Harpole A, Willcox D, Cavecchi Y and Katz M P 2020 *Astrophysical Journal* **894** 6 (*Preprint* 1912.04956)

[8] Harpole A, Ford N M, Eiden K, Zingale M, Willcox D E, Cavecchi Y and Katz M P 2021 *Astrophysical Journal* **912** 36 (*Preprint* 2102.00051)

[9] Goodwin A J, Heger A, Chambers F R N, Watts A L and Cavecchi Y 2021 *Monthly Notices of the Royal Astronomical Society* **505** 5530–5542 ISSN 0035-8711 (*Preprint* https://academic.oup.com/mnras/article-pdf/505/4/5530/38858828/stab1659.pdf) URL https://doi.org/10.1093/mnras/stab1659

[10] Zingale M, Eiden K and Katz M 2023 *Astrophysical Journal* **952** 160 (*Preprint* 2303.17077) URL https://dx.doi.org/10.3847/1538-4357/ace04e

[11] Strohmayer T and Bildsten L 2006 *Compact Stellar X-ray Sources* ed Lewin W H G and van der Klis M (Cambridge University Press) pp 113–156 URL https://doi.org/10.1017/cbo9780511536281.004

[12] Almgren A S, Beckner V E, Bell J B, Day M S, Howell L H, Joggerst C C, Lijewski M J, Nonaka A, Singer M and Zingale M 2010 *ApJ* **715** 1221–1238 ISSN 0004-637X, 1538-4357 (*Preprint* 1005.0114) URL https://doi.org/10.1088/0004-637x/715/2/1221

[13] Zhang W, Almgren A, Beckner V, Bell J, Blaschke J, Chan C, Day M, Friesen B, Gott K, Graves D, Katz M, Myers A, Nguyen T, Nonaka A, Rosso M, Williams S and Zingale M 2019 *JOSS* **4** 1370 ISSN 2475-9066 URL https://doi.org/10.21105/joss.01370

[14] Timmes F X and Swesty F D 2000 *ASTROPHYS J SUPPL S* **126** 501–516 ISSN 0067-0049, 1538-4365 source code obtained from http://cococubed.asu.edu/codes/eos.shtml/helmholtz.tbz URL https://doi.org/10.1086/313304

[15] Timmes F X 2000 *ApJ* **528** 913–945 ISSN 0004-637X, 1538-4357 URL https://doi.org/10.1086/308203

[16] Wallace R K and Woosley S E 1981 *ApJS* **45** 389 ISSN 0067-0049, 1538-4365 URL https://doi.org/10.1086/190717

[17] Malone C M, Zingale M, Nonaka A, Almgren A S and Bell J B 2014 *ApJ* **788** 115 ISSN 0004-637X, 1538-4357 URL https://doi.org/10.1088/0004-637x/788/2/115

[18] E Willcox D and Zingale M 2018 *JOSS* **3** 588 ISSN 2475-9066 URL https://doi.org/10.21105/joss.00588

[19] Smith A I, Johnson E T, Chen Z, Eiden K, Willcox D E, Boyd B, Cao L, DeGrendele C J and Zingale M 2023 *The Astrophysical Journal* **947** 65 URL https://dx.doi.org/10.3847/1538-4357/acbaff

[20] Guichandut S and Cumming A 2023 *Astrophysical Journal* **954** 54 (*Preprint* 2301.08769)

[21] Almgren A, Sazo M B, Bell J, Harpole A, Katz M, Sexton J, Willcox D, Zhang W and Zingale M 2020 *Journal of Open Source Software* **5** 2513 URL https://doi.org/10.21105/joss.02513

[22] Tange O 2018 *GNU Parallel 2018* (Ole Tange) ISBN 9781387509881 URL https://doi.org/10.5281/zenodo.1146014

[23] Hunter J D 2007 *Comput. Sci. Eng.* **9** 90–95 ISSN 1521-9615 URL https://doi.org/10.1109/mcse.2007.55

[24] Oliphant T E 2007 *Comput. Sci. Eng.* **9** 10–20 ISSN 1521-9615 URL https://doi.org/10.1109/mcse.2007.58

[25] van der Walt S, Colbert S C and Varoquaux G 2011 *Comput. Sci. Eng.* **13** 22–30 ISSN 1521-9615 (*Preprint* https://aip.scitation.org/doi/pdf/10.1109/MCSE.2011.37) URL https://doi.org/10.1109/mcse.2011.37

[26] Nethercote N and Seward J 2007 *Proceedings of the 2007 ACM SIGPLAN conference on Programming language design and implementation - PLDI '07* PLDI '07 (New York, NY, USA: ACM Press) p 89–100 ISBN 978-1-59593-633-2 URL https://doi.org/10.1145/1250734.1250746

[27] Brown P N, Byrne G D and Hindmarsh A C 1989 *SIAM J. Sci. and Stat. Comput.* **10** 1038–1051 ISSN 0196-5204, 2168-3417 URL https://doi.org/10.1137/0910062

[28] Turk M J, Smith B D, Oishi J S, Skory S, Skillman S W, Abel T and Norman M L 2010 *ApJS* **192** 9 ISSN 0067-0049, 1538-4365 (*Preprint* 1011.3514) URL https://doi.org/10.1088/0067-0049/192/1/9

Journal of Physics: Conference Series **2742** (2024) 012006 doi:10.1088/1742-6596/2742/1/012006

Semi-global Simulations of Star Formation in Nuclear Rings of Barred Galaxies

Woong-Tae Kim[1,2] **& Sanghyuk Moon**[3]

[1] Department of Physics & Astronomy, Seoul National University, Seoul 08826, Republic of Korea

[2] SNU Astronomy Research Center, Seoul National University, Seoul 08826, Republic of Korea

[3] Department of Astrophysical Sciences, Princeton University, Princeton, NJ 08544, USA

E-mail: `unitree@snu.ac.kr`

Abstract. Nuclear rings at the centers of barred galaxies are active in star formation. To understand what determines the star formation rate (SFR) and structure of nuclear rings, we conduct semi-global, magnetohydrodynamic simulations of nuclear rings subject to various mass inflow rates with and without magnetic fields. We adopt the TIGRESS framework of Kim & Ostriker to handle radiative heating and cooling, star formation, and related supernova feedback. Our findings suggest that supernova feedback cannot destroy the nuclear ring completely or halt star formation within it, while both the mass inflow rate and supernova feedback affect the ring star formation rate. The supernova feedback is responsible for small-amplitude SFR fluctuations with a timescale of less than 40 million years, while the SFR variations over longer timescales are due to changes in the mass inflow rates. Magnetic fields seeded by the inflows are amplified in the ring due to rotational shear and supernova feedback, greatly reducing the SFR at late times. Strong magnetic tension in the ring drives radially inward accretion flows from the ring to form a circumnuclear disk in the central region, which is absent in the unmagnetized model.

1. Introduction

Barred galaxies are common in the universe. Including weak bars, about two thirds of disk galaxies possess a bar at their centers [1, 2, 3, 4, 5, 6, 7, 8]. A stellar bar exerts a gravitational torque to gas to produce a pair of dust lanes at the leading side of the bar, which are shocks in the gas flows [9, 10, 11, 12, 13, 14, 15]. When gas in orbital motion hits the dust lanes, it loses angular momentum to move radially inward and fuels a star-forming nuclear ring at the location where the centrifugal force balances the gravity [14, 16]. By making CO observations for the gas densities and line-of-sight velocities, one can infer the mass inflow rate $\dot{M}_{\rm in}$ along the dust lanes to the ring. In the case of NGC 1097, the estimated mass inflow rate was found to be $\dot{M}_{\rm in} = 3.0 \pm 2.1\ M_\odot\ {\rm yr}^{-1}$ [17]. The condition of mass conservations requires that $\dot{M}_{\rm in} = \dot{M}_{\rm SF} + \dot{M}_{\rm out} + \dot{M}_{\rm ring}$, where $\dot{M}_{\rm SF}$ is the star formation rate (SFR), $\dot{M}_{\rm out}$ is the outflow rate from the galaxy as a wind, $\dot{M}_{\rm ring}$ is the rate of change in the ring mass. For NGC 1097, $\dot{M}_{\rm SF} \simeq 1.8 - 2\ M_\odot\ {\rm yr}^{-1}$ [18, 19] and $\dot{M}_{\rm out} = 0.6\ M_\odot\ {\rm yr}^{-1}$ [20]. This indicates that most of the inflowing gas is currently being consumed by star formation in the ring of NGC 1097.

Our own Galaxy also possesses a nuclear ring, called the central molecular zone (CMZ), with the current SFR of $\dot{M}_{\rm SF} = 0.04$–$0.1\ M_\odot\ {\rm yr}^{-1}$ [21, 22, 23], which is about 10 times smaller than the prediction of the Kennicutt-Schmidt relation for the observed CMZ mass [22, 24],

Table 1. Model Parameters

model	$R_{\rm ring}$ (pc)	$\dot{M}_{\rm in}$ ($M_\odot$ yr^{-1})	$\Delta\tau_{\rm in}$ (Myr)	$\beta_{\rm in}$
L0	600	0.125	∞	∞
L1	600	0.5	∞	∞
L2	600	2.0	∞	∞
L3	600	8.0	∞	∞
constant	600	0.5	∞	∞
P15	600	0.5	15	∞
P50	600	0.5	50	∞
P100	600	0.5	100	∞
Binf	500	1.0	∞	∞
B100	500	1.0	∞	100
B30	500	1.0	∞	30
B10	500	1.0	∞	10

possibly by solenoidal turbulence driving associated with strong velocity shear [25]. To explain the low SFR of the CMZ, [26] proposed a scenario of episodic star formation in which the ring becomes gradually massive as the gas driven by the large-scale bar accumulates, and undergoes gravitational instability and starburst when it becomes sufficiently massive. The associated strong supernova feedback disperses the ring material, making the ring quiescent temporarily until mass accumulates again in the ring. According to this scenario, the CMZ is currently in the quiescent phase. One-dimensional idealized simulations of [27] found that the ring SFR indeed exhibits quasi-periodic oscillations with a period of $\sim$ 20 Myr.

However, three-dimensional global simulations with more realistic treatment of star formation and feedback show different results. For example, the simulations of [28] found that ring star formation is episodic, but the associated period is about 20 times longer than that in the one-dimensional simulations of [27]. [15] found that $\dot{M}_{\rm SF}$ is very similar to the mass inflow rate to the ring, while it is maintained almost constant in the simulations of [29]. These diverse results from the three-dimensional global simulations imply that there is no consensus as to what controls the ring SFR. In these global simulations, the mass inflow rate to a nuclear ring is self-consistently determined by dynamical interactions between gas and a stellar bar, and thus highly time dependent. In such situations, the ring size and mass vary considerably with time, making it difficult to isolate key factors that determine the ring SFR.

Another important ingredient that governs gas dynamics and star formation in the central parts of barred galaxies is magnetic fields. Observations of radio synchrotron emission indicate that magnetic fields are mainly parallel to the dust lanes with a strength of $\sim$ 30 μG in the bar regions, while they penetrate the ring with a pitch angle of $\sim 40^\circ$ and have a strength of $\sim$ 60 μG in the ring regions [30, 31]. More recent SOFIA observations of dust emission suggest that magnetic fields are dominated by the $m = 0$ mode, while the radio polarization is dominated by the $m = 2$ spiral mode [32]. Magnetic fields in the rings appear to suppress the ring star formation. [33] measured the magnetic field strength of 11 giant molecular clouds distributed along the nuclear ring of NGC 1097, for which the gas depletion time $\tau_{\rm dep}$ is obtained from [18]. They found that $\tau_{\rm dep}$ is higher for clouds with stronger magnetic fields, suggesting that magnetic fields suppress the ring star formation.

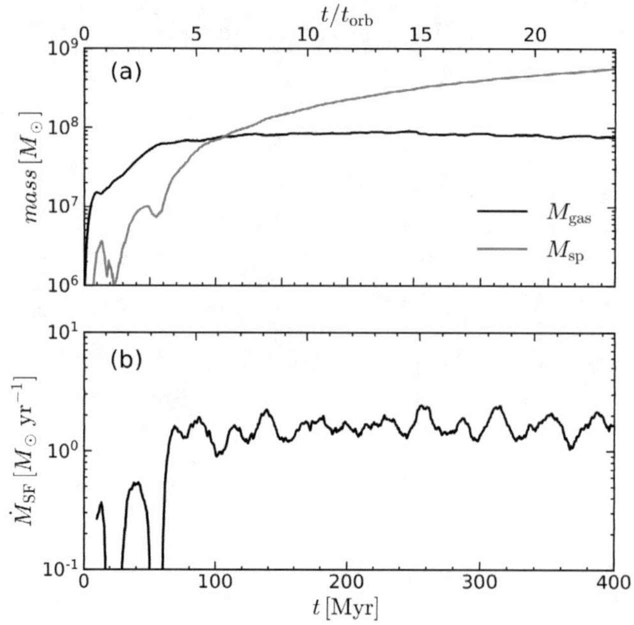

Figure 1. Temporal variations of (a) the gas mass $M_{\rm gas}$ in the ring as well as the mass $M_{\rm sp}$ of the newly formed stars and (b) the ring SFR $\dot{M}_{\rm SF}$ of the unmagnetized model with $\dot{M}_{\rm in} = 2\ M_\odot\ {\rm yr}^{-1}$.

To investigate what controls star formation in nuclear rings, we have run series of three-dimensional magnetohydrodynamic (MHD) simulations. Instead of running fully global simulations, we construct a semi-global model of a nuclear ring in which the bar-driven mass inflow rate $\dot{M}_{\rm in}$ is handled by the boundary conditions. The semi-global model resolves nuclear rings better than global models and allows us to directly control the mass inflow rate and ring radius. We first run unmagnetized models with various time-fixed $\dot{M}_{\rm in}$ [34], unmagnetized models with time-varying $\dot{M}_{\rm in}$ [35], and then magnetized models with fixed $\dot{M}_{\rm in}$ [36]. Here we highlight the main results of [34, 35, 36]. We refer the reader to these papers for more detailed and quantitative results and discussion.

2. Model and Method

As a simulation domain, we consider a cubic box with side length $L = 2048$ pc located at the center of a galaxy. The box is under the external gravitational potential from the central black hole and stellar bulge, corresponding to those in NGC 1097. The box rotates about the center at angular frequency $\Omega_p = 36$ km s^{-1} kpc^{-1}, equal to the bar pattern speed. Instead of modeling the bar explicitly, we incorporate the bar-driven mass inflows using gas streams through two nozzles located at the y-boundaries. To cover various galactic situations, we vary the inflow rate $\dot{M}_{\rm in}$ and inflow speed $v_{\rm in}$ of the gas in the nozzles, the latter of which is to set to give the ring size $R_{\rm ring} = 500$ or 600 pc. The box is initially filled with rarefied warm gas with temperature $T = 2\times10^4$ K and number density $n = 10^{-5}e^{-|z|/(50\ {\rm pc})}$ cm^{-3}, which is swept by the gas streams from the boundaries.

We use the Athena code [37] to integrate the basic MHD equations expanded in the rotating frame, and adopt the TIGRESS framework [38] to handle star formation and supernova feedback self-consistently after making it suitable for non-periodic boundary conditions of our semi-global

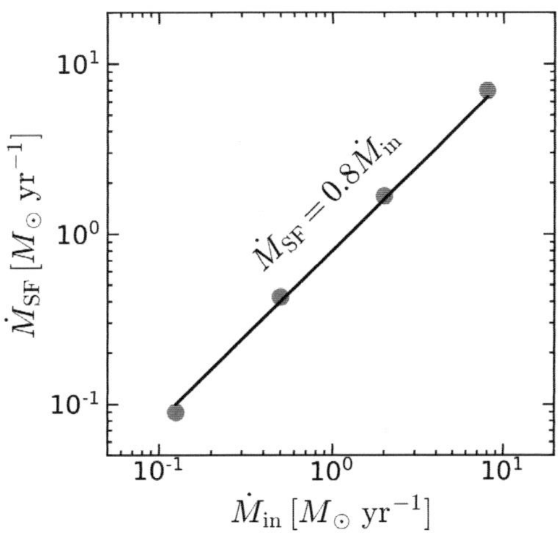

Figure 2. Star formation rate $\dot{M}_{SF}$ as a function of the mass inflow rate $\dot{M}_{in}$ for unmagnetized models with $R_{ring} = 600$ pc.

models. A sink particle (representing a star cluster) is created in a cell that satisfies the three conditions simultaneously: (1) the cell lies at a local potential minimum; (2) the gas density is greater than the Larson-Penston density threshold that depends on the resolution and the local temperature; (3) the velocity is converging in all three directions. We follow the orbits of sink particles by using the Boris algorithm [39] which conserves the momentum and energy of the particles almost exactly.

In what follows, we present three sets of models. In the first set, we run various models with differing $\dot{M}_{in} = 0.125$–$8\ M_\odot\ \mathrm{yr}^{-1}$ which is kept fixed over time. In the second set, we explore the models with a time-varying inflow rate

$$\dot{M}_{in}(t) = 0.3e^{-1.5\cos(2\pi t/\Delta\tau_{in})}\ M_\odot\ \mathrm{yr}^{-1}, \tag{1}$$

where $\Delta\tau_{in} = 15, 50, 100$ Myr is the oscillation period of the mass inflow rate. Finally in the third set, we make the inflow streams threaded by magnetic fields parallel to the inflow velocity by imposing the boundary condition

$$\mathbf{B}_{in} = \frac{B_{in}}{v_{in}}\mathbf{v}_{in}, \tag{2}$$

with strength B_{in} corresponding to the plasma parameter $\beta_{in} = 10, 30, 100$. Table 1 lists the model parameters presented in this article.

3. Results

3.1. Models with Fixed Mass Inflows

Figure 1 plots the temporal changes of the gas mass M_{gas} in the ring and the mass M_{sp} of newly-formed star particles, as well as the star formation rate $\dot{M}_{SF}$ in the ring averaged over a 10 Myr interval. At early time when the gas streams injected from the nozzles collide each other and start to form a ring, M_{sp} and $\dot{M}_{SF}$ exhibit some fluctuations. At $t \sim 100$ Myr, however, the system reaches a quasi-steady state in which the ring morphology and various statistical quantities such as the ring mass and SFR do not vary much over time. The mass of star particles keeps increasing over time due to continued star formation. In our models, star formation occurs stochastically rather than synchronously in different parts of the ring, leading

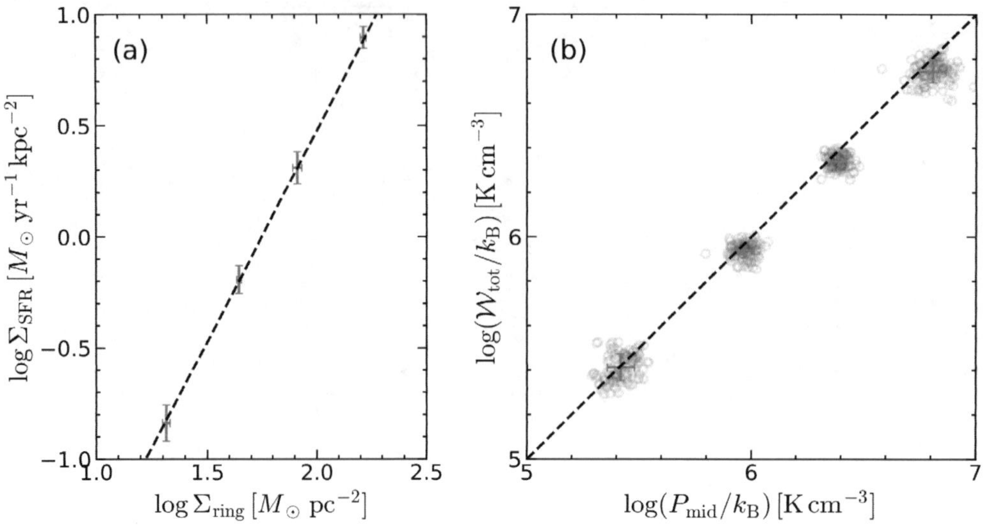

Figure 3. (a) SFR surface density $\Sigma_{\rm SFR}$ as a function of the gas surface density $\Sigma_{\rm ring}$ in the ring and (b) the total gas weight $\mathcal{W}$ against the total midplane pressure $P_{\rm mid}$ for unmagnetized models with $R_{\rm ring} = 600$ pc.

to an almost constant SFR when averaged over the entire ring. Small-amplitude fluctuations in $\dot{M}_{\rm SF}$ is caused by turbulence driven by supernova feedback.

We run various models for the ring radius of $R_{\rm ring} = 600$ pc with differing $\dot{M}_{\rm in}$. Figure 2 plots $\dot{M}_{\rm SF}$ against $\dot{M}_{\rm in}$, showing that ~ 80 % of the inflowing gas turn to stars in our models. This suggests that the ring SFR is determined primarily by the mass inflow rate. Remaining ~ 20 % escapes the simulations domain as galactic winds.

Figure 3(a) plots the SFR surface density $\Sigma_{\rm SFR} = \dot{M}_{\rm SF}/A_{\rm ring}$ in the ring as a function of the gas surface density $\Sigma_{\rm SFR}$ of the ring, where $A_{\rm ring}$ denotes the area covered by the ring. The best fit to the numerical results yields

$$\Sigma_{\rm SFR} = 3.0\ M_\odot\ {\rm yr}^{-1}\ {\rm kpc}^{-2} \left(\frac{\Sigma_{\rm ring}}{10^2\ M_\odot\ {\rm pc}^{-2}}\right)^{1.9}, \tag{3}$$

for our unmagnetized models. In our model, Equation (3) is a consequence of vertical dynamical equilibrium of the ring. That is, the total midplane pressure

$$P_{\rm mid} = \frac{1}{2\Delta z A_{\rm ring}} \int_{-\Delta z}^{\Delta z} \iint_{A_{\rm ring}} (P + \rho v_z^2) dx dy dz, \tag{4}$$

balances the weight

$$\mathcal{W}_{\rm tot} = \frac{1}{A_{\rm ring}} \iint_{A_{\rm ring}} \left(\int_0^\infty \rho \frac{\partial \Phi_{\rm tot}}{\partial z} dz \right) dx dy, \tag{5}$$

of the overlying gas, as Figure 3(b) displays. Here, Δz is the grid spacing in the vertical direction and $\Phi_{\rm tot}$ is the total (external and self) gravitational potential. The midplane pressure is dominated by the turbulent pressures (ρv_z^2) due to supernova feedback, with $P_{\rm mid} \propto \Sigma_{\rm SFR}^{0.9}$. Since the vertical weight $\mathcal{W}_{\rm tot}$ is the product of the gas surface density $\Sigma_{\rm ring}$ and the vertical gravity g_z, this together with Equation (3) and Figure 3(b) suggests that $g_z \propto \Sigma_{\rm ring}^{0.8}$. Given that g_z is independent of $\Sigma_{\rm ring}$ if the external gravity dominates, and $g_z \propto \Sigma_{\rm ring}$ if self-gravity

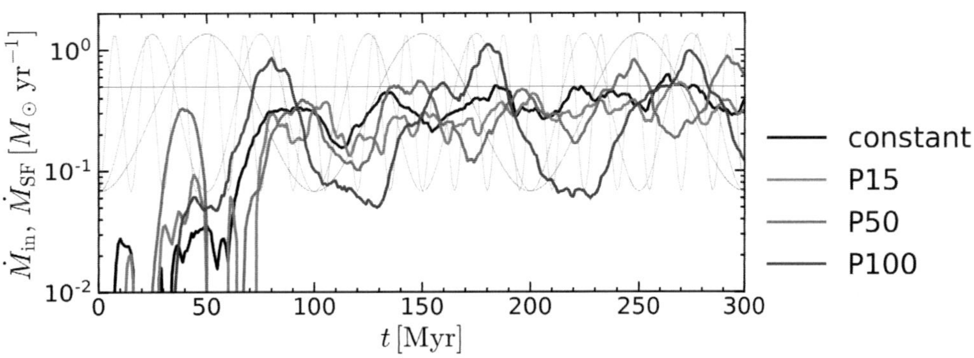

Figure 4. Temporal changes in the SFR $\dot{M}_{\rm SF}$ (thick lines) and the mass inflow rate $\dot{M}_{\rm in}$ (thin lines) for unmagnetized models with time-varying mass inflow rates

dominates, our results show that both external gravity and self-gravity (including the gravity of star particles) are important in determining the vertical dynamical equilibrium.

3.2. Models with Time-varying Mass Inflows

Now we turn to models with time-varying mass inflow rate $\dot{M}_{\rm in}$. Figure 4 plots the mass inflow rate $\dot{M}_{\rm in}$ (thin lines) and the ring SFR $\dot{M}_{\rm SF}$ (thick lines) for models with $\Delta\tau_{\rm in} = \infty$ (`constant`), 15 (`P15`), 50 (`P50`), 100 Myr (`P100`). The SFR is measured using the star particles that formed in the past 10 Myr. Note that When $\Delta\tau_{\rm in} \geq 50$ Myr, an oscillating inflow rate causes large-amplitude (a factor of ~ 5), quasi-periodic variations of the SFR. This is because the increased mass inflows first enhance the ring density to promote star formation, and the associated increase in supernova feedback then inflates the ring vertically, increasing the vertical scale height of the ring and thus reducing the midplane ring density. In models with $\Delta\tau_{\rm in} \leq 15$ Myr, in contrast, the SFR shows only modest (less than a factor of 2) fluctuations with timescale < 40 Myr, due to star formation feedback. This indicates that the long-period, large-amplitude fluctuations of the SFR is caused by the temporal change in the mass inflows rate, while the short-period, low-amplitude fluctuations of the SFR is due to supernova feedback.

3.3. Models with Magnetic Fields

We also use the same semi-global framework and introduce weak magnetic fields parallel to the velocity fields at the nozzles to explore the effect of magnetic fields on the ring star formation. We consider four models with different field strength, characterized by the plasma beta $\beta_{\rm in} = \infty, 100, 30, 10$. The $\beta_{\rm in} = 100$ model has the mean field strength of 0.76 μG at the nozzles. For all magnetized models, the mass inflow rate is fixed to $\dot{M}_{\rm in} = 1\ M_\odot\ {\rm yr}^{-1}$. Figure 5 plots the temporal changes in the strength of the regular and turbulent components of magnetic fields in the ring. Figure 6 plots the temporal changes in the ratios of the magnetic energy $E_{\rm mag}$ to the turbulent kinetic energy $E_{\rm trb}$ and to the orbital kinetic energy $E_{\rm orb}$ of the rings.

Early evolution of the magnetized models is similar to that of the unmagnetized counterpart. The gas streams from the nozzles at the opposite boundaries collide with each other, driving strong shocks. The streams lose their orbital kinetic energy as they pass through the shocks multiple times, and form a nuclear ring with $R_{\rm ring} = 500$ pc after about three orbital times. As the streams form a ring, magnetic fields become predominantly toroidal in the ring, although vertical expansions of superbubbles created by supernova explosions drag the toroidal fields in the vertical direction to produce poloidal fields in high-altitude regions.

Journal of Physics: Conference Series **2742** (2024) 012006 doi:10.1088/1742-6596/2742/1/012006

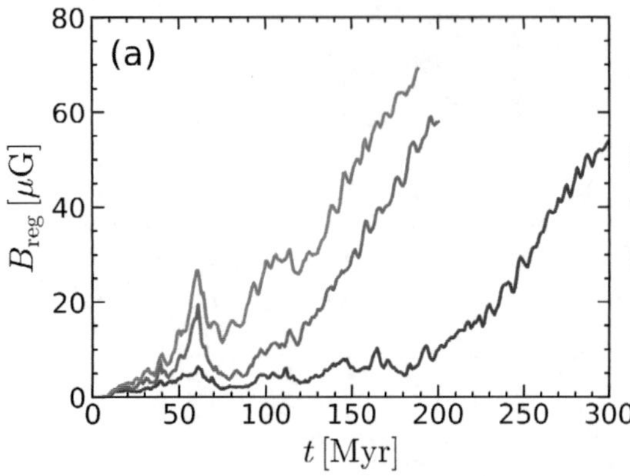

Figure 5. Temporal variations in the strength of (a) the regular component $B_{\rm reg}$ and (b) the turbulent component $B_{\rm trb}$ of magnetic fields in magnetized models with $\dot{M}_{\rm in} = 1\ M_\odot\ {\rm yr}^{-1}$.

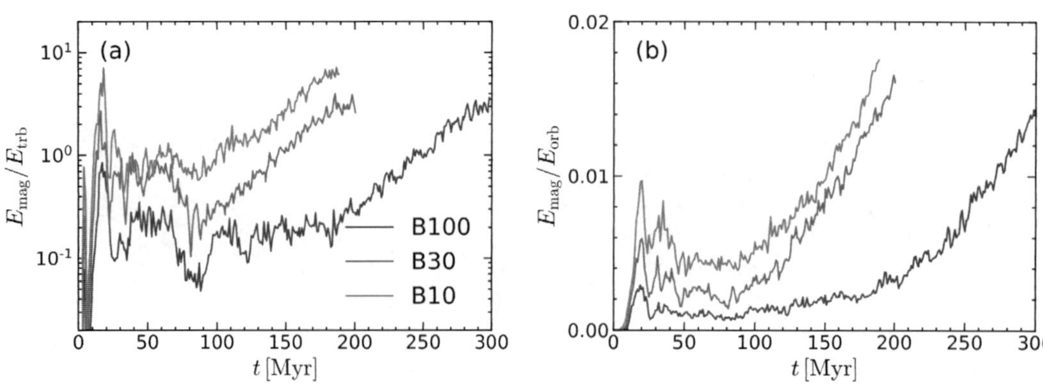

Figure 6. Temporal variations in the ratios of the magnetic energy $E_{\rm mag}$ to (a) the turbulent kinetic energy $E_{\rm trb}$ and (b) the orbital kinetic energy $E_{\rm orb}$ for the models shown in Figure 5.

Driven by rotational shear and supernova feedback, the magnetic fields in the ring become stronger over time, due to actions of both small- and large-scale dynamos. Figure 5 shows that the regular fields keep increasing with time, achieving strength of $\sim$ 50–70 μG presumably due to large-scale α–Ω type dynamo [40, 41]. On the other hand, the turbulence driven by supernova feedback may amplify the turbulent magnetic fields that saturate at $\sim$ 30–40 μG, probably balanced by turbulent dissipation [42]. Note that $E_{\rm mag}/E_{\rm trb}$ in each model rises rapidly as the ring forms at early time, saturates over the time span when $B_{\rm trb}$ dominates, and increases again at late time primarily due to the Ω effect. In all models, the saturation values of $E_{\rm mag}/E_{\rm trb}$ exceed 0.1, which can be achieved mainly by solenoidal, low-Mach-number driving in a uniform non-rotating medium [43]. It would be interesting to study how rotation and shear conspire with turbulent dynamo to set up the saturated level of $E_{\rm mag}/E_{\rm trb}$. As in the unmagetized cases, the magnetized ring maintains vertical dynamical equilibrium instantaneously, with the total mid-plane pressure equal to the weight of the overlying gas. Strong magnetic fields provide magnetic stress which reduces the demand for supernova feedback to produce turbulent and thermal pressures.

Figure 7(a,b) plots $\dot{M}_{\rm SF}$ and the gas depletion time $t_{\rm dep} \equiv M_{\rm gas}/\dot{M}_{\rm SF}$ for all models. At $t \sim 120$ Myr, the magnetic fields in the ring become strong enough to decrease $\dot{M}_{\rm SF}$ and

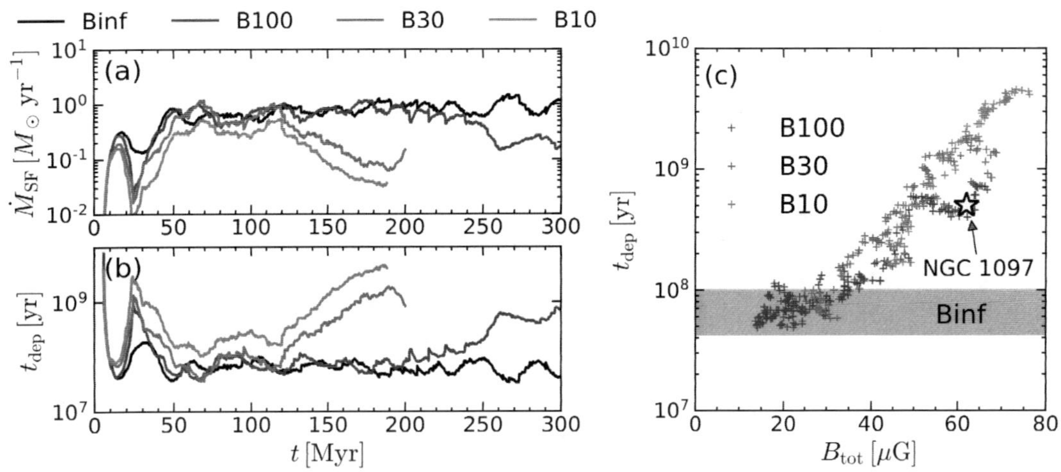

Figure 7. Temporal variations of (a) the ring SFR $\dot{M}_{\rm SF}$ and (b) the gas depletion time $t_{\rm dep}$, and (c) $t_{\rm dep}$ as a function of the total magnetic field strength $B_{\rm tot}$ for all magnetized models with $\dot{M}_{\rm in} = 1\ M_\odot\ {\rm yr}^{-1}$. The yellow star symbol in (c) marks the observational results of the nuclear ring in NGC 1097.

increase $t_{\rm dep}$. Figure 7(c) plots $t_{\rm dep}$ as a function of the total magnetic field strength $B_{\rm tot}$ for all magnetized models, showing that $t_{\rm dep}$ is a increasing function of $B_{\rm tot} \gtrsim 30\ \mu$G. The star symbol in Figure 7(c) denotes the observed conditions of the nuclear rings in NGC 1097 [33, 44], indicating that the ring of this galaxy is in the regime where the magnetic support against gravity is dynamically important.

Strong magnetic fields in the nuclear ring also exerts magnetic stress, transporting the angular momentum of rotating gas radially outward, and thus causing gas inflow. In a quasi-steady state, the mass accretion rate is given by

$$\dot{M}_{\rm acc} \approx \dot{M}_M + \dot{M}_R, \tag{6}$$

where $\dot{M}_M$ and $\dot{M}_R$ are the mass accretion rate due to the maxwell stress $\mathcal{T}_{R\phi} = -B_R B_\phi/(4\pi)$ and the Reynolds stress $\mathcal{R}_{R\phi} = \rho v_R v_\phi$ defined as

$$\dot{M}_M \equiv 2\pi \left[\frac{\partial(Rv_c)}{\partial R}\right]^{-1} \frac{\partial(R^2\mathcal{T}_{R\phi})}{\partial R}, \qquad \dot{M}_R = 2\pi \left[\frac{\partial(Rv_c)}{\partial R}\right]^{-1} \frac{\partial(R^2\mathcal{R}_{R\phi})}{\partial R}, \tag{7}$$

respectively, with $v_c = (\Omega - \Omega_p)R$ being the background circular velocity in the rotating frame. Figure 8(a) plots the numerically measured $\dot{M}_{\rm acc}$ at $R = 100$ pc (solid lines) as a function of time, in comparison with the predicted $\dot{M}_M$ (dashed lines) and $\dot{M}_R$ (dotted lines). In the unmagetized model, $\dot{M}_{\rm acc} \sim 4 \times 10^{-4}\ M_\odot\ {\rm yr}^{-1}$ on average, indicating that there is almost no gas accretion without magnetic fields. Although $\dot{M}_{\rm acc}$ remains small at early time, it increases quite rapidly to $\dot{M}_{\rm acc} \sim 0.02$–$0.03\ M_\odot\ {\rm yr}^{-1}$ at late time in the magnetized models. Note that $\dot{M}_{\rm acc} \simeq \dot{M}_M$ when magnetic fields are strong, suggesting that magnetic tension is the major driver of mass accretion in our simulations. Figure 8(b) plots the radial distributions of $\dot{M}_{\rm acc}$ for a few selected epochs for the model with $\beta_{\rm in} = 100$. Note that $\dot{M}_{\rm acc}$ is not constant in radius but decreases toward small R, with a slope decreasing with time. This implies that there is a sustained mass deposition at $R < R_{\rm ring}$ mediated by magnetic tension, eventually forming a nuclear disk at the galaxy center.

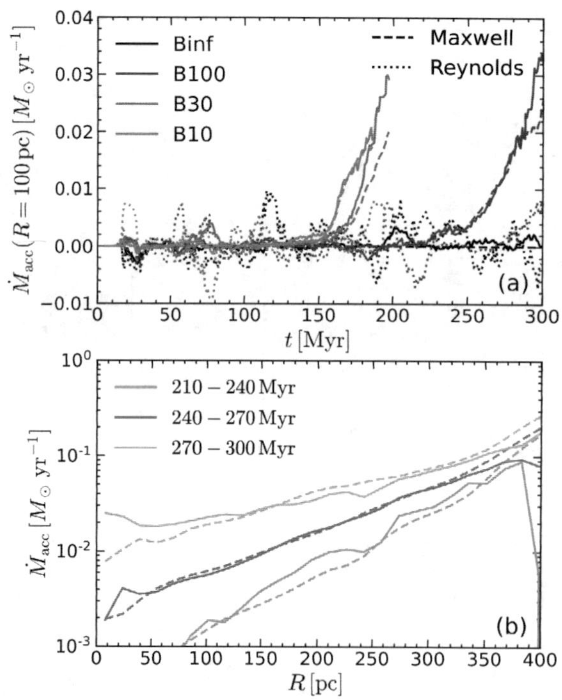

Figure 8. (*a*) Temporal changes in the accretion rate $\dot{M}_{\rm acc}$ (solid lines) measured at $R = 100$ pc for models with $\beta_{\rm in} = \infty$ (blakc), 100 (blue), 30 (green), and 10 (red). The dashed and dotted lines correspond to $\dot{M}_R$ and $\dot{M}_M$, respectively. (*b*) Radial distributions of $\dot{M}_{\rm acc}$ (solid lines) measured at different epoches for the model with $\beta_{\rm in} = 100$. The dashed line plots the predicted accretion rate $\dot{M}_M$ in Equation (7) due to the Maxwell stress.

4. Summary

Barred galaxies host a nuclear ring with intense star formation. We have presented the results of three-dimensional, semi-global, MHD simulations of star formation in nuclear rings. We find that the star formation rate $\dot{M}_{\rm SF}$ in nuclear ring is controlled primarily by the mass inflow rate $\dot{M}_{\rm in}$. Supernova feedback is responsible for small-amplitude fluctuations in $\dot{M}_{\rm SF}$ with a timescale less than 40 million years, while the SFR variations over longer timescales are due to changes in the mass inflow rates. When $\dot{M}_{\rm in}$ is constant, the ring SFR ($= 0.8\dot{M}_{\rm in}$) remains quite steady. While the time-varying $\dot{M}_{\rm in}$ can cause long-timescale ($\Delta\tau_{\rm in} \gtrsim 50$ Myr), large-amplitude SFR variations, short-timescale ($\Delta\tau_{\rm in} \sim 20$ Myr) fluctuations in the SFR are likely due to supernova feedback rather than $\dot{M}_{\rm in}$. The combination of supernova-driven turbulence and rotational shear amplifies both turbulent and regular magnetic fields. The breakout of superbubbles creates a toroidal-to-poloidal transition in high-altitude regions. Strong toroidal magnetic fields not only suppress star formation in the ring at late times but also induce gas accretion toward to the galaxy center, potentially forming a circumnuclear disk near the galaxy center.

Acknowledgments

The work of W.-T. Kim was supported by the grant of National Research Foundation of Korea (2022R1A2C1004810). Computational resources for this work were provided by Princeton Research Computing, a consortium including PICSciE and OIT at Princeton University, and by the Supercomputing Center/Korea Institute of Science and Technology Information with supercomputing resources including technical support (KSC-2021-CRE-0025).

References

[1] de Vaucouleurs G 1963, *ApJ* **138** 934
[2] Sellwood J A and Wilkinson A 1993 *Rep. on Prog. in Physics* **56** 173
[3] Knapen J H, Shlosman, I and Peletier, R F 2000 *ApJ* **529** 93
[4] Whyte L F, Abraham R G, Merrifield M R, et al. 2002 *MNRAS* **336** 1281
[5] Laurikainen E, Salo H and Buta R 2004 *ApJ* **607** 103
[6] Buta R J, Sheth K, Athanassoula E, et al. 2015 *ApJS* **217** 32
[7] Díaz-García S, Salo H, Laurikainen E and Herrera-Endoqui M 2016, *A&A* **587** A160
[8] Díaz-García S, Salo H, Knapen J H and Herrera-Endoqui M 2019 *A&A* **631** A94
[9] Sanders R H and Huntley J M 1976 *ApJ* **209** 53
[10] Athanassoula E 1992 *MNRAS* **259** 345
[11] Martini P, Regan M R, Mulchaey J S and Pogge R W 2003 *ApJS* **146** 353
[12] Martini P, Regan M R, Mulchaey J S and Pogge R W 2003 *ApJ* **589** 774
[13] Kim W-T, Seo W-Y, Stone J M, Yoon D and Teuben P J 2012 *ApJ* **747** 60
[14] Kim W-T, Seo W-Y and Kim Y 2012 *ApJ* **758** 14
[15] Seo W-Y, Kim W-T, Kwak S, et al. 2019 *ApJ* **872** 5
[16] Li Z, Shen J and Kim W-T 2015 *ApJ* **806** 150
[17] Sormani M C, Barnes A. T., Sun J, et al. 2023 *MNRAS* **523** 2918
[18] Hsieh P-Y, Matsushita S, Liu G, et al. 2011 *ApJ* **736** 129
[19] Song Y, Linden S T, Evans A S, et al. 2021 *ApJ* **916** 73
[20] Stuber S K, Saito T, Schinnerer E, et al. 2021 *A&A* **653** A172
[21] Morris M and Serabyn E 1996 *ARA&A* **34** 645
[22] Longmore S N, Bally J, Testi L, et al. 2013 *MNRAS* **429** 987
[23] Koepferl C M, Robitaille, T P, Morales E F E and Johnston K G 2015 *ApJ* **799** 53
[24] Tsuboi M, Handa T and Ukita N 1999 *ApJS* **120** 1
[25] Federrath C, Rathborne J M, Nongmore S N, et al. 2016 *ApJ* **832** 143
[26] Kruijssen J M D, Longmore S N, Elmegreen B G, et al. 2014, *MNRAS* **440** 3370
[27] Krumholz M R, Kruijssen, J M D and Crocker, R M 2017 *MNRAS* **466** 1213
[28] Armillotta L, Krumholz, M R, Di Teodoro, E M and McClure-Griffiths N M 2019 *MNRAS* **490** 4401
[29] Sormani M C, Tress R G, Glover S C O, et al. 2020 *MNRAS* **497** 5024
[30] Beck R, Ehle M, Shoutenkov V, Shukurov A and Sokoloff D. 1999, *Nature* **397** 324
[31] Beck R, Fletcher A, Shukurov A, et al. 2005 *A&A* **444** 739
[32] Lopez-Rodriguez E, Beck R, Clark S E, et al. 2021 *ApJ* **923** 150
[33] Tabatabaei F S, Minguez P, Prieto M A and Fernández-Ontiveros J A 2018 *Nature Astronomy* **2** 83
[34] Moon S, Kim W-T, Kim C-G and Ostriker E C 2021 *ApJ* **914** 9
[35] Moon S, Kim W-T, Kim C-G and Ostriker E C 2022 *ApJ* **925** 99
[36] Moon S, Kim W-T, Kim C-G and Ostriker E C 2023 *ApJ* **946** 114
[37] Stone J M, Gardiner T A, Teuben P, Hawley J F and Simon J B 2008 *ApJS* **178** 137
[38] Kim C-G and Ostriker E C 2017 *ApJ* **846** 133
[39] Boris J P 1970 *Proc. Fourth Conf. on Numerical Simulation of Plasmas* ed. J P Boris and R A Shanny R A (Washington, DC: Naval Research Laboratory) p3
[40] Brandenburg A and Subramanian K 2005 *Physics Reports* **417** 1
[41] Brandenburg A and Ntormousi E 2023 *Annu. Rev. Astron. Astrophys.* **61** 561
[42] Kim C-G and Ostriker E C 2015 *ApJ* **815** 67
[43] Federrath C, Chabrier G, Schober J, et al. 2011 *PRL* **107** 114504
[44] Prieto M A, Fernandez-Ontiveros J A, Bruzual G, et al. 2019 *MNRAS* **485** 3264

Journal of Physics: Conference Series **2742** (2024) 012007 doi:10.1088/1742-6596/2742/1/012007

Simulating Wind-Blown Nebulae from Single and Binary Massive Stars

Jonathan Mackey

Dublin Institute for Advanced Studies, Astronomy & Astrophysics Section, DIAS Dunsink Observatory, Dublin, D15 XR2R, Ireland

E-mail: jmackey@cp.dias.ie

Abstract. Winds from massive stars expand supersonically into their surroundings, creating dynamic and fascinating nebulae that can give us insight into physical processes in interstellar plasma, and into the evolutionary history of the stars. Around single stars, parsec-scale bubbles such as bow shocks and ring nebulae are formed, whereas in colliding-wind binary (CWB) systems the high wind density produces intense time- and space-dependent emission across the electromagnetic spectrum from radio to gamma-rays. This contribution summarizes some recent results from 3D MHD modelling of bow shocks around runaway stars such as ζ Oph, and of the wind-collision zone of the CWB systems WR 140 and WR 21a. A resolution study of 3D simulations of bow shocks shows that X-ray emission from the shocked wind is time-variable and that converged results can be obtained once the Kelvin-Helmholtz instability at the contact discontinuity is resolved. Simulations of the CWB system WR140 show that inverse-Compton cooling of the shocked plasma can trigger runaway cooling when the orbit is near periastron, producing strong compression and dynamical instabilities. This sharply reduces the hard-X-ray emission around periastron, in agreement with observations. Scaling tests of the simulation software PION are also presented for a model of the CWB system WR21a run on up to 8192 cores using the HPC system *Karolina*.

1. Introduction

Massive stars drive strong winds into the interstellar medium (ISM), generating shockwaves and often producing bright nebulae including both thermal and non-thermal emission. The nebulae can be used to probe shock physics as well as to constrain mass-loss rates of stars [1], the evolutionary history of post-main-sequence stars [2, 3], and the effects of radiative heating on winds [4]. While most massive stars form in binary or triple systems within star clusters, some of these are ejected and disperse through the Galaxy [5].

Bow shocks form around massive stars when the relative motion between star and interstellar medium (ISM) exceeds the ISM sound speed, because of the hydrodynamic interaction between stellar wind and ISM. A two-shock structure forms, divided by a contact discontinuity and swept into a bow shape by the dynamical pressure of the relative motion [6]. For hot stars the (inner) reverse shock is usually adiabatic whereas the (outer) forward shock is often radiative [7, 8]. Magnetic fields can impact the structure and stability of bow shocks by limiting the compressibility of the gas and by inhibiting instability growth at the contact discontinuity [9–11]. Direct detection of the shocked wind through thermal X-ray emission has long been predicted [12, 13] but, despite many searches [14, 15], has only been detected in the bow shock of the closest

Journal of Physics: Conference Series **2742** (2024) 012007 doi:10.1088/1742-6596/2742/1/012007

O star to Earth, ζ Oph [11, 16]. This detection is important for measuring the energy content of wind bubbles, and to constrain effectiveness of physical processes such as thermal conduction and instability-driven turbulent mixing. These are the primary energy loss mechanisms for the hot phase of the ISM [17–19].

Colliding-wind binaries (CWBs) consist of two massive stars, each of which drives a strong wind, and the wind-wind collision between the two stars emits radiation at a level detectable above the stellar emission (usually at radio, optical, X-ray and sometimes γ-ray energies) [e.g. 20, 21]. With typical orbital periods of days to a few years, they provide ideal laboratories to study shocks with predictable and time-dependent properties [22], particle acceleration [23] and dust formation [24].

The work presented here uses the radiation-magnetohydrodynamics (MHD) software PION [25], with some upgrades described in [26]. PION solves the Euler or ideal-MHD equations on a uniform mesh using the finite-volume method with an integration scheme accurate to 2nd order in time and space. Static mesh-refinement (*nested grids*) is used to focus resolution and computing resources on regions of interest; this method is quite efficient when such regions can be predicted ahead of time [27]. Parallelisation is via a hybrid MPI+OpenMP scheme using MPI for domain decomposition and OpenMP threads operating on 1D columns of cells in each sub-domain. The stellar wind boundary condition consists of a sphere of radius at least 7 cells on the coarsest level containing the boundary. Mass, momentum and energy flux into the domain from this sphere are calculated based on the adopted mass-loss rate and stellar-wind velocity of the star, with an option for wind acceleration according to a commonly used β-law prescription [26]. For binary systems a simple N-body integrator calculates the orbital motion of the stars and the wind boundary is updated whenever the star moves by 0.1 cells.

2. Bow shocks around massive stars

The direct detection of the shocked stellar wind in bow shocks around hot stars has proven difficult because of the low density of the emitting gas. So far the only detection is of the hot plasma around the closest O star to Earth, ζ Oph, using *Chandra* observations [11, 16]. The observed soft, thermal X-ray emission has a temperature ($T \sim 0.2\,\mathrm{keV}$) well below that of the shocked wind, and is thought to arise from gas near the contact discontinuity (CD) with temperature intermediate between the shocked wind and ISM. This layer may be produced by thermal conduction [18] and/or turbulent mixing [7, 17] at the wind-ISM interface. Radiative cooling in this X-ray emitting layer is thought to be the primary energy-loss mechanism for the hot phase of the ISM [e.g. 19], and so characterising its properties are important for our understanding of the multi-phase ISM.

Predictions for X-ray emission in the Bubble Nebula (NGC 7635) using 2D hydrodynamical simulations of this bow shock [13] proved to be overoptimistic (no emission was detected by [15]) for reasons that are still not clear. Conversely, [11] made 3D MHD simulations of the bow shock of ζ Oph, obtaining synthetic emission maps that show weaker thermal X-ray emission than is actually observed. The 2D hydrodynamic simulations [13] showed vigorous mixing of wind and ISM gas driven by the Kelvin-Helmholtz (KH) instability at the CD, whereas the 3D MHD simulations showed a smooth CD. [11] speculated that the combination of MHD inhibition of the KH instability [28], together with the diffusive HLL solver used for MHD, resulted in much weaker mixing and therefore fainter X-ray emission. One should distinguish between the former (physical) mechanism and the latter (purely numerical) effect in order to understand the conflicting results from modelling NGC 7635 and ζ Oph.

To this end, we performed a numerical study of a bow shock around a runaway massive star using different sets of equations (Euler, ideal MHD), solvers (HLL, HLLD), dimensionality (2D, 3D) and spatial resolution (spanning a factor of 8 in cell diameter). The full results will be presented in a paper currently in preparation; here we show some initial results of the study.

Journal of Physics: Conference Series **2742** (2024) 012007 doi:10.1088/1742-6596/2742/1/012007

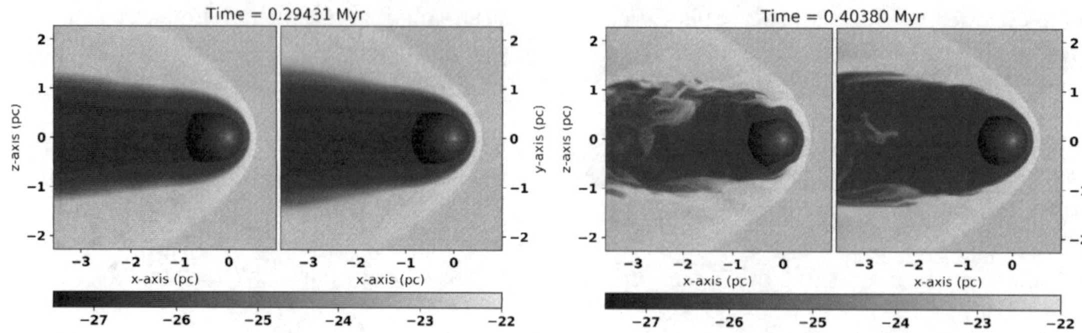

Figure 1. Slices of $\log_{10}\rho/(\mathrm{g\,cm^{-3}})$ through the planes $y = 0$ and $z = 0$ for low-resolution simulation 1 (left two panels) and high-resolution simulation 3 (right two panels) at the times indicated. The star is at the origin, and the reference frame is such that the star is stationary on the grid, with ISM flowing from right to left at speed $v_\star$. The stellar wind is mostly low density (blue) whereas ISM is high density (green/yellow), with a sharp interface at the CD.

A simulation is set up of a massive star moving at $v_\star = 30\,\mathrm{km\,s^{-1}}$ in the $\hat{x}$ direction through a uniform ISM with density $\rho = 10^{-23}\,\mathrm{g\,cm^{-3}}$. The mass-loss rate of the star is $\dot{M} = 10^{-7}\,\mathrm{M_\odot\,yr^{-1}}$ and wind terminal velocity $v_\infty = 2000\,\mathrm{km\,s^{-1}}$; the star has surface rotation velocity $v_{\rm rot} = 100\,\mathrm{km\,s^{-1}}$, a radius of $10\,\mathrm{R_\odot}$ and a surface split-monopole magnetic field strength of 1 G. We assume the star emits enough ionizing photons to keep the domain photoionized, leading to a minimum equilibrium temperature around 8000 K in the ISM. For the MHD simulations the ISM magnetic field is set to a uniform value of $\mathbf{B}_0 = (1, 4, 0)\,\mu$G, oriented almost perpendicular to $\mathbf{v}_\star$. A hybrid HLLD/HLL Riemann solver is used, such that HLL is used at strong shocks and when the density ratio between left and right states is more than 5 [26]. A nested grid is set up with 3 levels of refinement focused on the bow shock, such that the finest level contains both the star and the apex of the bow shock. Simulation 1 has resolution 128^3 grid cells per level, 2 has 256^3 and 3 has 384^3.

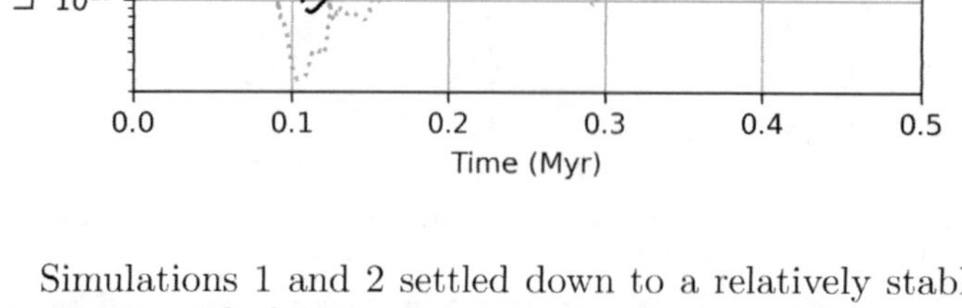

Figure 2. Soft X-ray (0.3-2 keV) luminosity of simulations 1 (blue solid line), 2 (red dashed line) and 3 (black dot-dashed line) as a function of simulation time. A high-resolution 2D simulation is overplotted (cyan dotted line) for comparison (with $\mathbf{B}_0 \parallel \mathbf{v}_\star$). The data at $t < 0.12$ Myr show effects of the initial conditions.

Simulations 1 and 2 settled down to a relatively stable state with a smooth CD, similar to the results of [11] (simulation 1 is shown in the left 2 panels of Fig. 1). In contrast, the (highest-resolution) simulation 3 shows the development of KH instability at the CD, with growing amplitude waves seeded near the apex of the bow shock and moving downstream. This entrains significant quantities of ISM gas into the hot shocked wind, reducing its temperature, increasing

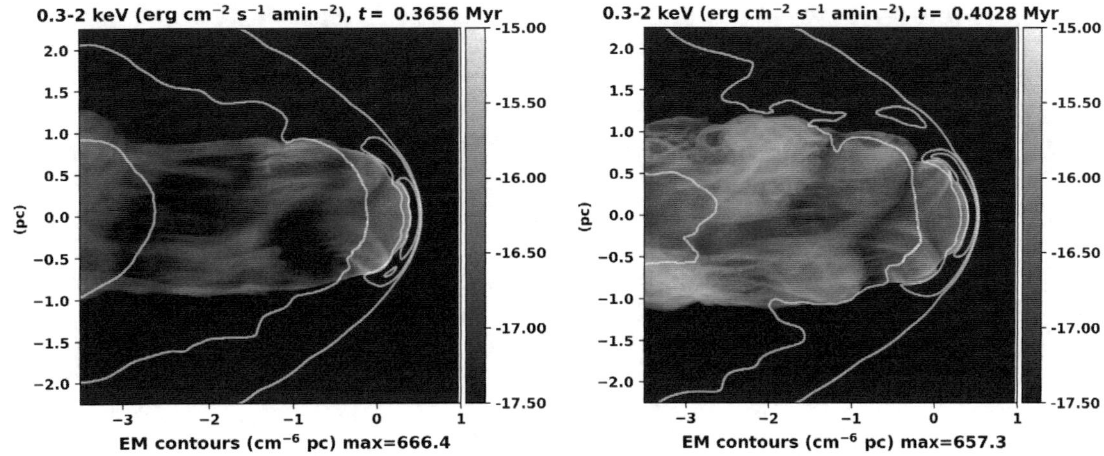

Figure 3. Projected soft-X-ray emission from simulation 3 at simulation times corresponding to weak (left) and strong (right) X-ray emission, separated by 37 000 years. The projection is along the $\hat{y}$ direction, and the surface brightness is plotted in the units indicated in the plot title. Contours of radio emission measure (integral of electron density squared along the line of sight) are overplotted on a linear scale from the peak value to 10% of the peak, tracing the shocked ISM.

its density, and hence increasing the X-ray emission. The symmetry axis of the termination shock is noticeably tilted with respect to $\mathbf{v}_\star$ in the x-z plane; this is a temporary feature driven by the KH vortices and at other times it is tilted in the opposite sense. As well as mixing wind and ISM gas, the turbulent flow also inhibits flow of shocked wind through the $-\hat{x}$ boundary, increasing the pressure (and density) of the shocked-wind region in the downstream. This forces the wind termination shock closer to the star in the downstream direction than in simulation 1.

Fig. 2 shows the time evolution of the X-ray luminosity from the full simulation domain for simulations 1-3. Simulation 1 has soft-X-ray luminosity of $L_\mathrm{X} \sim 3 \times 10^{29}\,\mathrm{erg\,s^{-1}}$ with weak fluctuations on $\sim 0.1\,\mathrm{Myr}$ timescales. Simulation 2 has somewhat stronger fluctuations, up to a factor of 2 in L_X, centred around the same value as simulation 1. Simulation 3 shows even larger fluctuations, up to a factor of 5 from maximum to minimum, driven by the periodic growth of KH vortices that induce strong mixing in the turbulent wake downstream from the star. Compared with a higher-resolution 2D simulation (cyan dotted line) the amplitude and timescale of the X-ray variability is quite consistent, albeit the simulations are not identical because in 2D the ISM magnetic field is initially parallel to $\mathbf{v}_\star$. The contrast between maximum and minimum emission is shown in Fig. 3, where we plot the X-ray surface brightness, projected along the $\hat{z}$ direction, at two different times. In the right panel the regions of bright emission correspond to vortices that have entrained ISM gas into the wind bubble, producing soft-X-ray emitting plasma. One may expect that adding time-dependence to the ISM properties (e.g., turbulence and overdense clouds) will also induce further variability in observable properties [e.g. 29].

We can draw two main conclusions from these simulations:

(i) The simulations of [11] had insufficient spatial resolution to capture the turbulent mixing induced by KH instability in the wake behind the bow shock of ζ Oph, and this is most likely the reason why they predict weaker X-ray emission than is observed.

(ii) Even in the absence of ISM inhomogeneity and turbulent flows, the X-ray emission of the shocked wind is time-variable, with fluctuations of up to a factor of 5-10 on a timescale of

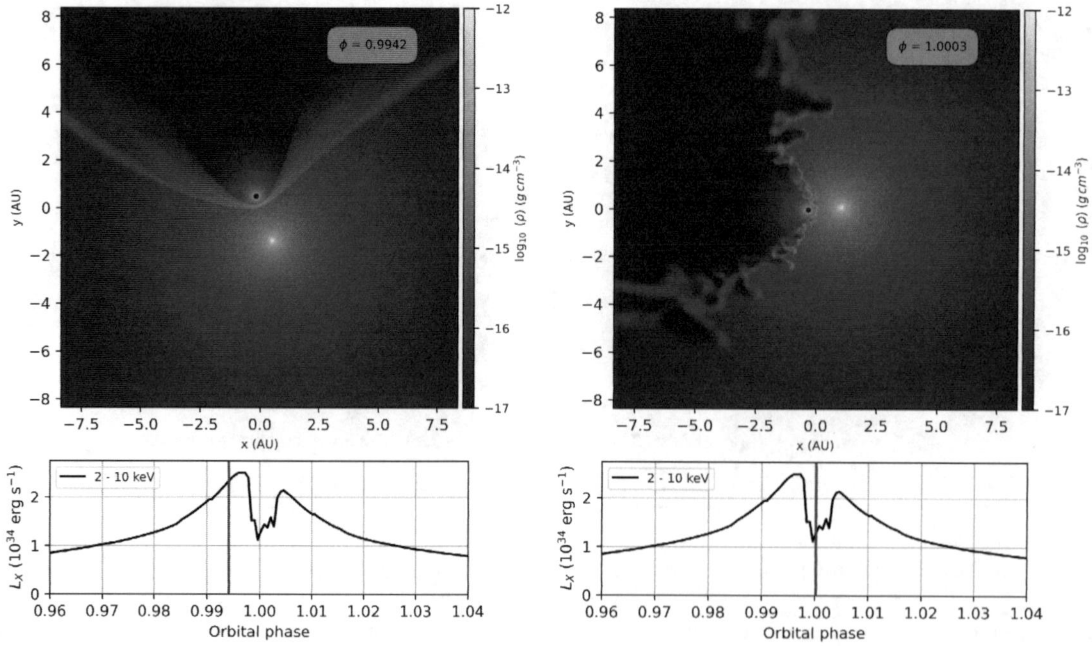

Figure 4. Snapshots of gas density shortly before (left) and at (right) periastron for the simulation of the WR 140 system. Gas density in the orbital plane $z = 0$ is shown on a logarithmic scale with the WR star being bright yellow and the O star a black circle. The hard-X-ray lightcurve is also plotted in the lower panels, and the vertical line shows the time of the snapshot.

~ 0.05 Myr. Comparison of simulations with observations therefore requires a statistical sample of bow shocks for conclusions to be drawn about the effectiveness of this turbulent mixing process in nature as compared with simulations.

3. Colliding-wind binaries

In contrast to the large but relatively faint bow shocks around runaway massive stars, CWBs are typically point sources (except for interferometric observations) with intense multiwavelength emission, both thermal and non-thermal. The wind-wind interaction shares many characteristics with bow shocks, including the bow shape that arises when the winds have unequal momentum density. The gas density in the shocked plasma is orders of magnitude larger, and characteristic size orders of magnitude smaller, than bow shocks, and the shocked region is swept into a spiral shape by orbital motion. Radiative processes are very similar between the two classes of object, but in CWBs the radiation and matter densities are usually large enough that local absorption plays an important role in the observed emission.

In [26] we investigated the archetypal CWB WR 140, consisting of a carbon-type Wolf-Rayet star (WC) and a main-sequence O star in an eccentric 7.9 year orbit that has separation ranging from 1.4 au at periastron to 10 times that at apastron. The wind-wind interaction was simulated for about 300 days centred on periastron, to model the strong dip observed in the X-ray lightcurve around periastron [22]. Using 3D MHD simulations with 7 levels of static mesh-refinement, we showed that such a dip is not produced by a standard calculation because the shocked wind of both stars remains largely adiabatic. The shocked layers remain dynamically stable and the X-ray luminosity through periastron follows the expectations for an adiabatic wind-collision zone, modified slightly to take account of wind acceleration.

Some time ago it was proposed that inverse-Compton cooling of the thermal plasma could be important in CWBs [30], but this cooling process has been largely ignored in modelling efforts for the past 30 years. We calculated that this should be important in WR 140 around periastron, using analytic arguments and 2D simulations of wind-wind collisions. This extra radiative cooling, that is effective when the shocks are within a few stellar radii of a luminous star, is enough in the WR 140 system to trigger runaway radiative cooling in the shocked layer, and the associated dynamical instability seen in radiative shocks.

Fig. 4 shows two snapshots of a 3D MHD simulation of the WR 140 system about 17 days before periastron (left panel) and at periastron (right panel). Gas density is plotted in the upper panels, and 2-10 keV X-ray luminosity in the lower panels. In the left-hand panel the separation of the two stars is such that neither IC nor collisional cooling processes are effective, and so the bow shock appears smooth and dynamically stable. Once the separation of the stars crosses a critical threshold, where the cooling timescale becomes shorter than the advection timescale (the time for shocked gas to leave the wind-collision zone), then the shocked layer undergoes runaway cooling, losing pressure support and collapsing to a thin sheet that is dynamically unstable to the thin shell instability [31]. While we do not spatially resolve the physical sheet thickness, the vigorous instability nevertheless induces growing oscillations that deform the shock fronts. This means that much of the shock experiences a normal flow velocity reduced by the cosine of the deformation angle, resulting in a lower post-shock temperature and weaker hard-X-ray emission.

This dip is one of the most striking features of the observed hard-X-ray lightcurve seen over at least 3 orbits [22]. Our results can qualitatively match the data in terms of the sharpness and asymmetry of the dip, but compared with observations the onset of the increase is too late in the simulations, and the decrease in emission is not deep enough. Further work is required to investigate these discrepencies, although the qualitative agreement is very encouraging. In particular, radiative transfer calculations to take account of the variable local absorption in the WR wind would be very valuable, because the correction of the observed X-ray flux for absorption only considers a single value of the column density (the source is unresolved).

4. Parallel scaling: WR 21a as a case study

The binary system WR 21a is one of the most massive in the Galaxy, with component masses of 93 $M_\odot$ and 53 $M_\odot$ for the WR and O star, respectively [32]. It is also one of the most X-ray luminous [33], a close binary with orbital period only 31.7 days and significant eccentricity, $e = 0.695$. We used this binary system as as the problem setup for testing the scaling of PION on the IT4I HPC system *Karolina* through a *EuroHPC* benchmarking project[1].

A 3D hydrodynamics simulation was set up with 8 refinement levels, each containing $320^2 \times 160$ grid cells, with a domain size $[8.192, 8.192, 4.096] \times 10^{14}$ cm on the coarsest level, centred on the origin. The system centre-of-mass is at the origin, the orbital plane is $z = 0$, and the stellar separation at periastron is 4.7×10^{12} cm. The simulation was integrated from periastron to close to apastron and a snapshot was saved. Starting from this snapshot, the simulation was then restarted using 1, 2, 4, 8, 16, 32, and 64 nodes, each of which was run with 1, 2, 4, and 8 OpenMP threads per MPI process, for total core counts of 128 - 8192 cores (*Karolina* has 128 cores per node). The simulation was run in each case for 199 coarse-level timesteps (25 472 on the finest level) to test the strong scaling.

Furthermore, lower-resolution simulations with the same setup were run to the same point in time, using $128^2 \times 64$, $160^2 \times 80$, $192^2 \times 96$ and $256^2 \times 128$ grid cells per level. These were run

[1] European High Performance Computing Joint Undertaking (EuroHPC JU) project `EHPC-BEN-2023B05-029`, run on Karolina, operated by the IT4Innovations National Supercomputing Center of the Czech Republic.

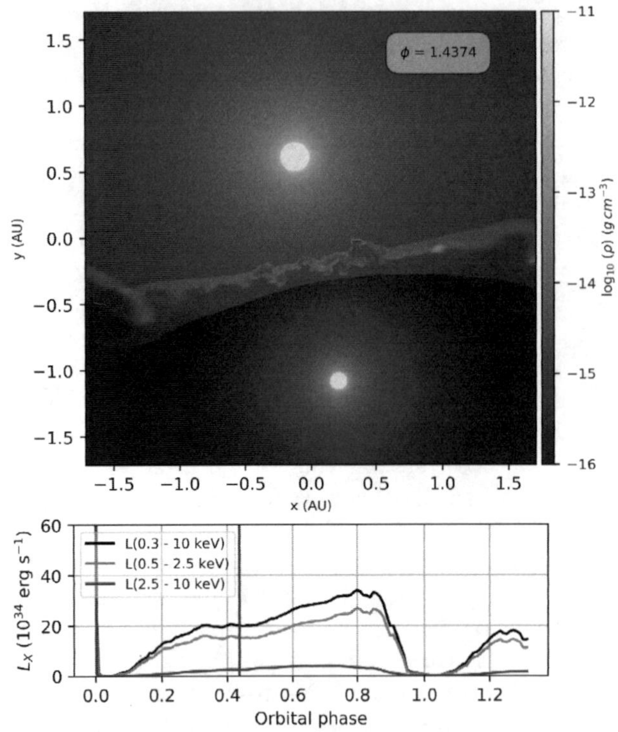

Figure 5. Snapshot showing a slice of 3D simulation of the CWB system WR 21a close to apastron, in the orbital plane $z = 0$. Log of gas density is plotted on a logarithmic scale. The lower panel shows the X-ray luminosity of the wind-collision zone as a function of orbital phase, and the vertical line shows the current phase. The system centre of mass is at the origin, the stars are orbiting counter-clockwise, and not all of the simulation domain is shown.

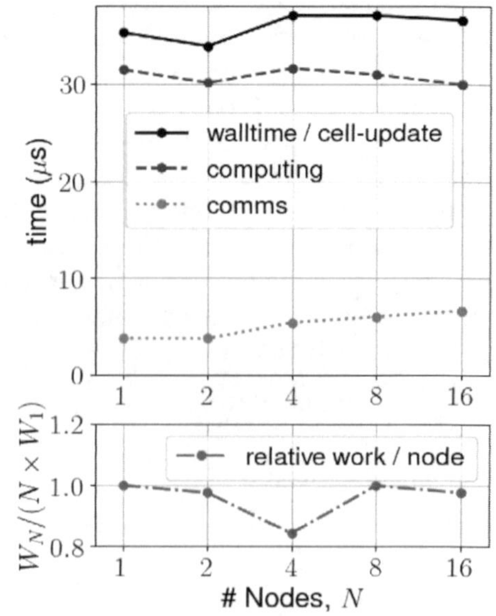

Figure 6. Weak scaling of PION for simulations of the WR 21a system. The upper panel shows the time required per cell-update per core as a function of the number of nodes. The time required for computation and communication is also plotted. The lower panel shows the amount of work per core for the simulations (it was not possible to make this exactly equal in all cases because the number of cells in each dimension should be divisible by 32).

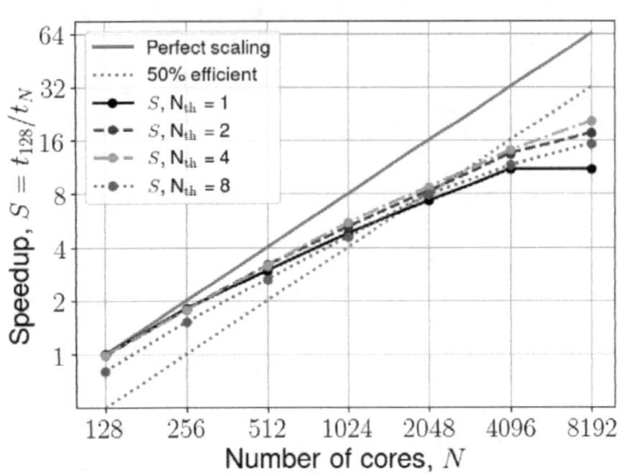

Figure 7. Strong scaling of PION for simulating the CWB system WR 21a over 199 coarse-level timesteps with $320^2 \times 160$ grid cells per level and 8 refinement levels. Speedup, S, is shown relative to a simulation using 1 node with 128 MPI processes and 1 OpenMP thread per process. Curves are shown for $N_{\rm th} = 1$, 2, 4, and 8 threads per MPI process. For $N_{\rm th} = 4$ the number of ghost cells exceeds the number of grid cells for $N > 2048$ cores.

Journal of Physics: Conference Series **2742** (2024) 012007 doi:10.1088/1742-6596/2742/1/012007

with 1, 2, 4, and 8 nodes, and compared with the highest-resolution simulation run on 16 nodes, in all cases using 8 OpenMP threads per MPI process, i.e., 16 MPI processes per node. These simulations were used to test the weak scaling of PION. It is not a perfect weak-scaling test because the amount of work per core/node was not constant (generally the cell counts in PION should be divisible by 32 in each dimension), but it gives a good baseline measurement with which we can predict the performance at even higher resolution on a similar problem, which is the main aim of the calculation.

The starting snapshot for the strong-scaling test, and the weak-scaling test on 16 nodes, is plotted in Fig. 5, where gas density is shown in the orbital plane. The larger and more massive WR star is towards the top, and the O star with the weaker wind is below. The wind-collision region is close to the stability limit, with weak KH instability along the contact discontinuity providing the seed for overdensities that locally undergo runaway cooling, although the bulk of the shocked plasma behaves adiabatically. The lower-resolution simulations look similar but with less instability as resolution degrades.

Results for weak scaling are plotted in Fig. 6, where the upper panel shows the time required per cell-update per core, as well as the time spent in communication and in computation. The lower panel shows the relative amount of work per node/core for the different simulations, which is approximately constant. The upper panel shows that performance is consistent from 1 node (128 cores) up to 16 nodes (2048 cores), with wall time and compute time basically constant, and a weak trend of increasing time on communication with increasing node count, from 10% of the total runtime on 1 node to 20% on 16 nodes. An extrapolation of this performance would imply that higher resolution simulations could run on up to 8192 cores with $< 30\%$ of runtime on communications.

Strong scaling results are plotted in Fig. 7, where we show the speedup obtained relative to a simulation with 1 node (128 cores), 128 MPI processes and 1 OpenMP thread per process. The efficiency remains above 50% from 1 to 16 nodes (2048 cores), which is the point at which the number of ghost cells equals the number of grid cells for each MPI process (when run with 4 OpenMP threads per process). For 32 and 64 nodes there is more ghost-cell calculation than grid-cell calculation, and so we don't expect high efficiency.

The strong-scaling plot shows that running PION is most efficient with 4 OpenMP threads per MPI process: for lower thread counts the scaling drops off faster whereas for larger thread counts the overall speed is lower. After some investigation we discovered that the communication routines themselves are the bottleneck for large OpenMP thread counts, because the boundary communication is not yet multithreaded. The next step is to implement this multithreading so that PION can run efficiently on 8-20 threads per MPI process.

5. Summary

Computational research on wind-driven nebulae around single and binary massive stars is advancing rapidly, enabled by mature and robust simulation codes and increasing computational resources. This contribution summarizes published and ongoing work from my research group on bow shocks around massive stars, wind-wind collisions in binary systems, and code development and testing. A resolution study of 3D simulations of bow shocks shows that X-ray emission from the shocked wind is strongly time-variable and that results consistent with higher-resolution 2D simulations can be obtained once the KH instability at the contact discontinuity is resolved. This may contribute to resolving discrepencies between previous simulations and X-ray observations: the overprediction of X-rays from simulations of the Bubble Nebula [13, 15] and underprediction from simulations of the bow shock of ζ Oph [11, 16]. With considerably increased computational cost, similar simulations could be run including diffusive processes such as thermal conduction [34] to attempt to resolve the physical diffusion length-scale and achieve numerically converged results.

Simulations of the CWB system WR 140 show that inverse-Compton cooling of the shocked plasma can trigger runaway cooling when the orbit is near periastron, producing strong compression and dynamical instabilities. This sharply reduces the hard-X-ray emission around periastron, in agreement with observations.

Scaling tests of PION were also presented for a model of the CWB system WR 21a run on up to 8192 cores, showing good strong scaling up to the point where the number of ghost cells equals the number of grid cells. Weak scaling was tested from 128 to 2048 cores and performance is consistent across this range. These scaling results show that PION can run efficiently for simulations of CWB systems on Tier-0 supercomputing systems.

Acknowledgments
I gratefully acknowledge support from: a Royal Society-Science Foundation Ireland University Research Fellowship; a Royal Society Research Fellows Enhancement Award 2021; an Irish Research Council (IRC) Starting Laureate Award; the Irish Centre for High-End Computing (ICHEC) for provision of computing facilities and support (projects `eurocc-af-2`, `dsast028b`, `dsast029c`); and the European High Performance Computing Joint Undertaking (EuroHPC JU) and the IT4Innovations National Supercomputing Center for a grant of computing time on Karolina (project `EHPC-BEN-2023B05-029`). I am especially grateful to my collaborators and current and former group members for their contributions to the results presented here.

References

[1] Gvaramadze V V, Langer N and Mackey J 2012 *MNRAS* **427** L50–L54 (*Preprint* 1209.0455)

[2] Mohamed S, Mackey J and Langer N 2012 *A&A* **541** A1 (*Preprint* 1109.1555)

[3] Mackey J, Mohamed S, Neilson H R, Langer N and Meyer D M A 2012 *ApJL* **751** L10 (*Preprint* 1204.3925)

[4] Mackey J, Mohamed S, Gvaramadze V V, Kotak R, Langer N, Meyer D M A, Moriya T J and Neilson H R 2014 *Nature* **512** 282–285 (*Preprint* 1408.2522)

[5] Gies D R 1987 *ApJS* **64** 545–563

[6] van Buren D and McCray R 1988 *ApJL* **329** L93–L96

[7] Mackey J, Gvaramadze V V, Mohamed S and Langer N 2015 *A&A* **573** A10 (*Preprint* 1410.0019)

[8] Henney W J and Arthur S J 2019 *MNRAS* **486** 3423–3433 (*Preprint* 1903.03737)

[9] van Marle A J, Decin L and Meliani Z 2014 *A&A* **561** A152 (*Preprint* 1312.5877)

[10] Meyer D M A, Mignone A, Petrov M, Scherer K, Velázquez P F and Boumis P 2021 *MNRAS* **506** 5170–5189 (*Preprint* 2107.05513)

[11] Green S, Mackey J, Kavanagh P, Haworth T J, Moutzouri M and Gvaramadze V V 2022 *A&A* **665** A35 (*Preprint* 2203.06331)

[12] Weaver R, McCray R, Castor J, Shapiro P and Moore R 1977 *ApJ* **218** 377–395

[13] Green S, Mackey J, Haworth T J, Gvaramadze V V and Duffy P 2019 *A&A* **625** A4 (*Preprint* 1903.05505)

[14] Toalá J A, Oskinova L M and Ignace R 2017 *ApJL* **838** L19 (*Preprint* 1703.04059)

[15] Toalá J A, Guerrero M A, Todt H, Sabin L, Oskinova L M, Chu Y H, Ramos-Larios G and Gómez-González V M A 2020 *MNRAS* (*Preprint* 2003.06371)

[16] Toalá J A, Oskinova L M, González-Galán A, Guerrero M A, Ignace R and Pohl M 2016 *ApJ* **821** 79 (*Preprint* 1602.07805)

[17] Slavin J D, Shull J M and Begelman M C 1993 *ApJ* **407** 83

[18] Comerón F and Kaper L 1998 *A&A* **338** 273–291

[19] Sasaki M, Ponti G and Mackey J 2022 *Handbook of X-ray and Gamma-ray Astrophysics* p 24 (*Preprint* 2207.01263)

[20] Dougherty S M, Beasley A J, Claussen M J, Zauderer B A and Bolingbroke N J 2005 *ApJ* **623** 447–459 (*Preprint* astro-ph/0501391)

[21] Rauw G 2022 *X-Ray Emission of Massive Stars and Their Winds* (Singapore: Springer Nature Singapore) pp 1–31 ISBN 978-981-16-4544-0 (*Preprint* 2203.16842)

[22] Pollock A M T, Corcoran M F, Stevens I R, Russell C M P, Hamaguchi K, Williams P M, Moffat A F J, Weigelt G, Shenavrin V, Richardson N D, Espinoza D and Drake S A 2021 *ApJ* **923** 191 (*Preprint* 2109.10350)

[23] White R, Breuhaus M, Konno R, Ohm S, Reville B and Hinton J A 2020 *A&A* **635** A144 (*Preprint* 1911.01079)

[24] Lau R M, Eldridge J J, Hankins M J, Lamberts A, Sakon I and Williams P M 2020 *ApJ* **898** 74 (*Preprint* 2006.08695)

[25] Mackey J, Green S, Moutzouri M, Haworth T J, Kavanagh R D, Zargaryan D and Celeste M 2021 *MNRAS* **504** 983–1008 (*Preprint* 2103.07555)

[26] Mackey J, Jones T A K, Brose R, Grassitelli L, Reville B and Mathew A 2023 *MNRAS* **526** 3099–3114 (*Preprint* 2301.13716)

[27] Freyer T, Hensler G and Yorke H W 2003 *ApJ* **594** 888–910 (*Preprint* arXiv:astro-ph/0306541)

[28] Frank A, Jones T W, Ryu D and Gaalaas J B 1996 *ApJ* **460** 777 (*Preprint* astro-ph/9510115)

[29] Baalmann L R, Scherer K, Kleimann J, Fichtner H, Bomans D J and Weis K 2021 *A&A* **650** A36 (*Preprint* 2104.03748)

[30] Cherepashchuk A M 1976 *Soviet Astronomy Letters* **2** 138

[31] Vishniac E T 1994 *ApJ* **428** 186–208 (*Preprint* astro-ph/9306025)

[32] Barbá R H, Gamen R C, Martín-Ravelo P, Arias J I and Morrell N I 2022 *MNRAS* **516** 1149–1157 (*Preprint* 2109.06311)

[33] Gosset E and Nazé Y 2016 *A&A* **590** A113 (*Preprint* 1604.01536)

[34] Meyer D M A, Mackey J, Langer N, Gvaramadze V V, Mignone A, Izzard R G and Kaper L 2014 *MNRAS* **444** 2754–2775 (*Preprint* 1408.2828)

Diffusive shock acceleration in relativistic, oblique shocks

Allard Jan van Marle[1], Artem Bohdan[2,3], Anabella Araudo[1,4], Fabien Casse[5], Alexandre Marcowith[4]

[1] Extreme Light Infrastructure ERIC, ELI Beamlines Facility, Za Radnici 835, CZ-25241 Dolni Brezany, Czech Republic
[2] Max-Planck-Institut für Plasmaphysik, Boltzmannstr. 2, DE-85748 Garching, Germany
[3] Excellence Cluster ORIGINS, Boltzmannstr. 2, DE-85748 Garching, Germany
[4] Laboratoire Univers et Particules de Montpellier (LUPM) Université Montpellier, CNRS/IN2P3, CC72, Place Eugéne Bataillon, F-34095 Montpellier Cedex5, France
[5] Université de Paris, AstroParticle & Cosmologie, CNRS CEA, Observatoire de Paris, PSL Research University, CNES, F-75013 Paris, France

E-mail: ajvanmarle@gmail.com

Abstract. Cosmic rays are charged particles that are accelerated to relativistic speeds by astrophysical shocks. Numerical models have been successful in confirming the acceleration process for (quasi-)parallel shocks, which have the magnetic field aligned with the direction of the shock motion. However, the process is less clear when it comes to (quasi-)perpendicular shocks, where the field makes a large angle with the shock-normal. For such shocks, the angle between the magnetic field and flow ensures that only highly energetic particles can travel upstream at all, reducing the upstream current. This process is further inhibited for relativistic shocks, since the shock can become superluminal when the required particle velocity exceeds the speed of light, effectively inhibiting any upstream particle flow. In order to determine whether such shocks can accelerate particles, we use the particle-in-cell (PIC) method to determine what fraction of particles gets reflected initially at the shock. We then use this as input for a new simulation that combines the PIC method with grid-based magnetohydrodynamics to follow the acceleration (if any) of the particles over a larger time-period in a two-dimensional grid. We find that quasi-perpendicular, relativistic shocks are capable of accelerating particles through the DSA process, provided that the shock has a sufficiently high Alfvénic Mach number.

1. Introduction

Cosmic rays are charged particles that have been accelerated to relativistic speeds. It is generally accepted that the acceleration process occurs at astrophysical shocks, with the particles gaining energy through repeated shock crossings. One of the main processes involved is that of Diffusive Shock Acceleration (DSA) [1], also known as Fermi 1 acceleration. This involves the deflection of charged particles by the upstream and downstream magnetic field, which requires magnetic instabilities. One of the sources of such instabilities is the presence of the charged particles themselves. The current generated by these particles as they move relative to the thermal gas can trigger the "streaming instability", which twists the magnetic field [2,3]. Because the wavelength of this instability scales with the current, it can continue to interact with the particles as they are accelerated, making the process both self-induced and self-sustaining.

Journal of Physics: Conference Series **2742** (2024) 012008 doi:10.1088/1742-6596/2742/1/012008

1.1. Diffusive Shock Acceleration in oblique shocks

The interaction between charged, non-thermal particles and a magnetic field and the DSA that follows has been studied in considerable detail in the case of a non-relativistic parallel shock, where the magnetic field is aligned with the shock-normal [4,5,6] and the motion of the gas does not approach the speed of light. However, the situation is less clear in the case of oblique, relativistic shocks. [4,5] found that at an angle of more than approximately 60°, the upstream current became too weak to trigger instabilities and therefore DSA could no longer occur. By contrast, [6] did find DSA. However, these simulations assumed that both parallel and oblique shocks would inject non-thermal particles into the upstream medium at the same rate.

In [7], the problem was further investigated by combining both particle-in-cell (PIC) and PIC-MHD (a combination of PIC and grid-based magnetohydrodynamics). Using a PIC code to model the shock itself, combined with PIC-MHD to model large-scale, long-term behaviour, it was determined that the occurrence of streaming instability, and the DSA that it causes, depends on the Alfvénic Mach number of the shock. For non-thermal particles to trigger an instability, their energy density has to exceed the energy density of the magnetic field. In the upstream medium, this is a problem. To travel upstream, the particle has to exceed the velocity at which the upstream medium advects toward the shock. Particles follow the magnetic field lines, which, for an oblique shock, means that the particle velocity has to exceed $v_s/\tan\theta_B$, where v_s is the velocity of the upstream medium in the shock rest-frame; and θ_B is the angle between the magnetic field and the shock normal. The larger θ_B, the higher the required velocity, which means that only the fastest particles -a small fraction of the non-thermal particles that are generated at the shock-will be able to travel upstream. As a result, they can only perturb the magnetic field if that field is sufficiently weak, which translates to a high Alfvénic Mach number. [7] showed that for $\theta_B = 60^o$ the upstream Alfvénic Mach number has to be at least $\simeq 30$ for even partial DSA to take place, and only at Alfvénic Mach numbers exceeding 100 does the process become efficient. Furthermore, this dependence is extremely strong: for $\theta_B = 70^o$, the minimum Alfvénic Mach number would be around 1000.

1.2. Can DSA occur in oblique, relativistic shocks?

The model conclusions drawn in the previous section only apply to non-relativistic shocks. If the shock speed approaches the speed of light, a second obstacle to DSA appears: While particles have to move at a speed larger than $v_s/\tan\theta_B$ to travel into the upstream medium, at the same time, they still need to obey $v < c$, with c the light velocity. So, for any shock where $v_s/\tan\theta_B \geq c$, (also referred to as super-luminal shocks) no particle will be able to move upstream, thereby precluding the possibility of DSA.

Therefore, DSA in oblique, relativistic shocks can only occur within a relatively narrow parameter space, where the shock's Mach number is high enough that upstream particles can trigger instabilities, and the obliquity is not so large that the shock becomes super-luminal. Here it should also be noted that at relativistic speeds, the Lorentz contraction causes the obliquity in the shock rest-frame to be larger than in the co-moving frame, and it is the obliquity in the shock rest-frame that determines whether the shock is super-luminal. We will now show the preliminary results of our simulations of such shocks.

2. PIC simulations

For our simulations, we repeat the process described in [7], which involves two sets of simulations. In the first, PIC is used to model the structure of the shock and determine the injection rate of non-thermal particles at the shock front. This data is then used in PIC-MHD models to determine the evolution of the plasma and its magnetic field in a long-term, 2-D simulation.

The PIC simulations are performed with a fully relativistic electromagnetic 2-D PIC code based on the TRISTAN code [8,9]. Shocks are initialized using the reflecting-wall setup, where a beam of plasma is aimed at a reflective boundary, and the resulting shock is traced as it moves back into the upstream medium. For these simulations, we choose to use an out-of-plane magnetic field configuration, so the magnetic field lies in the x-z, plane, whereas the simulation box is in the x-y plane. We use an upstream Lorentz-factor $\Gamma = 1.86$ and vary θ_B from 0 to 33 in the upstream rest-frame, which puts it at 0 to 60 degrees in the shock rest-frame. This covers the full parameter space from parallel to superluminal shocks The Alfvénic and sonic Mach numbers are 20 and 22 respectively to give us a plasma-$\beta = 1$. The results of these simulations were presented in [10], and showed the dependence of the particle injection rate in the upstream medium as a function of the shock obliquity, starting high for e parallel shock, but decreasing with an increased obliquity. For the superluminal shock, no particles were injected into the upstream medium.

3. PIC-MHD simulations

From the injection rate obtained in the PIC model, we start our PIC-MHD model. The PIC-MHD method, which was described in [11], assumes that a plasma can be described as a combination of two components: a thermal plasma, which can be treated as a fluid through traditional grid-based magnetohydrodynamics (MHD) and a non-thermal component that is treated as a collection of particles, modelled through PIC. These two components exist in the same time and space and interact with each other through the Lorentz force (which describes the force, exerted by the electro-magnetic field and the particles on each other), and a modified version of Ohm's law that takes into account an additional component of the electric field, generated by the non-thermal current. This method was implemented in the existing MPI-AMRVAC code [12], as described in [6]. Although this PIC-MHD code was initially limited to non-relativistic MHD, it has since been extended to include relativistic magneto-hydrodynamics. By combining MHD and PIC, we can model the interaction between non-thermal particles and an electromagnetic field at a lower computational cost than if we used only PIC. This allows us to use a larger simulation box and run the simulation for a longer period of time than would otherwise be possible. However, because the shock itself is treated as a discontinuity in the thermal plasma, the transition from thermal to non-thermal particles is not part of the model and needs to be prescribed based on information obtained from PIC simulations.

For this simulation we use a 2-D box-size with $180 \times 30\, R_L$; R_L being the Larmour radius defined by the velocity of the injected particles and the upstream magnetic field. This box is covered by a fixed grid with individual cells of $0.15\, R_L$, which guarantees that the gyro motion of the individual particles is adequately resolved. Unlike the PIC simulations, we run this model in the reference frame of the shock, starting from an analytical solution for a standing shock with an upstream $\theta_B = 30^o$ derived from the Rankine-Hugoniot conditions. The flow moves from the inner x-boundary toward the outer x-boundary. Once the simulation starts, we inject protons at the shock with a rate of two percent of the mass that is passing through the shock front, matching the results shown in PIC simulations as shown in [10]. We neglect non-thermal electrons. This allows us to scale the resolution of the grid cells to the proton-gyro radius without the need to resolve the much shorter electron gyro-radius.

At injection, the particles are given a Lorentz factor that is three times that of the upstream flow, with the velocity isotropically distributed in the frame of the shock.

We repeat the simulation with an upstream flow that has a Lorentz factor of 3.0 to test the code's ability to perform simulations for relativistic shocks. This also obliges us to reduce θ_B to ten degrees, because a $\Gamma = 3$ shock would become superluminal at an angle of thirty degrees.

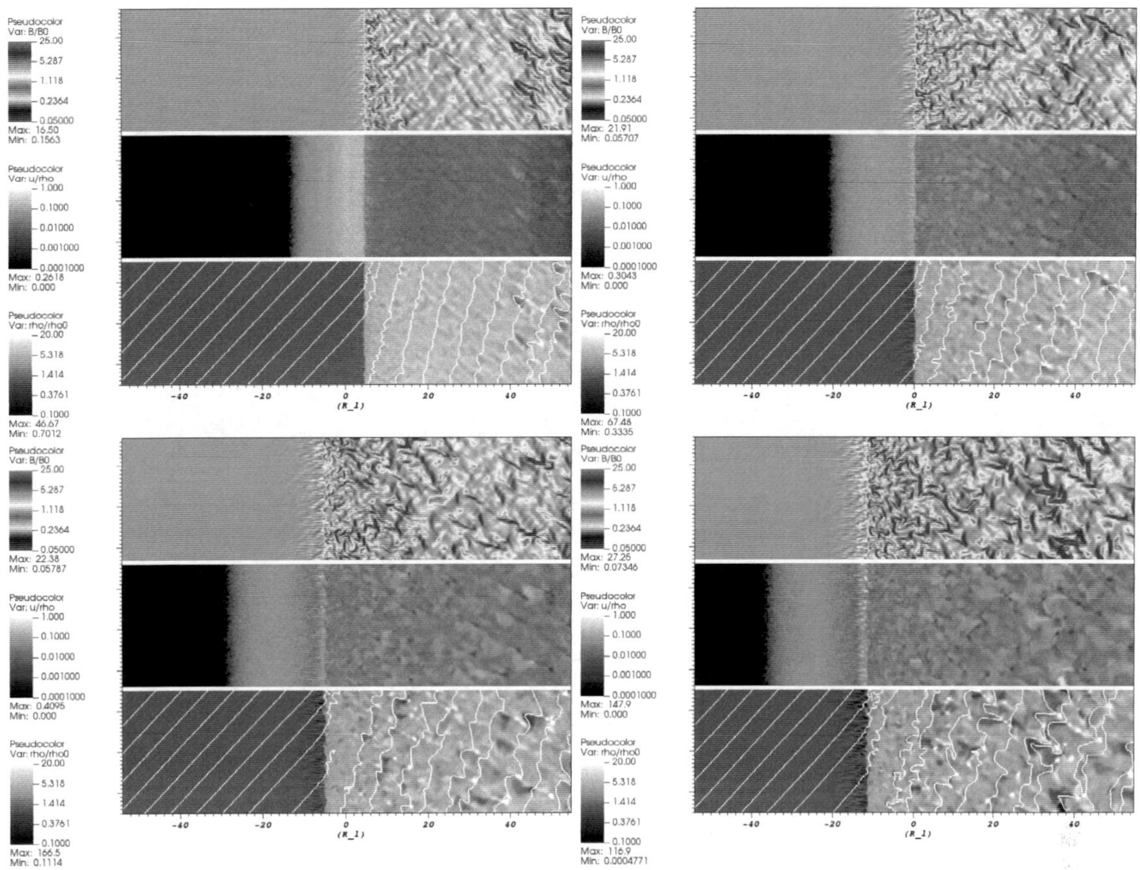

Figure 1. Simulation results for a shock with $\Gamma = 1.86$, $\theta_B = 30^o$. From top to bottom, the figures show the magnetic field relative to the initial upstream field, the ratio of non-thermal to thermal particles, and the thermal gas density relative to the initial upstream density with the magnetic field lines for $t = 200\ R_L/c$ (upper left), $300\ R_L/c$ (upper right),$400\ R_L/c$ (lower left) and $500\ R_L/c$ (lower right).

4. Results

4.1. Simulation results for $\Gamma = 1.86$

Figure 1 demonstrates the result of our simulation with $\Gamma = 1.86$ and $\theta_B = 30$. Here we show two snapshots in time at $t = 200\ R_L/c$ (upper left panel) through $500\ R_L/c$ (lower right panel). Each panel shows the magnetic field relative to the initial upstream field (top), the non-thermal particle density relative to the thermal gas particle density (centre), and the thermal gas density relative to the initial upstream thermal gas density (bottom) as well as the magnetic field lines. In the early stage (upper left), the upstream magnetic field remains largely unperturbed. In any case, the particles have had no time to move far upstream. Downstream, the field lines show deviation from the initial condition, but the changes are still small. Over time, this changes. The upstream field shows signs of perturbation, which follows the expected form for a streaming instability though the deviations are relatively small, particularly when compared to non-relativistic results such as shown in [4,5,6]. In all of this, we should keep in mind that we do not include the effect of non-thermal electrons. As a result, we don't see electron-induced instabilities such as the Whistler and fire-hose instabilities.

The particle acceleration is demonstrated in fig. 2, which shows the spectral energy

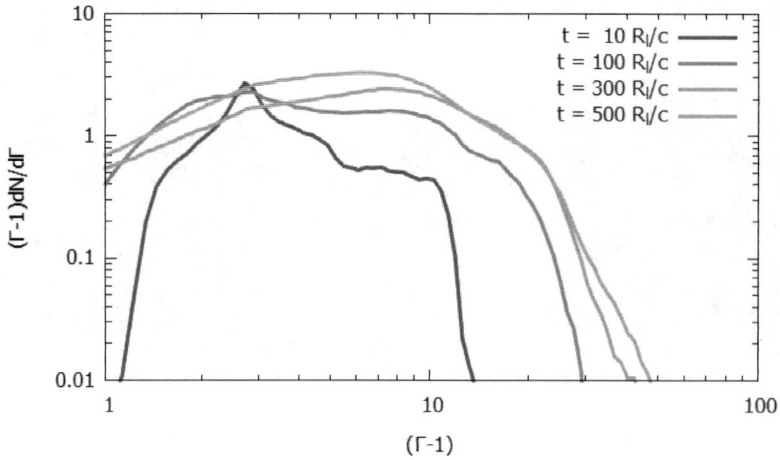

Figure 2. Time-evolution of the spectral energy distribution of non-thermal particles for a shock with $\Gamma = 1.86$, $\theta_B = 30^o$.

distribution (SED) of the non-thermal particles as a function of time.

To understand the evolution of the particle energies, we have to differentiate between two acceleration mechanisms. The first, DSA, was described previously. However, there is a second mechanism: Shock Drift Acceleration (SDA). Unlike DSA, which involves particles travelling large distances up- and down-stream before being reflected by local instabilities in the magnetic field, SDA is a localized phenomenon, which involves particles moving along the shock front, repeatedly transitioning between the upstream and downstream medium. It is effective in giving the particles an initial boost in energy. However, once a particle escapes from the vicinity of the shock, the process comes to an end, which limits the maximum energy that can be obtained. DSA, which involves a much larger volume of space, can continue as the particle gains energy as long as some instability catches the particle and reflects it toward the shock.

Initially, the SED shows only limited growth, ranging from non-relativistic to $\Gamma \simeq 15$. In this stage, there is no DSA, but some particles are accelerated through SDA)]. SDA, which does not require large-scale instability in the upstream magnetic field, can occur because particles can repeatedly cross the shock as they circle the magnetic field lines. This process was shown in detail in [6]. While it would not be significant for a $\theta_B = 30^o$ shock in a non-relativistic case, the Lorentz compression increases the effective angle of the upstream magnetic field in the shock rest frame, allowing for a more efficient SDA. Over time, the high-energy tail of the SED increases until it approaches $\Gamma = 80$, demonstrating that the particles are being accelerated efficiently. We also see the initial stage of a power-law distribution, indicating the onset of DSA.

4.2. Simulation results for $\Gamma = 3$

The results for our model with $\Gamma = 3$ and $\theta_B = 10^o$ are shown in Fig. 3, which shows the same variables as Fig. 1 at $t = 100R_L/c$ (left) and $t = 200R_L/c$ (right). Compared to the model with $\Gamma = 1.86$, the instabilities develop very quickly, with significant upstream distortion of the magnetic field clearly visible at $t = 200R_L/c$. The SED for this simulation is shown in Fig. 4, which demonstrates that the acceleration is even more efficient than for the slower model, with particles exceeding $\Gamma = 100$ within a short ($t = 200R_L/c$) timeframe.

Journal of Physics: Conference Series **2742** (2024) 012008 doi:10.1088/1742-6596/2742/1/012008

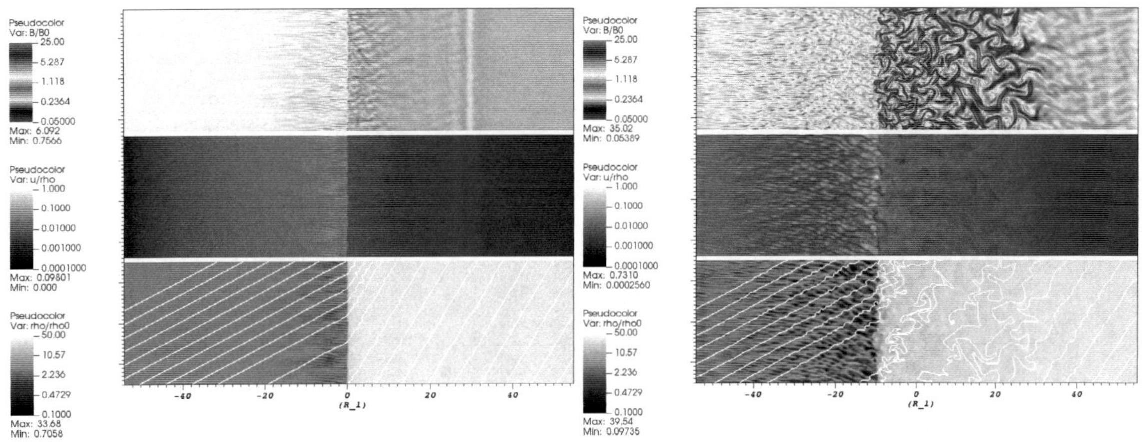

Figure 3. Similar to Fig. 1 but for a shock with $\Gamma = 3$, $\theta_B = 10^o$ for $t = 100\ R_L/c$ (left), $200\,R_L/c$ (right).

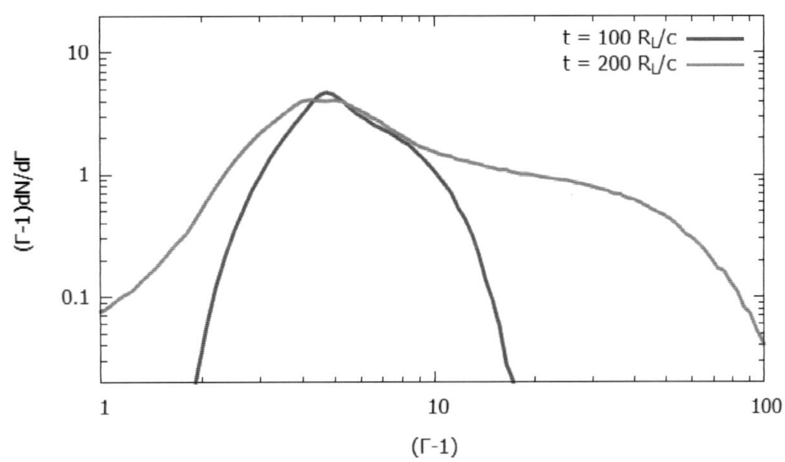

Figure 4. Time-evolution of the spectral energy distribution of non-thermal particles for a shock with $\Gamma = 3$, $\theta_B = 10^o$.

5. Conclusions

By combining PIC and PIC-MHD simulations, we can create a model of an oblique, relativistic shock and investigate its ability to accelerate particles to the speed observed in cosmic rays. We find that for those shocks that are not superluminal, the injection rate is high. This allows the particles to create perturbations of the magnetic field in the upstream medium and trigger the non-resonant streaming instability. As a result, particles are accelerated, and we begin to observe the formation of a power-law SED. In fact, our results show that oblique, relativistic shocks can be highly efficient accelerators because of the combination of a high compression rate across the shock and the fact that the Lorentz contraction allows even mildly oblique shocks to cause SDA which can effectively combine with DSA. From this, we can conclude that these shocks can contribute to the observed cosmic-ray spectrum.

In the future, we intend to extend our parameter space to higher relativistic shocks ($\Gamma \geq 10$), and, potentially, extend the model to 3-D.

Acknowledgments
A. B. was supported by the German Research Foundation (DFG) as part of the Excellence Strategy of the federal and state governments - EXC 2094 - 390783311. Computations were performed on the HPC system Raven at the Max Planck Computing and Data Facility.

References
[1] Drury, L.O'.C., 1983, *Rep. Prog. Phys.*, **46**, 973
[2] Bell, A. R. 2004, *MNRAS*, **353**, 550
[3] Bell, A. R. 2005,*MNRAS*, **358**, 181
[4] Caprioli D. & Spitkovski A. 2014 *ApJ* **783** 91
[5] Caprioli D. & Spitkovski A. 2014 *ApJ* **794** 46
[6] van Marle, A. J., Casse, F., Marcowith, A. 2018 *MNRAS*, **473** 3394
[7] van Marle, A. J., Bohdan, A., Morris, P. J., et al. 2022, *ApJ*, **929**, 7.
[8] Buneman, O., *Computer Space Plasma Physics: Simulation Techniques and Software Eds.*: H. Matsumoto & Y. Omura, Tokyo: *Terra Scientific*, 1993
[9] Niemiec, J., Pohl, M., Stroman, T., et al. 2008, *ApJ*, **684**, 1174
[10] Bohdan, A., Araudo, A., van marle, A. J. et al. 2023, *ArXiv:2308.12721*
[11] Bai, X-N., Caprioli, D., Sironi, L., & Spitkovsky, A. 2015, *ApJ*, **809**, 55
[12] van der Holst, B., Keppens, R., & meliani, Z. 2008, *CPC*, **179**, 617

Secular Outflows from Long-Lived Neutron Star Merger Remnants

David Radice

Institute for Gravitation & the Cosmos, The Pennsylvania State University, University Park, PA 16802
Department of Physics, The Pennsylvania State University, University Park, PA 16802
Department of Astronomy & Astrophysics, The Pennsylvania State University,University Park, PA 16802

E-mail: david.radice@psu.edu

Sebastiano Bernuzzi

Theoretisch-Physikalisches Institut, Friedrich-Schiller-Universität Jena, 07743, Jena, Germany

Abstract. We study mass ejection from a binary neutron star merger producing a long-lived massive neutron star remnant with general-relativistic neutrino-radiation hydrodynamics simulations. In addition to outflows generated by shocks and tidal torques during and shortly after the merger, we observe the appearance of a wind driven by spiral density waves in the disk. This spiral-wave-driven outflow is predominantly located close to the disk orbital plane and have a broad distribution of electron fractions. At higher latitudes, a high electron-fraction wind is driven by neutrino radiation. The combined nucleosynthesis yields from all the ejecta components is in good agreement with Solar abundance measurements.

1. Introduction

Neutron star merger outflows are thought to be one of the main astrophysical sites of production of heavy r-process elements [1]. Strong evidence linking mergers and r-process nucleosynthesis is provided by AT2017gfo, the kilonova associated with the neutron star merger GW170817. In particular, models of the light-curve and spectra of AT2017gfo require the presence of rare earth elements in the ejecta [2, 3, 4, 5, 6, 7]. More recently, kilonova-like signals were observed following the long gamma-ray bursts 211211A [8, 9, 10] and 230307A [11, 12, 13], which were also interpreted as binary neutron star mergers [14]. However, with the possible exception of strontium [15, 16] and tellurium [17], it has not been possible to identify specific heavy r-process elements in AT2017gfo, or other kilonovae. As such, nucleosynthesis yields need to be estimated from merger simulations.

A major difficulty for theoretical models is that the outflows in neutron star mergers are generated through a variety of mechanisms operating over different timescales [18]. Some of the ejecta is generated by tidal torques and shocks over a timescale of milliseconds during and shortly after the merger [19, 20, 21, 22, 23, 24, 25, 26, 27, 28, 29, 30, 31] the so-called dynamical ejecta. Winds from the remnant operate on a timescale of seconds, the so-called secular ejecta [32, 33, 34, 35, 36, 37, 38, 39, 40, 41, 42, 43, 44]. Most studies have focused on

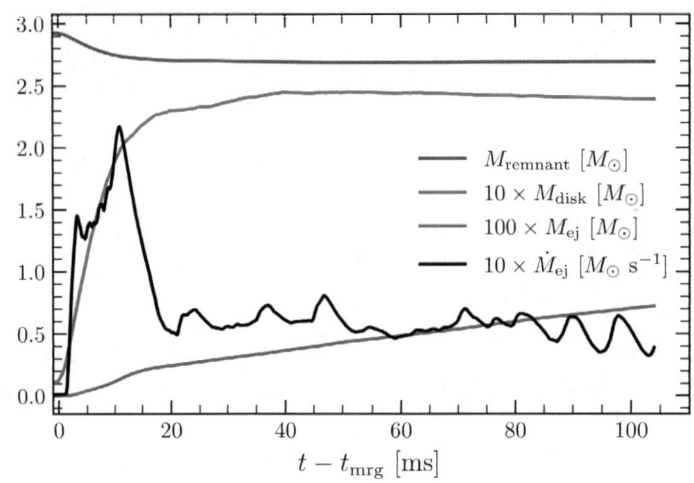

Figure 1. Baryonic rest mass of the remnant ($\rho > 10^{13}$ g cm^{-3}), of the disk ($\rho < 10^{13}$ g cm^{-3}), and of the ejecta ($-hu_t \geq h_{\rm min}$), and mass outflow rate as a function of time from merger. The disk is initially formed of material expelled from the remnant on a timescale of ~20 ms. It then evolves on a secular timescale under the effect of the spiral-wave instability. The disk mass evolution curve has been smoothed using a square window with width of 1 ms.

either the secular, or the dynamical ejecta. A notable exception are the recent works by Kiuchi and collaborators [45], considering the dynamical and secular ejecta from a binary neutron star system producing a black hole shortly after merger, and of Hayashi et al. [46], considering the long-term evolution of a neutron-star black hole merger remnant. However, such simulations are extremely challenging because, while fully general-relativistic codes are required to model the merger, the large computational costs of these simulations make them unsuitable for long-term integration. On the other hand, various groups have successfully modeled the secular ejecta, particularly from black-hole torus systems, while neglecting the back-reaction of the metric, or in Newtonian gravity. This approach is viable for remnants forming black-holes, but not for mergers producing long-lived massive neutron star remnants. In the latter case, a major outflow component is the spiral-wave wind, which is driven by hydrodynamic instabilities inside the remnant. Capturing such instabilities requires the same-kind of full-3D simulations with dynamical spacetime that are used to model the merger phase. Understanding such mergers is important, since it is believed that a substantial fraction of mergers produce long-lived remnants. GW170817 is also likely to have resulted in the formation of a meta-stable remnant [47, 48, 49, 50, 51, 52, 53, 44, 54, 55].

In Ref. [56], we have reported on the first ab-initio, general-relativistic neutrino-radiation hydrodynamics simulations of a long-lived neutron star merger remnant to secular timescales. We reported on the structure of the remnant and of the disk and on their evolution in the first 100 ms after the merger. Here, we study the mass ejection and nucleosynthesis yield from our models. We find that a massive outflow is launched by the spiral-wave wind mechanism, in agreement with previous studies with more approximate neutrino treatment. We discuss the geometry and composition of the outflows. Finally, we show that our model predicts an r-process nucleosynthesis pattern in good agreement with Solar abundance measurements.

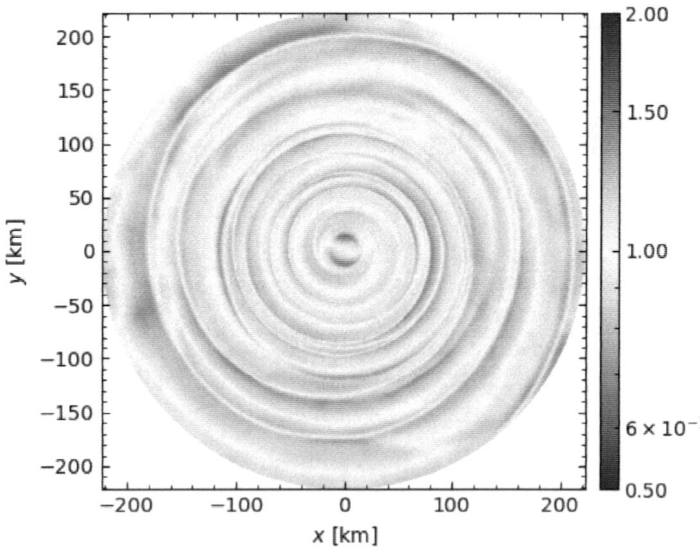

Figure 2. Density contrast in the disk 111.36 ms after merger. The spiral-wave in the disk is generated by the $m = 2$ deformation of the remnant (the so-called bar mode) and persists throughout our simulation.

2. Methods

We consider a binary with component masses 1.35 $M_\odot$ and 1.35 $M_\odot$. We adopt the HS(DD2) equation of state [57, 58], which predicts a maximum nonrotating neutron-star mass of 2.42 $M_\odot$ and the radius of a nonrotating 1.4 $M_\odot$ neutron star to be $R_{1.4} = 13.2$ km. We construct irrotational initial data with the `Lorene` pseudo-spectral code [59]. The initial separation is of 50 km. Evolutions are carried out using the gray moment-based (M1) general-relativistic neutrino-radiation hydrodynamics code `THC_M1` [60, 61, 62, 63, 64], which is based on the `Einstein Toolkit` [65, 66]. For the simulations discussed here, we use the `Carpet` adaptive mesh-refinement (AMR) driver [67, 68], which implements the Berger-Oilger scheme with refluxing [69, 70], and evolve the spacetime geometry using the `CTGamma` code [71, 72], which solves the Z4c formulation of Einstein's equations [73, 74]. The simulations reported here neglect magnetic fields. On the one hand, it is known that if the stars are originally endowed with magnetar-level fields, the magnetic stresses can qualitatively affect the outflow dynamics [44]. On the other hand, simulations starting with realistic initial fields and employing a sophisticated treatment to capture unresolved turbulence and dynamo action show that the magnetic fields are unlikely to be dynamically significant over the timescales considered here [75, 76]. In particular, the spiral-wave wind is expected to dominate the angular momentum transport in the disk, for the values of turbulent viscosity inferred by those simulations [51]. The results of additional simulations, including with more equations of state and with the inclusion of angular momentum transport due to turbulent viscosity will be reported elsewhere.

The evolution grid employs 7 levels of AMR. Our simulation is performed at two resolutions with the finest grid having finest grid spacing of $h = 0.167\ GM_\odot/c^2 \simeq 247$ m and $h = 0.125\ GM_\odot/c^2 \simeq 185$ m, respectively denoted as LR and SR setups. Here, we report the results from the higher resolution simulations. There lower-resolution data is in good quantitative and qualitative agreement.

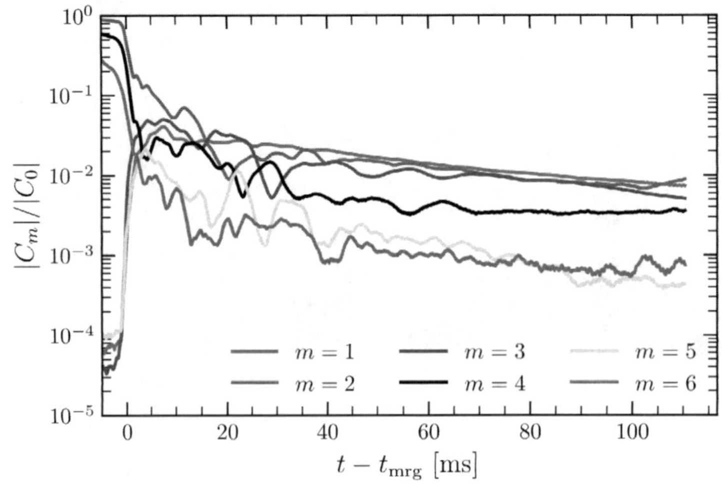

Figure 3. Normalized density modes on the xy-plane (see Eq. 2). The $m = 1, 2, 3$ density modes are the dominant modes. They are driven by the low-$T/|W|$ instability after merger. The curves have been smoothed using a square windows with width of 1 ms.

3. Results

The binary considered in this study results in the formation of a remnant that is dynamically stable, as documented in, e.g. [77]. Figure 1 shows the evolution of the baryon mass density inside the remnant massive neutron star and in the disk. Following a standard convention, we flag material as belonging to the remnant massive neutron star if its density is larger than 10^{13} g cm^{-3}. Material with lower density and situated within a cubical box of half-diameter 160 $GM_\odot/c^2 \simeq 236$ km is consider as belonging to the disk. The latter is formed of matter that is squeezed out of the collisional interface between the neutron stars during and shortly after merger [26, 78]. After ~20–30 ms mass ejection from the central remnant terminates. The subsequent disk mass evolution is primarily driven by mass loss to outflows with negligible accretion onto the central object.

The ejecta are extracted on a sphere of coordinate radius $R = 300\ GM_\odot/c^2 \simeq 443$ km, only material with outgoing radial velocity and that is unbound according to the Bernoulli condition[1] is considered in our analysis. Our simulations show two distinct mass ejection components: a first pulse of material (~$3 \times 10^{-3}\ M_\odot$) component ejected during and shortly after the merger, followed by a steady wind with $\dot{M} \simeq 0.05\ M_\odot$ s^{-1}. The first ejecta component comprises the tidal tails of the stars and the shocked ejecta launched shortly after merger [26]. The second component is dominated by the spiral-wave wind [51]. However, we also observe the emergence of a neutrino-driven wind at high latitudes [28, 64].

The spiral-wave wind dominates the overall mass outflow rate in the postmerger. This wind is the result of hydrodynamic and gravitational torques exerted by the remnant on the disk. The former is deformed as a result of the so-called low-$T/|W|$ instability [80, 81, 82, 83, 84, 85]. To diagnose this we show in Fig. 2 the density contrast on the orbital plane of the disk:

$$\frac{\rho - \langle \rho \rangle}{\langle \rho \rangle}, \tag{1}$$

where ρ is the fluid rest-mass density and $\langle \cdot \rangle$ denotes average in the azimuthal direction. The

[1] That is with $-hu_t \geq h_{\rm min}$, where h is the specific enthalpy, $h_{\rm min}$ is the minimum specific enthalpy of the equation of state, and u is the fluid four-velocity [79].

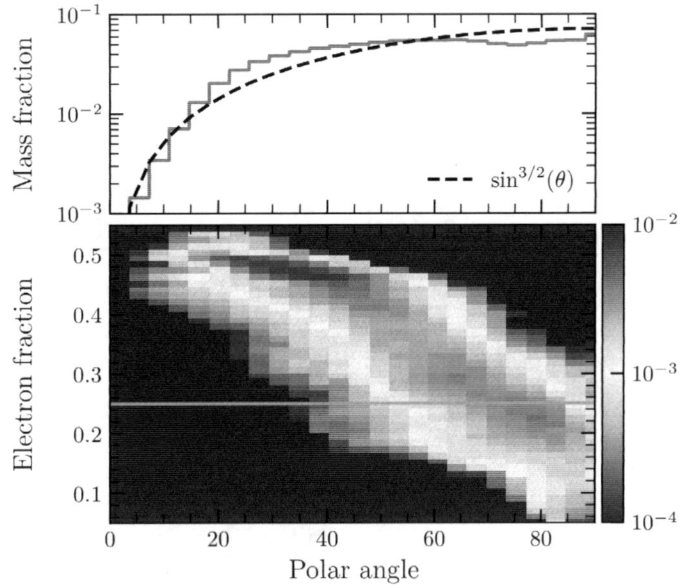

Figure 4. Angular distribution and electron fraction of the ejecta. Most of the ejecta is equatorial and has a broad range of electron fractions. The polar ejecta is less massive and characterized by significantly larger electron fraction $Y_e \gtrsim 0.4$.

$m = 2$ pattern in the remnant is clearly visible at the center, as are the spiral waves. These waves carry angular momentum outwards and drive the mass outflow, as anticipated by previous long-term post-merger simulations with simplified neutrino physics [51, 28]. This spiral wave pattern forms shortly after merger and persists for the entire duration of our simulations. The spiral-wave wind is predominantly emitted close to the disk orbital plane and has a broad distribution of electron fractions (see Fig. 4 below).

Figure 3 shows the evolution of the density modes on the orbital plane,

$$C_m = \int_{z=0} e^{-\mathrm{i}m\varphi} \rho W \sqrt{\gamma}\, \mathrm{d}^2 x, \tag{2}$$

as a function of time. The odd-m modes are initially zero, because equal mass binaries have a π-symmetry[2]. However, the symmetry is broken by turbulence during merger and subsequently the odd-m modes are excited by the low-$T/|W|$ instability [84, 86]. The even m modes are present from the beginning, particularly the $m = 2$ mode. In absence of the low-$T/|W|$ instability these modes would be efficiently damped by gravitational radiation [87]. However, in our simulation they are continuously driven by the instability and saturate at a roughly constant amplitude or decay slowly. The one-arm ($m = 1$ mode) and the triaxial ($m = 3$ mode) instabilities dominate the remnant dynamics from ~ 20 ms to $\sim$60 ms postmerger. Afterwards, the $m = 1, 2, 3$ modes equally contribute to the remnant dynamics. This has been observed in both isolated neutron star [81, 88] and binary mergers [84] simulations, although the details of the evolution are dependent on the equation of state, the compactness and possibly the microphysics [86]. The gravitational waves associated to these nonaxisymmetric modes carry the imprinting of the equation of state but, even under optimistic assumption, an observation will likely require third-generation detectors and a source at distance of $\sim$30 Mpc [84]. We caution the reader that, because our code uses a Cartesian grid, the $m = 4$ mode might be artificially enhanced.

[2] Meaning that they are symmetric under a rotation of 180° about the z-axis.

Journal of Physics: Conference Series **2742** (2024) 012009 doi:10.1088/1742-6596/2742/1/012009

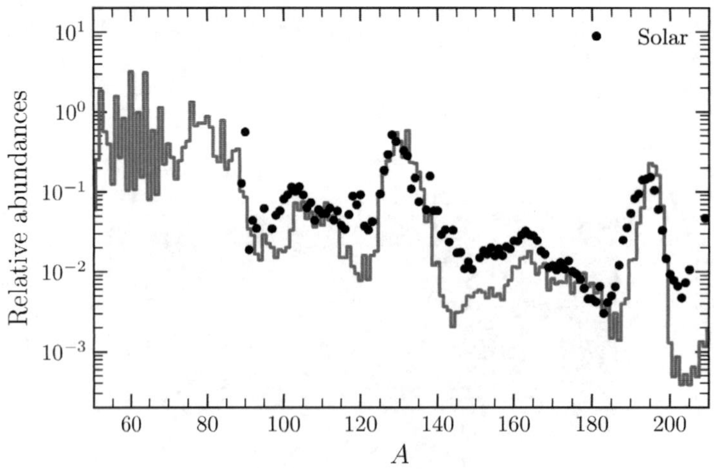

Figure 5. Nucleosynthesis yields of the ejecta (solid line) and Solar r-process abundances from Ref. [92] (points). The spiral-wave wind produces a full r-process yield.

While the structure of the wind close to the equator is similar to that observed in simulations with simplified M0 neutrino transport we have published in the past [51, 28, 89], the wind at high latitude is substantially enhanced in our new M1 simulations, as we have also documented for a lower resolution simulation in [64]. The angular structure of the ejecta is shown in Fig. 4, upper panel. This figure should be contrasted with Fig. 1 of [6], where the same diagnostics is shown from our older simulations using M0 neutrino transport. In contrast to the M0 simulations, which had angular profiles well fitted by $\sin^2(\theta)$ – θ being the angle from the equatorial plane – the new M1 simulations show a shallower profile, closer to $\sin^{3/2}(\theta)$. The composition of the ejecta is also quite different: the M0 simulations predict ejecta with electron fraction $Y_e < 0.4$ at all latitudes, while the newer M1 simulations predict a broader distribution with $Y_e \lesssim 0.5$. These differences arise from the fact that the M0 scheme artificially suppresses neutrino heating in the polar region of the remnant, suppressing the wind [64].

Figure 5 shows the combined nucleosynthesis yields for all ejecta components. These have been obtained using the `SkyNet` code [90] with the procedure discussed in Ref. [26]. We find that our model well reproduces the Solar r-process pattern. The only significant discrepancy is in the region with $A \sim 140$, however those can be ascribed to the particular nuclear-mass model used in our nucleosynthesis calculations [91].

4. Conclusions

We have performed long-term simulations of a binary neutron star merger producing a long-lived remnant with a general-relativistic neutrino-radiation hydrodynamics code. We find that such events drive massive outflows through a combination of tidal torques and shocks, during and shortly after the merger, and spiral-wave and neutrino-driven winds after the merger. The dynamics of the spiral-wave wind is in good agreement with predictions from simulations with lower-fidelity neutrino transport. However, there are substantial differences in the dynamics and composition of the neutrino-driven wind. In particular, our previous studies underestimated both the mass and the electron fraction of this component of the ejecta. The combined nucleosynthesis yield of all ejecta components is in good agreement with the Solar r-process abundance pattern, indicating that neutron star mergers producing long-lived remnants provide a natural environment for the formation of these elements.

Acknowledgments

DR acknowledges funding from the U.S. Department of Energy, Office of Science, Division of Nuclear Physics under Award Number(s) DE-SC0021177 and from the National Science Foundation under Grants No. PHY-2011725, PHY-2020275, PHY-2116686, and AST-2108467. This research used resources of the National Energy Research Scientific Computing Center, a DOE Office of Science User Facility supported by the Office of Science of the U.S. Department of Energy under Contract No. DE-AC02-05CH11231. SB knowledges funding from the EU Horizon under ERC Consolidator Grant, no. InspiReM-101043372 and from the Deutsche Forschungsgemeinschaft, DFG, project MEMI number BE 6301/2-1. The authors acknowledge the Gauss Centre for Supercomputing e.V. (`www.gauss-centre.eu`) for funding this project by providing computing time on the GCS Supercomputer SuperMUC-NG at LRZ (allocation `pn36ge` and `pn36jo`).

References

[1] Thielemann F K, Eichler M, Panov I and Wehmeyer B 2017 *Ann.Rev.Nucl.Part.Sci.* **67** 253 (*Preprint* 1710.02142)
[2] Coulter D, Foley R, Kilpatrick C, Drout M, Piro A *et al.* 2017 *Science* **358** 1556 (*Preprint* 1710.05452)
[3] Kasliwal M, Nakar E, Singer L, Kaplan D, Cook D *et al.* 2017 *Science* **358** 1559 (*Preprint* 1710.05436)
[4] Chornock R, Berger E, Kasen D, Cowperthwaite P, Nicholl M *et al.* 2017 *Astrophys.J.Lett.* **848** L19 (*Preprint* 1710.05454)
[5] Villar V A, Guillochon J, Berger E, Metzger B D, Cowperthwaite P S *et al.* 2017 *Astrophys.J.Lett.* **851** L21 (*Preprint* 1710.11576)
[6] Perego A, Radice D and Bernuzzi S 2017 *Astrophys.J.Lett.* **850** L37 (*Preprint* 1711.03982)
[7] Kawaguchi K, Shibata M and Tanaka M 2018 *Astrophys.J.Lett.* **865** L21 (*Preprint* 1806.04088)
[8] Rastinejad J C, Gompertz B P, Levan A J, Fong W f, Nicholl M *et al.* 2022 *Nature* **612** 223 (*Preprint* 2204.10864)
[9] Troja E, Fryer C, O'Connor B, Ryan G, Dichiara S *et al.* 2022 *Nature* **612** 228 (*Preprint* 2209.03363)
[10] Yang J, Ai S, Zhang B B, Zhang B, Liu Z K *et al.* 2022 *Nature* **612** 232 (*Preprint* 2204.12771)
[11] Sun H, Wang C W, Yang J, Zhang B B, Xiong S L *et al.* 2023 *arXiv:2307.05689* (*Preprint* 2307.05689)
[12] Levan A, Gompertz B P, Salafia O S, Bulla M, Burns E *et al.* (JWST) 2023 *arXiv:2307.02098* (*Preprint* 2307.02098)
[13] Yang Y H, Troja E, O'Connor B, Fryer C L, Im M *et al.* 2023 *arXiv:2308.00638* (*Preprint* 2308.00638)
[14] Gottlieb O, Metzger B D, Quataert E, Issa D and Foucart F 2023 *arXiv:2309.00038* (*Preprint* 2309.00038)
[15] Watson D, Hansen C J, Selsing J, Koch A, Malesani D B *et al.* 2019 *Nature* **574** 497 (*Preprint* 1910.10510)
[16] Perego A, Vescovi D, Fiore A, Chiesa L, Vogl C *et al.* 2022 *Astrophys.J.* **925** 22 (*Preprint* 2009.08988)
[17] Hotokezaka K, Tanaka M, Kato D and Gaigalas G 2023 *arXiv:2307.00988* (*Preprint* 2307.00988)
[18] Shibata M and Hotokezaka K 2019 *Ann.Rev.Nucl.Part.Sci.* **69** 41 (*Preprint* 1908.02350)
[19] Hotokezaka K, Kiuchi K, Kyutoku K, Okawa H, Sekiguchi Y i *et al.* 2013 *Phys.Rev.D* **87** (*Preprint* 1212.0905)
[20] Korobkin O, Rosswog S, Arcones A and Winteler C 2012 *Mon.Not.Roy.Astron.Soc.* **426** 1940 (*Preprint* 1206.2379)
[21] Bauswein A, Goriely S and Janka H T 2013 *Astrophys.J.* **773** 78 (*Preprint* 1302.6530)
[22] Sekiguchi Y, Kiuchi K, Kyutoku K and Shibata M 2015 *Phys.Rev.D* **91** (*Preprint* 1502.06660)
[23] Radice D, Galeazzi F, Lippuner J, Roberts L F, Ott C D *et al.* 2016 *Mon.Not.Roy.Astron.Soc.* **460** 3255 (*Preprint* 1601.02426)
[24] Lehner L, Liebling S L, Palenzuela C, Caballero O, O'Connor E *et al.* 2016 *Class.Quant.Grav.* **33** (*Preprint* 1603.00501)
[25] Dietrich T, Ujevic M, Tichy W, Bernuzzi S and Bruegmann B 2017 *Phys.Rev.D* **95** (*Preprint* 1607.06636)
[26] Radice D, Perego A, Hotokezaka K, Fromm S A, Bernuzzi S *et al.* 2018 *Astrophys.J.* **869** 130 (*Preprint* 1809.11161)
[27] Vincent T, Foucart F, Duez M D, Haas R, Kidder L E *et al.* 2020 *Phys.Rev.D* **101** (*Preprint* 1908.00655)
[28] Nedora V, Bernuzzi S, Radice D, Daszuta B, Endrizzi A *et al.* 2021 *Astrophys.J.* **906** 98 (*Preprint* 2008.04333)
[29] Foucart F, Duez M D, Hebert F, Kidder L E, Pfeiffer H P *et al.* 2020 *Astrophys.J.Lett.* **902** L27 (*Preprint* 2008.08089)
[30] Combi L and Siegel D M 2023 *Astrophys.J.* **944** 28 (*Preprint* 2206.03618)
[31] Schianchi F, Gieg H, Nedora V, Neuweiler A, Ujevic M *et al.* 2023 *arXiv:2307.04572* (*Preprint* 2307.04572)

[32] Perego A, Rosswog S, Cabezón R M, Korobkin O, Käppeli R *et al.* 2014 *Mon.Not.Roy.Astron.Soc.* **443** 3134 (*Preprint* 1405.6730)
[33] Martin D, Perego A, Arcones A, Thielemann F K, Korobkin O *et al.* 2015 *Astrophys.J.* **813** 2 (*Preprint* 1506.05048)
[34] Fernández R and Metzger B D 2013 *Mon.Not.Roy.Astron.Soc.* **435** 502 (*Preprint* 1304.6720)
[35] Fujibayashi S, Kiuchi K, Nishimura N, Sekiguchi Y and Shibata M 2018 *Astrophys.J.* **860** 64 (*Preprint* 1711.02093)
[36] Siegel D M and Metzger B D 2018 *Astrophys.J.* **858** 52 (*Preprint* 1711.00868)
[37] Fernández R, Tchekhovskoy A, Quataert E, Foucart F and Kasen D 2019 *Mon.Not.Roy.Astron.Soc.* **482** 3373 (*Preprint* 1808.00461)
[38] Miller J M, Ryan B R, Dolence J C, Burrows A, Fontes C J *et al.* 2019 *Phys.Rev.D* **100** (*Preprint* 1905.07477)
[39] Most E R, Papenfort L J, Tootle S D and Rezzolla L 2021 *Mon.Not.Roy.Astron.Soc.* **506** 3511 (*Preprint* 2106.06391)
[40] Curtis S, Miller J M, Frohlich C, Sprouse T, Lloyd-Ronning N *et al.* 2023 *Astrophys.J.Lett.* **945** L13 (*Preprint* 2212.10691)
[41] Fahlman S and Fernández R 2022 *Mon.Not.Roy.Astron.Soc.* **513** 2689 (*Preprint* 2204.03005)
[42] Sprouse T, Lund K, Miller J, McLaughlin G and Mumpower M 2023 *arXiv:2309.07966* (*Preprint* 2309.07966)
[43] Just O, Vijayan V, Xiong Z, Goriely S, Soultanis T *et al.* 2023 *Astrophys.J.Lett.* **951** L12 (*Preprint* 2302.10928)
[44] Combi L and Siegel D M 2023 *arXiv:2303.12284* (*Preprint* 2303.12284)
[45] Kiuchi K, Fujibayashi S, Hayashi K, Kyutoku K, Sekiguchi Y *et al.* 2023 *Phys.Rev.Lett.* **131** (*Preprint* 2211.07637)
[46] Hayashi K, Fujibayashi S, Kiuchi K, Kyutoku K, Sekiguchi Y *et al.* 2022 *Phys.Rev.D* **106** (*Preprint* 2111.04621)
[47] Rezzolla L, Most E R and Weih L R 2018 *Astrophys.J.Lett.* **852** L25 (*Preprint* 1711.00314)
[48] Li S Z, Liu L D, Yu Y W and Zhang B 2018 *Astrophys.J.Lett.* **861** L12 (*Preprint* 1804.06597)
[49] Metzger B D, Thompson T A and Quataert E 2018 *Astrophys.J.* **856** 101 (*Preprint* 1801.04286)
[50] Ai S, Gao H, Dai Z G, Wu X F, Li A *et al.* 2018 *Astrophys.J.* **860** 57 (*Preprint* 1802.00571)
[51] Nedora V, Bernuzzi S, Radice D, Perego A, Endrizzi A *et al.* 2019 *Astrophys.J.Lett.* **886** L30 (*Preprint* 1907.04872)
[52] Ciolfi R and Kalinani J V 2020 *Astrophys.J.Lett.* **900** L35 (*Preprint* 2004.11298)
[53] Mösta P, Radice D, Haas R, Schnetter E and Bernuzzi S 2020 *Astrophys.J.Lett.* **901** L37 (*Preprint* 2003.06043)
[54] Curtis S, Bosch P, Mösta P, Radice D, Bernuzzi S *et al.* 2023 *arXiv:2305.07738* (*Preprint* 2305.07738)
[55] Kawaguchi K, Fujibayashi S, Domoto N, Kiuchi K, Shibata M *et al.* 2023 *arXiv:2306.06961* (*Preprint* 2306.06961)
[56] Radice D and Bernuzzi S 2023 *arXiv:2306.13709* (*Preprint* 2306.13709)
[57] Typel S, Ropke G, Klahn T, Blaschke D and Wolter H 2010 *Phys.Rev.C* **81** (*Preprint* 0908.2344)
[58] Hempel M and Schaffner-Bielich J 2010 *Nucl.Phys.A* **837** 210 (*Preprint* 0911.4073)
[59] Gourgoulhon E, Grandclement P, Taniguchi K, Marck J A and Bonazzola S 2001 *Phys.Rev.D* **63** (*Preprint* gr-qc/0007028)
[60] Radice D and Rezzolla L 2012 *Astron.Astrophys.* **547** A26 (*Preprint* 1206.6502)
[61] Radice D, Rezzolla L and Galeazzi F 2014 *Mon.Not.Roy.Astron.Soc.* **437** L46 (*Preprint* 1306.6052)
[62] Radice D, Rezzolla L and Galeazzi F 2014 *Class.Quant.Grav.* **31** (*Preprint* 1312.5004)
[63] Radice D, Rezzolla L and Galeazzi F 2015 *ASP Conf.Ser.* **498** 121 (*Preprint* 1502.00551)
[64] Radice D, Bernuzzi S, Perego A and Haas R 2022 *Mon.Not.Roy.Astron.Soc.* **512** 1499 (*Preprint* 2111.14858)
[65] Loffler F, Faber J, Bentivegna E, Bode T, Diener P *et al.* 2012 *Class.Quant.Grav.* **29** (*Preprint* 1111.3344)
[66] Werneck L, Cupp S, Assumpção T, Brandt S R, Cheng C H *et al.* 2023 The einstein toolkit to find out more, visit http://einsteintoolkit.org URL https://doi.org/10.5281/zenodo.7942541
[67] Schnetter E, Hawley S H and Hawke I 2004 *Class.Quant.Grav.* **21** 1465 (*Preprint* gr-qc/0310042)
[68] Reisswig C, Haas R, Ott C, Abdikamalov E, Mösta P *et al.* 2013 *Phys.Rev.D* **87** (*Preprint* 1212.1191)
[69] Berger M J and Oliger J 1984 *J.Comput.Phys.* **53** 484
[70] Berger M J and Colella P 1989 *Journal of Computational Physics* **82** 64–84
[71] Pollney D, Reisswig C, Schnetter E, Dorband N and Diener P 2011 *Phys.Rev.D* **83** (*Preprint* 0910.3803)
[72] Reisswig C, Ott C, Abdikamalov E, Haas R, Moesta P *et al.* 2013 *Phys.Rev.Lett.* **111** (*Preprint* 1304.7787)
[73] Bernuzzi S and Hilditch D 2010 *Phys.Rev.D* **81** (*Preprint* 0912.2920)
[74] Hilditch D, Bernuzzi S, Thierfelder M, Cao Z, Tichy W *et al.* 2013 *Phys.Rev.D* **88** (*Preprint* 1212.2901)
[75] Palenzuela C, Liebling S and Miñano B 2022 *Phys.Rev.D* **105** (*Preprint* 2204.02721)
[76] Aguilera-Miret R, Palenzuela C, Carrasco F and Viganò D 2023 *arXiv:2307.04837* (*Preprint* 2307.04837)

[77] Radice D, Perego A, Bernuzzi S and Zhang B 2018 *Mon.Not.Roy.Astron.Soc.* **481** 3670 (*Preprint* 1803.10865)
[78] Zenati Y, Krolik J H, Werneck L R, Murguia-Berthier A, Etienne Z B *et al.* 2023 *arXiv:2306.09464* (*Preprint* 2306.09464)
[79] Foucart F, Moesta P, Ramirez T, Wright A J, Darbha S *et al.* 2021 *Phys.Rev.D* **104** (*Preprint* 2109.00565)
[80] Rampp M, Mueller E and Ruffert M 1998 *Astron.Astrophys.* **332** 969 (*Preprint* astro-ph/9711122)
[81] Ou S and Tohline J 2006 *Astrophys.J.* **651** 1068 (*Preprint* astro-ph/0604099)
[82] Corvino G, Rezzolla L, Bernuzzi S, De Pietri R and Giacomazzo B 2010 *Class.Quant.Grav.* **27** (*Preprint* 1001.5281)
[83] Paschalidis V, East W E, Pretorius F and Shapiro S L 2015 *Phys.Rev.D* **92** (*Preprint* 1510.03432)
[84] Radice D, Bernuzzi S and Ott C D 2016 *Phys.Rev.D* **94** (*Preprint* 1603.05726)
[85] Longo Micchi L F, Radice D and Chirenti C 2023 *Mon.Not.Roy.Astron.Soc.* **525** 6359 (*Preprint* 2306.04711)
[86] Espino P L, Prakash A, Radice D and Logoteta D 2023 *arXiv:2301.03619* (*Preprint* 2301.03619)
[87] Bernuzzi S, Radice D, Ott C D, Roberts L F, Moesta P *et al.* 2016 *Phys.Rev.D* **94** (*Preprint* 1512.06397)
[88] Manca G M, Baiotti L, De Pietri R and Rezzolla L 2007 *Class.Quant.Grav.* **24** S171 (*Preprint* 0705.1826)
[89] Zappa F, Bernuzzi S, Radice D and Perego A 2022 *arXiv:2210.11491* (*Preprint* 2210.11491)
[90] Lippuner J and Roberts L F 2017 *Astrophys.J.Suppl.* **233** 18 (*Preprint* 1706.06198)
[91] Vassh N, McLaughlin G C, Mumpower M R and Surman R 2022 *arXiv:2202.09437* (*Preprint* 2202.09437)
[92] Arlandini C, Kaeppeler F, Wisshak K, Gallino R, Lugaro M *et al.* 1999 *Astrophys.J.* **525** 886 (*Preprint* astro-ph/9906266)

Three-dimensional Modeling of the Solar Wind and Local Interstellar Medium Interaction with Pickup Ions in the Presence of Heliospheric Current Sheet

R. K. Bera

Center for Space Plasma and Aeronomic Research, The University of Alabama in Huntsville, Huntsville, AL 35899, USA

F. Fraternale

Center for Space Plasma and Aeronomic Research, The University of Alabama in Huntsville, Huntsville, AL 35899, USA

N. V. Pogorelov

Center for Space Plasma and Aeronomic Research, The University of Alabama in Huntsville, Huntsville, AL 35899, USA
Department of Space Science, The University of Alabama in Huntsville, Huntsville, AL 35899, USA

E-mail: rkb0019@uah.edu

Abstract.
Our three-dimensional, time-dependent, multi-fluid model has been used to investigate the solar wind (SW)–local interstellar medium (LISM) interaction with pickup ions (PUIs) treated as a separate fluid. A non-zero, but fixed, angle between the Sun's magnetic and rotation axis is adopted. The flow of the plasma mixture (thermal SW protons, PUIs, and electrons), is described by the system of ideal magnetohydrodynamic equations with the source terms responsible for charge exchange between ions and neutral atoms. Different populations of neutral atoms are governed by the individual sets of the Euler equations. As the standard Rankine–Hugoniot relations are not appropriate to describe the anisotropic behavior of PUIs at the termination shock, we use a kinetically-derived set of boundary conditions at it. We extend our previous work [1] and perform these new simulations on a Cartesian grid. This approach allows us to maintain a uniform grid resolution in all directions, without compromising resolution, at large distances from the Sun. The possibility of transition of the SW flow to a stochastic regime in the region between the termination shock and heliopause is further investigated.

1. Introduction

Interaction of the solar wind (SW) with the local interstellar medium (LISM) is an exciting area of space physics research which encompasses very different time and length scales. Collision of these plasma flows creates the heliosphere. The dynamical properties of the heliosphere are significantly influenced by the (time-dependent) properties of the SW and of the LISM. From the gas dynamic (or magnetohydrodynamic, MHD) perspective, the boundary between the SW and LISM plasmas is a tangential discontinuity known as the heliopause (HP). Deceleration of the supersonic SW by the HP leads to the formation of a heliospheric termination shock (TS). Both the TS and the HP were identified in situ by the Voyager 1 (V1) and Voyager 2 (V2) spacecraft. The region between TS and HP is commonly referred to as the inner heliosheath (IHS), or simply the heliosheath. The presence of the so-called bow shock (BS) in front of the HP depends on the flow parameters in the LISM and was discussed in [2–6].

Interstellar neutral (ISN) atoms cross the HP and penetrate deeply into the heliosphere. Some of them undergo ionization through charge exchange with SW ions, solar radiation, and electron impact, giving birth to new populations of ions known as pickup ions (PUIs). Once created, PUIs respond to the motional electric field almost instantaneously and start to move outwards with the SW, acquiring its bulk speed [7]. Although their number density is low, high temperature of PUIs ($\sim 10^7$ K) makes their pressure dominant as compared both with the thermal SW and magnetic pressure in the outer heliosphere beyond ~ 20 AU. This was confirmed by Voyager 2 and New Horizons (NH) observations [8–12], and demonstrated by theoretical models and simulations [1, 13–21]. Due to the large mean free path of Coulomb collisions between PUIs and thermal SW ions, or electrons, PUIs are never in thermodynamic equilibrium with the surrounding plasma. They constitute a distinct suprathermal ion component within the SW plasma [17, 22].

Numerical simulations are essential for comprehending the three-dimensional characteristics of the SW–LISM interaction and interpreting in situ measurements by the Voyagers and NH missions, as well as remote observations made by the Interstellar Boundary Explorer (IBEX) [23–26]. It is not unusual to describe the SW–LISM interaction with multi-fluid models that are based on the combination of the ideal MHD equations and multiple sets of the Euler gas dynamics equations [1, 19, 27, 28] or kinetically [29, 30]. In recent years, significant progress has been made in incorporating PUIs into global models of the SW–LISM interaction. As was first demonstrated in [22], PUIs must be treated separately from the thermal ions. As a consequence, a number of fluid models have been developed where PUIs are treated as a separate fluid [19, 28, 31]. While PUIs away from the TS can be treated as having an isotropic distribution function, it is not so at shocks. Crossing of a collisionless shock, such as the TS, by PUIs is associated with complex kinetic processes [16, 32–37], which make the PUI distribution function in the TS vicinity highly anisotropic [16, 38, 39] and the applicability of the fluid approach questionable. While the traditional MHD Rankine—Hugoniot relations, which express the conservation of mass, momentum, energy, and magnetic flux at shocks in the mixture of charged particles, are valid at collisionless shocks on the MHD scale, it is not obvious how to derive the jumps in the PUI pressure and temperature across the TS. On the basis of extensive hybrid simulations of the PUI behavior across collisionless shocks [40], a set of boundary conditions (b.c.'s) applicable for global MHD modeling has been derived in [1].

In [1], the simulations were performed on spherical grids. The SW was assumed spherically-symmetric and the heliospheric magnetic field (HMF) was unipolar at the inner boundary of the simulation domain. Since the LISM flow is uniform and time-independent, only the stationary solutions were considered. The assumption of a unipolar HMF has been criticized earlier in [41, 42] as producing an unrealistically strong magnetic field in the IHS. In this work, we extend our model to Cartesian grids and present simulations in a more complex and computationally challenging statement, where the HMF is bipolar and a heliospheric current sheet (HCS) separates the regions of opposite polarity. These regions of positive and negative polarities are separated by the HCS and referred to as magnetic sectors. Typically, it takes a few hundred years for a steady state to be reached with our choice of initial conditions [47]. In the current simulation, the SW at the inner boundary is steady, except for the HMF polarity variations due to the Sun's rotation and a tilt between the Sun's rotation and magnetic axes. This is why, the steady state can be sought only in a statistical sense.

One advantage of Cartesian grids is that SW-LISM interaction simulations can be performed in large space domains ($\gtrsim$ 3000 AU) without compromising grid resolution at large distances. This is especially important for the SW–LISM interaction in the heliotail. Furthermore, Cartesian grids make it possible to maintain a uniform grid resolution in all directions.

Due to the non-zero tilt between the Sun's rotation and magnetic-dipole axis, the HCS naturally forms. The HCS moves with the plasma flow outward from the Sun, creating a shape that initially looks like a "ballerina skirt", although this similarity disappears with distance. It was shown in, e.g., [27, 43–45] that the HCS presence may create time-dependent features in the IHS. Moreover, it was suggested in [43] that a stochastic behavior of magnetic field can be obtained numerically when the sector regions are not resolved.

The sector width is proportional to the SW speed, so no grid is sufficient to resolve the sectors in the regions where the velocity component perpendicular to the HP is small. In the earlier simulations of this kind, PUIs were not treated as a separate fluid. Instead, an immediate assimilation of PUIs into a single Maxwellian distribution with the thermal SW was assumed. This paper supplements and extends the results of those simulations.

In Section 2, we briefly describe our physical and mathematical models. The simulation setup and boundary conditions are discussed in Section 3. The implementation of b.c's for the PUIs crossing the TS is described in Section 4. The simulation results are presented in Section 5, which is followed by the conclusions in Section 6.

2. Physical Model

We build our model on the multi-fluid approach described in [19, 46, 47] and further elaborated in [1]. We solve a system of equations governing the flow of the whole mixture of charged particles, neutral atoms, and PUIs. The plasma mixture consists of three co-moving populations of charged particles: thermal protons, electrons, and PUIs. This flow is described by the ideal MHD equations with the source terms responsible for charge exchange. We also distinguish three populations of neutral hydrogen (H) atoms: population 1 represents H atoms of interstellar origin, while populations 2 and 3 are the neutral atoms born by charge exchange in the IHS and in the supersonic SW, respectively. Each of these atom populations originates from thermodynamically distinct regions of the heliosphere (we call the Regions 1, 2, and 3), thus necessitating the use of separate sets of Euler equations to describe their behavior. PUIs are treated as a separate co-moving fluid by solving the corresponding continuity and pressure equations. The complete system of equations is given in [1, 19].

The analytical expressions for the source terms are derived assuming Maxwellian distributions for ions and neutral atoms.

The electron density is calculated assuming the quasi-neutrality condition, $n_{\rm e} = n_{\rm p,th} + n_{\rm PUI} = n$, where $n_{\rm e}$, $n_{\rm p,th}$, and $n_{\rm PUI}$ are the electron, thermal proton, and PUI number densities, respectively. The electron temperature is assumed to be equal to the temperature of thermal protons, $T_{\rm e} = T_{\rm p,th}$. The thermal pressure of each species is determined using the equation of state, $p_s = n_s k_{\rm B} T_s$, where p_s, n_s and T_s are the pressure, number density, and temperature of species s, and $k_{\rm B}$ is the Boltzmann constant. There is no need to solve separate equations for the thermal protons. Their properties can be derived from those of the mixture and PUIs.

A significant fraction of the thermal proton heating in the distant SW is caused by PUI-generated turbulence [31, 33, 48–50]. Here we adopt a simplified approach [1], where the source term in the PUI pressure equation is decreased by 5%. This automatically transfers some amount of energy from PUIs to thermal protons in our model. The fraction of energy to be transferred is adjusted to make the temperature of the thermal SW closer to that in V2 data. This modification of the source term is done only in the supersonic SW.

In addition to Eqs. (1)–(9) from [1], we solve three level-set equations, as described in [45]. The first level set is used to correctly identify the position of the HP by tracking the time

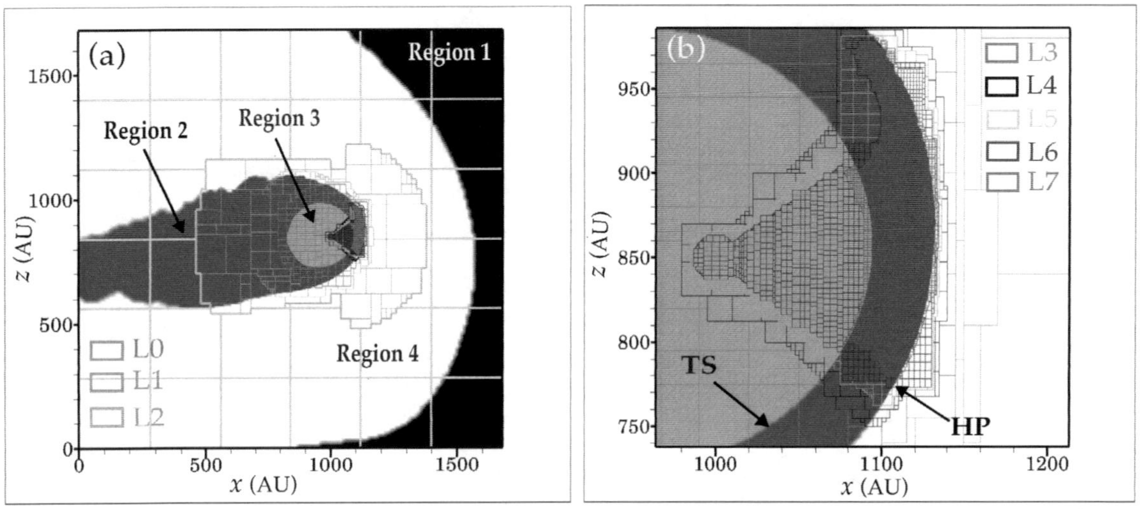

Figure 1. The numerical grid used in our simulation. The left panel shows the grid patches corresponding to all seven levels of refinements (L0-L7, color-coded) in the meridional plane. The right panel shows a zoomed view of the most refined regions.

evolution of its surface starting from the initial position at time $t = 0$. The other two level-set equations are introduced to identify the HCS and the boundary of the region covered by the magnetic field sectors.

3. Simulation Setup and Boundary Conditions

We solve the system of equations using the Multi-Scale Fluid-Kinetic Simulation Suite (MS-FLUKSS), which takes advantage of the Chombo adaptive mesh refinement (AMR) framework for efficient data choreography [1, 19, 21, 23, 29, 45, 47, 51, 52]. The simulation is performed on a Cartesian grid (x, y, z). The simulation domain is a cube with each side equal to 1680 AU. The z-axis is set to be parallel to the Sun's rotation axis. The x-axis belongs to the plane containing the z-axis and the velocity vector in the unperturbed LISM, $\mathbf{V}_{\rm LISM}$, and is oriented upstream into the LISM. The y-axis completes the right coordinate system. The Sun's position is at $(x_S, y_S, z_S) = (1000, 850, 850)$ AU. The objective of this paper is to test the model and assess the performance of our new code while addressing a challenging problem previously analyzed in [53], i.e., the SW–LISM interaction in the presence of an HCS under the assumption of a fixed tilt of the Sun's magnetic axis to its rotation axis. This is a computationally challenging problem. Due to the deceleration of the SW flow initially at the TS and further by the HP, the sector width decreases substantially approaching zero at the HP (see the discussion in Section 5). Therefore, from the perspective of global MHD simulations, it is practically impossible to resolve the sector structure in the HP vicinity.

The base grid size in our simulations is $\Delta x = \Delta y = \Delta z = 10$ AU. Seven levels of refinements with a refinement ratio equal to 2 at each level are applied. This creates the finest resolution of ≈ 0.07 AU cubed. Figure 1 shows the boundary of grid patches corresponding to the different levels as seen in the meridional plane (x-z plane), labeled from L0 to L7. The finest level (L7) is used in the IHS to cover the latitudinal extent of the sector region (the region containing the HCS) ($\pm 60°$ from the equatorial plane). In the azimuthal direction, the grid extends from $-15°$ to $35°$ from the x-axis to cover a broad region the direction towards the SW stagnation point. L7 starts in the supersonic SW a few AU ahead of the TS and extends to a few AU into the LISM, to cover the HP. The total number of cells is approximately 3×10^9.

Table 1 summarizes the inner and outer boundary conditions used in our simulations. The

Table 1. The boundary conditions used at 1 AU and at outer boundary.

Parameters at 1 AU	Value
Thermal proton density	9 cm^{-3}
SW speed	420 km s^{-1}
SW temperature	90000 K
Radial HMF component	39 μG
PUI density at 10 AU	1.5×10^{-3} cm^{-3}
PUI temperature at 10 AU	3.36×10^{6} K
LISM parameters	**Value**
Proton density	0.054 cm^{-3}
Flow speed	25.4 km s^{-1}
Plasma temperature	7500 K
ISMF strength	3.5 μG
Neutral H density	0.2 cm^{-3}
Velocity direction (HAE-J2000)	$(255^\circ.7, 5^\circ.1)$
ISMF (HAE-J2000)	$(224^\circ.46, 36^\circ.61)$

physical boundary conditions are enforced at $R_0 = 10$ AU. It is assumed that the SW is radial and its scalar properties are spherically symmetric at this boundary. Technically speaking, there is no mathematical inner boundary in the Cartesian simulation domain, i.e., we have the computational cells at distances smaller than 10 AU, but the quantities in those cells are substituted with the analytical, Parker's solution. However, only the ghost cells adjacent to the physical boundary are used. For the neutral fluids, there is no physical inner boundary

While the SW properties are known at the Earth orbit ($R = 1$ AU), we use the polytropic expressions to translate them from 1 AU to 10 AU. The choice of SW properties at 1 AU is based on the averaged OMNI data for the period from 2012 to 2022. The number density, velocity, and temperature of the SW at the inner boundary are stationary. The magnetic field vector at the inner boundary is assumed to be a Parker spiral [54]. The variation of the HCS location at the inner boundary, as a function of time due to the solar rotation with a period of 25 days, is given in [55]. As a consequence, the HMF strength, **B**, and the boundary between the regions of its opposite polarity, become functions of time. The PUI properties at 10 AU are specified based on the NH observations [12]. The ghost cells beyond the outer boundary are filled with the properties of the undisturbed LISM, characterized by time-independent values based on recently accepted parameters. Numerical fluxes through the outer boundary are obtained with the method described in [56, 57].

4. Treatment of PUIs at the Termination Shock

To describe the behavior of PUIs at the TS, we employ the shock relations recently described in [1]. They are obtained by summarizing the hybrid simulations from [40] in such a way that the *isotropic* temperature of PUIs well behind the TS is obtained in terms of the three local upstream shock parameters, i.e., the Alfvénic Mach number (M_A), plasma beta (β), and the angle between the shock normal and the upstream magnetic field (θ_{Bn}). The shock normal is calculated using the formulae given by [58]. Note that the upstream shock parameters are readily available in our global model of the plasma mixture. The idea is to substitute the solution to the PUI pressure equation behind the shock with the PUI pressure derived from the kinetic (hybrid) simulations. It is worth noting that we do not modify the plasma mixture explicitly, therefore it continues to satisfy the MHD conservation laws. A detailed derivation of the b.c.'s from the hybrid simulation results and their implementation in MS-FLUKSS are discussed in [1].

The application of such b.c.'s requires identification of the cells adjacent to the TS, both upstream and downstream. The algorithm designed for this purpose on spherical grids is developed in [1]. Then, it is relatively straightforward to choose the downstream and upstream cells for a given shock-containing cell. This is illustrated in Figure 1(a), where the shock-containing cells are colored red, with the upstream and downstream cells being of blue and green color, respectively. We label the shock cell and its corresponding upstream and downstream cells with S, U, and D. The shock has necessarily a numerical structure, with the typical thickness of a few cells along the normal direction. A shock cell, S, is identified as a cell that meets the following two simultaneous conditions: (i) it belongs in region 2, and (ii) it shares at least one neighboring cell in region 3. We found that the results are sensitive to the choice of the upstream cell and that it is essential to choose upstream cells carefully. Specifically, it is crucial to avoid (i) selecting a cell within the shock structure (indicated by the layer of white cells between the red and blue cells in panel (a)) and (ii) selecting a cell too far upstream because of the SW changing with the heliocentric distance. The second issue is partially mitigated when the grid resolution is very high.

The algorithm for selecting the upstream and downstream cells becomes more complicated for Cartesian grids. We illustrate the newly developed procedure in the right panel of Figure 2. The selection of shock cells (red cells in the figure) is based on the previously described criteria, i.e., we identify all cells that have at least one side adjacent to Region 3 while belonging to Region 2. For each of these cells, we then identify one appropriate upstream cell (blue) and one downstream cell (green). The coordinates of an ideal upstream point are obtained analytically using the direction of the flow (or the radial direction) and by setting a distance 2δ from the shock cell. Afterward, the cell closest to this point can be immediately found. For the reasons outlined above, we set $\delta = 1.2\Delta x$. The same procedure is used to select the downstream cell, in which the PUI pressure is modified, but the distance is set equal to δ in this case.

However, after this procedure is applied to all red cells, there remain a few more downstream cells that require modifications (yellow cells in the figure). We identify these cells and modify the PUI pressure in them by applying an appropriate interpolation procedure. Otherwise, the PUI pressure may become excessively low at these cells, creating spurious fluctuations in the heliosheath.

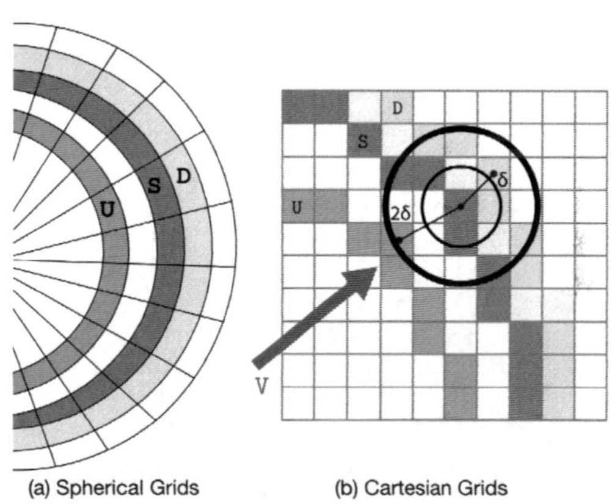

Figure 2. Schematic for the implementation of shock b.c.'s for PUIs in spherical and Cartesian grids.

5. Numerical results

We present solutions that have evolved beyond 300 years from the initial condition. Figure 3(a) shows the distribution of PUI pressure (p_{PUI}) in the meridional plane. To demonstrate the effect of the b.c.'s for PUIs at the TS, in Fig. 3(b) we compare the PUI pressure distributions along the x-axis obtained with the application of our b.c.'s (red curve) and without them (blue curve). Hereinafter these simulations as labeled as "TS-BC" and "no TS-BC." It is seen that the PUI pressure increases by approximately $40 - 50\%$ behind the shock when the TS-BC approach is used. It can also be seen that the IHS becomes narrower by a few AU. This is associated with cooling of the IHS plasma caused by the enhanced charge exchange. A detailed discussion of this effect can be found in [1]. Our simulation clearly shows the presence of the HCS, which

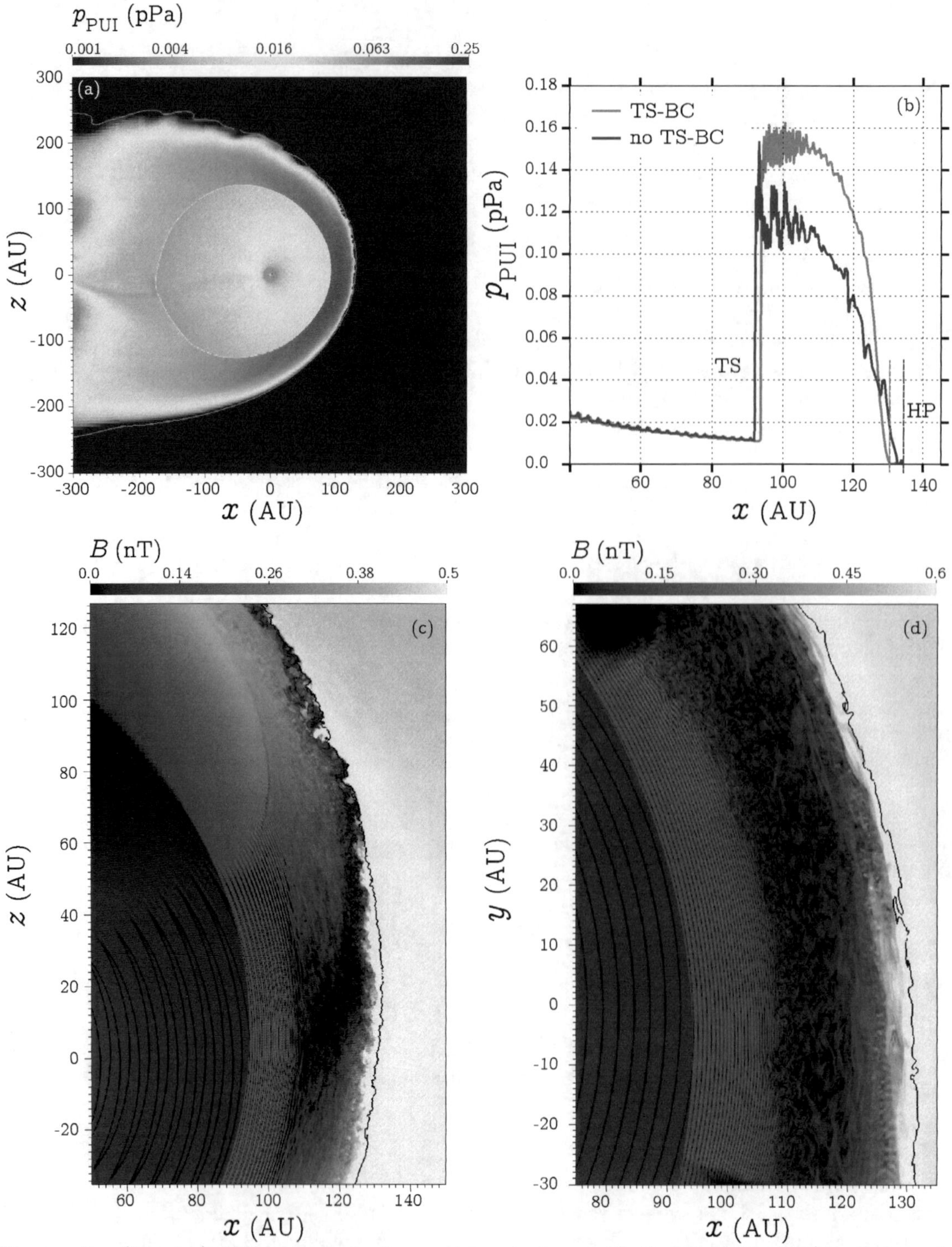

Figure 3. (Panel a) PUI pressure in the meridional plane with the shock b.c.'s; (panels b and c) PUI pressure along x-axis in the simulation with our b.c.'s at the TS (red curve) and without them (blue curve); and (panels c and d) magnetic field strength in the meridional and equatorial planes. The position of the HP is indicated by the white contour line in panel (a) and by the black contour lines in panels (c) and (d). Note that the grid boundaries can be seen in the equatorial plane.

is also seen in the PUI distributions. In particular, in the regions of well-resolved magnetic sectors, the PUI pressure increases inside the HCS, which results in periodic variations of about 15-20%. This occurs because of energy conservation, which is ensured by the MHD statement of the problem. Decreases in magnetic pressure are accompanied by increases in the thermal pressure of plasma, and vice versa. The numerical HCS structure inevitably contains a point where $B = 0$. Indeed, some, but not all, current sheet crossings identified by Voyagers exhibited small (below 0.02 nT) values of B. A typical width of these sheets was observed to be of the order of $\sim$ 15 PUI gyroscales ($\sim 5 \times 10^{-3}$ AU) [59–61]. This is way below our grid resolution. It is to be reminded that such a regular sector structure does not exist in the realistic SW. It was shown in [62] that fewer HCS crossings were observed by V2 in the IHS than it was expected on the basis of OMNI data at 1 AU.

Both solutions exhibit a wave-like behavior, but they occur on different scales (Figure 3b). The small-scale oscillations present in both curves are attributed to the physical presence of the HCS. However, it's noteworthy that the simulation without the TS-BC approach (blue curve) also exhibits larger-scale oscillations, which are identified as numerical artifacts. They arise because of the differences in numerical smearing of the plasma mixture and PUI pressure profiles across the TS. The smearing of the latter is stronger because the PUI pressure equation is being solved instead of the MHD equations in the conservation-law form applied to the mixture. Consequently, the actual numerical structure of the TS is wider for PUIs as compared with the plasma mixture. As they both are further used to calculate the analytic source terms, inappropriate, and inaccurate, values occasionally occur inside numerical shock structures. They, in turn, affect the source terms and eventually the PUI pressure. These issues are absent when the TS-BC approach is used.

The bottom panels of Figure 3 show the distributions of magnetic field magnitude in the meridional (panel c) and equatorial (panel d) planes. As in [53] (see Figure 14 therein), the sector region of the IHS gets compressed with the distance from the TS and bends as a whole following the plasma flow. Similarly to [53], a transition to turbulence, or stochasticity, is also seen in the IHS. This happens when the sectors become sufficiently compressed to trigger a tearing mode instability and consequent (numerical) magnetic reconnection. This behavior is in contrast with the solution presented in [44].

It is clear that this "turbulence" is underresolved and numerically dissipates on the grid. In the actual SW, the MHD turbulence cascade proceeds down to scales as small as a few thousand km [26, 63] and filamentary structures are created at the reconnection sites. These filamentary structures are roughly aligned with the z-axis. They evolve towards isotropic eddies, which eventually become elongated azimuthally as they are advected outward.

The distributions shown in Figure 3 are qualitatively similar to those obtained in [53]. The differences are due to the presence of PUIs treated as a separate fluid and may be affected by the differences in boundary conditions. In particular, [53] shows the the streamlines carrying magnetic field polarity on the unipolar region of the southern hemisphere turn upward into the northern hemisphere creating a unipolar "magnetic barrier" on the inner side of the HP surface. It is not entirely clear that this will ultimately happen in our solution if the SW stagnation point remains in the sector region, which would result in a layer of "turbulent" magnetic field spreading over the entire HP. Moreover, the wavenumber anisotropy of turbulence seems to differ from the previous calculations made on a spherical grid with the high computational cell aspect ratio.

6. Conclusions

We have extended the model presented in [1] to Cartesian grids. The new model utilizes a multi-fluid description of three-dimensional SW–LISM interaction with PUIs treated as a separate fluid and uses kinetically derived b.c.'s at the TS to describe their behavior. The model was used to investigate the SW-LISM interaction in the presence of a global HCS separating regions of opposite magnetic field polarity. The HCS presence affects the PUI properties by imposing oscillations on the length scale corresponding to the sector width. With a resolution of ≈ 0.07 AU in IHS, we were able to qualitatively reproduce the transition to turbulence in the IHS, as was first described in [53]. However, a number of noticeable differences between these two results were identified. The presence of MHD turbulence in the IHS and its impact on the properties of the heliosphere require further detailed investigation. It will be presented elsewhere. The new model is capable of investigating time-dependent SW–LISM interactions. In our future publications, we will further investigate the solar-cycle effects on the SW–LISM interaction. Data-driven simulations, similar to those in [64], are expected to be performed.

7. Acknowledgements

This work is supported by NASA grants 80NSSC18K1212, 80NSSC18K1649, 80NSSC24K0267, and NSF-BSF grants 2010450, 2010144 and 2019744. The authors acknowledge the Texas Advanced Computing Center (TACC) at The University of Texas at Austin for providing HPC resources on Frontera supported by NSF award CISE-OAC-2031611. Supercomputer time allocations were also provided by NASA High-End Computing Program award SMD-17-1537. FF and NP acknowledge support from the International Space Science Institute (ISSI) in Bern, through ISSI International Team project #574 "Shocks, Waves, Turbulence, and Suprathermal Electrons in the Very Local Interstellar Medium".

References

[1] Bera R K, Fraternale F, Pogorelov N V, Roytershteyn V, Gedalin M, McComas D J and Zank G P 2023 *ApJ* **954** 147
[2] Florinski V, Pogorelov N V, Zank G P, Wood B E and Cox D P 2004 *ApJ* **604** 700
[3] Pogorelov N V, Heerikhuisen J, Zank G P, Borovikov S N, Frisch P C and McComas D J 2011 *ApJ* **742** 104
[4] McComas D J, et al. 2012 **336** 1291–1293
[5] Zank G P, Heerikhuisen J, Pogorelov N V, Zirnstein E, Wood B E and McComas D J 2013 *ApJ* **763**
[6] Pogorelov N V, Heerikhuisen J, Roytershteyn V, Burlaga L F, Gurnett D A and Kurth W S 2017 *ApJ* **845** 9
[7] Parker E 1965 **13** 9–49
[8] Richardson J D, Paularena K I, Lazarus A J and Belcher J W 1995 *GRL* **22** 1469–1472
[9] Richardson J D and Smith C W 2003 *GRL* **30** 10
[10] Decker R B, Krimigis S M, Roelof E C, Hill M E, Armstrong T P, Gloeckler G, Hamilton D C and Lanzerotti L J 2008 *Nat* **454** 67–70
[11] McComas D J, et al. 2017 *ApJS* **233** 8
[12] McComas D J, et al. 2021 *ApJS* **254** 19
[13] Wallis M 1971 *NPhS* **233** 23–25
[14] Vasyliunas V M and Siscoe G L 1976 *JGRA* **81** 1247–1252
[15] Malama Y G, Izmodenov V V and Chalov S V 2006 *A&A* **445** 693–701
[16] Zank G P, Pauls H L, Cairns I H and Webb G M 1996 *JGRA* **101** 457–477
[17] Zank G P, Hunana P, Mostafavi P and Goldstein M L 2014 *ApJ* **797** 87
[18] Zank G P, Adhikari L, Zhao L L, Mostafavi P, Zirnstein E J and McComas D J 2018 *ApJ* **869** 23
[19] Pogorelov N V, Bedford M C, Kryukov I A and Zank G P 2016 *JPhCS* **767**
[20] Kim T K, Pogorelov N V, Zank G P, Elliott H A and McComas D J 2016 *ApJL* **832** 72
[21] Kim T, Kryukov I, Pogorelov N, Elliott H and Zank G P 2018 *ESSOAr* 1
[22] Isenberg P A 1986 *JGRA* **91** 9965–9972
[23] Pogorelov N V, et al. 2017 *SSRv* **212** 193–248
[24] Kleimann J, et al. 2022 *SSRv* **218** 36
[25] Sokół J M, et al. 2022 *SSRv* **218**
[26] Fraternale F, et al. 2022 *SSRv* **218** 50
[27] Pogorelov N V, Borovikov S N, Zank G P and Ogino T 2009 *ApJ* **696** 1478
[28] Opher M, Loeb A, Drake J and Toth G 2020 *NatAs* **4** 675–683
[29] Heerikhuisen J, Zirnstein E J, Pogorelov N V, Zank G P and Desai M 2019 *ApJ* **874** 76
[30] Fraternale F, Pogorelov N V and Bera R K 2023 *ApJ* **946** 97
[31] Usmanov A V, Goldstein M L and Matthaeus W H 2016 *ApJ* **820** 17
[32] Gedalin M, Balikhin M and Krasnosselskikh V 1995 *AdSpR* **15**(8-9)
[33] Richardson J D, Kasper J C, Wang C, Belcher J W and Lazarus A J 2008 *Nat* **454** 63–66
[34] Gedalin M, Pogorelov N V and Roytershteyn V 2021 *ApJ* **916** 57
[35] Zank G P, Heerikhuisen J, Pogorelov N V, Burrows R and McComas D 2010 *ApJ* **708** 1092–1106

[36] Chalov S V and Fahr H J 2000 *A&A* **360** 381–390
[37] Giacalone J, Nakanotani M, Zank G P, Kòta J, Opher M and Richardson J D 2021 *ApJ* **911** 27
[38] Chalov S V, Malama Y G, Alexashov D B and Izmodenov V V 2015 *MNRAS* **455** 431
[39] Gedalin M, Pogorelov N V and Roytershteyn V 2021 *ApJ* **910** 107
[40] Gedalin M, Roytershteyn V and Pogorelov N V 2023 *ApJ* **945** 50
[41] Pogorelov N V, et al. 2017 *SSRv* **212** 193–248
[42] Pogorelov N V, Fraternale F, Kim T K, Burlaga L F and Gurnett D A 2021 *ApJ* **917** L20
[43] Pogorelov N V, Borovikov S N, Bedford M C, Heerikhuisen J, Kim T K, Kryukov I A and Zank G P 2013 *Numerical modeling of space plasma flows: Astronum-2012* vol 474 ed Pogorelov N V, Audit E and Zank G P 165
[44] Opher M, Drake J F, Schoeffler K M, Richardson J D, Decker R B and Toth G 2011 *ApJ* **734** 71
[45] Borovikov S N, Pogorelov N V, Burlaga L F and Richardson J D 2011 *ApJL* **728** L21
[46] Pauls H L and Zank G P 1997 *JGRA* **102** 19779
[47] Pogorelov N V, Zank G P and Ogino T 2006 *ApJ* **644** 1299
[48] Williams L L, Zank G P and Matthaeus W H 1995 *JGRA* **100** 17059
[49] Matthaeus W H, Zank G P, Oughton S, Mullan D J and Dmitruk P 1999 *ApJ* **523** L93
[50] Isenberg P A, Smith C W, Matthaeus W H and Richardson J D 2010 *ApJ* **719** 716
[51] Pogorelov N V, Borovikov S, Heerikhuisen J, Kim T, Kryukov I and Zank G P 2014 *Proc. 2014 Ann. Conf. on Extreme Science and Engineering Discovery Environment* XSEDE'14
[52] Kryukov I A, Pogorelov N V, Zank G P and Borovikov S N 2012 *American Institute of Physics Conference Series* (*American Institute of Physics Conference Series* vol 1436) ed Heerikhuisen J, Li G, Pogorelov N and Zank G 48
[53] Pogorelov N V, Suess S T, Borovikov S N, Ebert R W, McComas D J and Zank G P 2013 *ApJ* **772** 2
[54] Parker E N 1961 *ApJ* **134** 20
[55] Pogorelov N V, Stone E C, Florinski V and Zank G P 2007 *ApJ* **668** 611
[56] Pogorelov N V and Semenov A Y 1997 *A&A* **321** 330
[57] Kulikovskii A G, Pogorelov N V and Semenov A Y 2000 *Mathematical Aspects of Numerical Solution of Hyperbolic Systems* Monographs and Surveys in Pure and Applied Mathematics (Chapman & Hall/CRC)
[58] Abraham-Shrauner B 1972 *JGRA* **77** 736
[59] Burlaga L F and Ness N F 2011 *JGRA* **116** A05102
[60] Burlaga L F and Ness N F 2014 *ApJ* **784** 146
[61] Burlaga L F, Ness N F and Richardson J D 2017 *ApJ* **841** 47
[62] Richardson J D, Burlaga L F, Drake J F, Hill M E and Opher M 2016 *ApJ* **831** 115
[63] Fraternale F, Pogorelov N V, Richardson J D and Tordella D 2019 *ApJ* **872** 40
[64] Kim T K, Pogorelov N V and Burlaga L F 2017 *ApJL* **843** L32

Journal of Physics: Conference Series **2742** (2024) 012011 doi:10.1088/1742-6596/2742/1/012011

Modeling the Time Dependent Interaction between the Solar Wind and the Local Interstellar Medium with Kinetic Neutral Hydrogen and Helium Atoms

Federico Fraternale

The University of Alabama in Huntsville, Center for Space Plasma and Aeronomic Research, Huntsville AL, 35899

Nikolai V. Pogorelov

Center for Space Plasma and Aeronomic Research, The University of Alabama in Huntsville, Huntsville, AL 35899, USA

Department of Space Science, The University of Alabama in Huntsville, Huntsville, AL 35899, USA

Ratan K. Bera

Center for Space Plasma and Aeronomic Research, The University of Alabama in Huntsville, Huntsville, AL 35899, USA

E-mail: federico.fraternale@uah.edu

Abstract.

We present recent advancements in our 3D modeling of the interaction between the solar wind and the local interstellar medium (LISM). The latest model results (Fraternale et al., ApJ, 2023) have raised a question about the electron density of the LISM near the heliopause. We have shown that the presence of helium ions leads to a significant underestimation of this parameter compared to the past simulations and Voyager 1 PWS observations. The latter observations, with over 12 years' worth of LISM data, offers a robust constraint on our models. Here we present additional simulations in support of the idea that the LISM proton density may need to be revised from approximately 0.054 cm^{-3} to values around 0.07 cm^{-3} or higher. Additionally, we have developed and successfully tested a new version of the kinetic code suitable for simulating time-dependent solutions.

1. Introduction

The interaction between the solar wind (SW) and the local interstellar medium (LISM) creates the heliosphere, serving as a natural laboratory for the investigation of various physical processes that occur under plasma conditions that are impossible to replicate in a terrestrial setting. Only a handful of missions provide data from the remote solar wind (SW) and the local interstellar medium (LISM), such as the still operational Voyager 1 and 2 (V1, V2), the Interstellar Boundary EXplorer (IBEX), and the New Horizons (NH) missions.

The LISM is influenced by the presence of the heliosphere in several ways. It extends over 1000 AU in the upwind direction and potentially spans tenfold that distance in the downwind direction exhibiting the heliotail. The term very local interstellar medium (VLISM) is employed here to refer to this perturbed medium.

Determining the properties of the LISM from remote and in situ observations is a challenging task. Primary interstellar neutral (ISN) atom fluxes observed at 1 AU are used to infer the properties of the unperturbed LISM. Energetic neutral atoms (ENAs) also enable remote studies of the global heliosphere [1, 2]. However, these tasks necessitate an extensive theoretical analysis and global modeling efforts. In fact, because of the filtration occurring in the outer heliosheath (OHS) and SW regions, the effects of ionization, gravity, radiation pressure, collisions, etc., the velocity distribution functions (VDFs) of primary atoms are heavily modified.

ISN helium is particularly valuable due to the smaller cross sections for charge exchange, as compared to hydrogen, and relatively high abundance. Nonetheless, even the helium distributions, including those of primary atoms, deviate from a Maxwellian [3–5], which makes kinetic modeling necessary. As number of questions regarding the properties of the unperturbed LISM remain unanswered. Recent studies, supported by the pickup ion (PUI) measurements made by NH, have been focused on reevaluating the estimates of LISM parameters, e.g., see [6–8]. A possibility also exists that the LISM in the solar neighborhood is composed of a mixture of the Local Interstellar Cloud (LIC) and the G-Cloud [9].

We have recently developed a fully 3D MHD-plasma/kinetic-neutrals global model, which, for the first time includes both neutral hydrogen (H) and helium (He) atoms self-consistently [4]. The most recent version also includes electrons treated as a separate, charge-neutralizing fluid [10, 11].

We have shown that the inclusion of He in a SW–LISM interaction model requires substantial reassessments of the H and H^+ distributions in the LISM with important consequences for the global heliosphere. A suggestion was made to revise also the LISM proton density to values higher than 0.054 cm^{-3} [10]. In addition, the heliospheric community accumulated enough data (22 years of Ulysses and IBEX observations combined, as well as 11 years of V1 sampling the VLISM in situ) to investigate the solar cycle effects on the neutral atom transport.

In this paper, we further elaborate on the discussion presented in [10] and provide updated versions of the model also suitable for time-dependent simulations.

2. Physical model and boundary conditions

Our physical model includes an MHD description of plasma and a kinetic description for neutral H and He atoms. The most recent version is described in [10, 11]. The plasma description is based on the solution of a set of the conservation laws for the mixture of thermal protons, pickup protons , electrons, and helium ions, hereinafter referred to as the plasma mixture:

$$\partial_t \rho + \boldsymbol{\nabla} \cdot (\rho \mathbf{u}) = \sum_s S_s^\rho, \tag{1}$$

$$\partial_t(\rho \mathbf{u}) + \boldsymbol{\nabla} \cdot \left[\rho \mathbf{u}\mathbf{u} + (p + \mathbf{B}^2/8\pi)\mathbf{I} - \frac{1}{4\pi}\mathbf{B}\mathbf{B}\right] = \sum_s \mathbf{S}_s^{\mathrm{m}}, \tag{2}$$

$$\partial_t E + \boldsymbol{\nabla} \cdot \left[(E + p + \mathbf{B}^2/8\pi)\mathbf{u} - \frac{1}{4\pi}(\mathbf{B} \cdot \mathbf{u})\mathbf{B}\right] = \sum_s S_s^E, \tag{3}$$

$$\partial_t \mathbf{B} + \nabla \cdot (\mathbf{u}\mathbf{B} - \mathbf{B}\mathbf{u}) = 0, \tag{4}$$

$$\partial_t \rho_{\mathrm{He}^z} + \boldsymbol{\nabla} \cdot (\mathbf{u}\ \rho_{\mathrm{He}^z}) = S^\rho_{\mathrm{He}^z}, \tag{5}$$

$$\partial_t p_{\mathrm{He}^z} + \boldsymbol{\nabla} \cdot (\mathbf{u}\ p_{\mathrm{He}^z}) = (1-\gamma)\ p_{\mathrm{He}^z} \nabla \cdot \mathbf{u} + S^p_{\mathrm{He}^z} + Q^{\mathrm{C}}_{\mathrm{He}^z}, \tag{6}$$

$$\partial_t p_{\rm e} + \boldsymbol{\nabla} \cdot (\mathbf{u}\, p_{\rm e}) = (1-\gamma)\, p_{\rm e} \nabla \cdot \mathbf{u} + S_{\rm e}^p + Q_{\rm e}^{\rm C}, \tag{7}$$

$$S_s^p = (\gamma - 1)\left(S_s^E - \mathbf{u} \cdot \mathbf{S}_s^{\rm m} + \frac{1}{2} u^2 S_s^\rho\right), \tag{8}$$

Here ρ_s, p_s, T_s, are the mass density, thermal pressure and temperature of each plasma component s, $\mathbf{u}$ and $\mathbf{B}$ are the bulk velocity and magnetic field vectors, respectively, and u and B are their magnitudes. The total energy density is $E = \rho(\epsilon + \mathbf{u}^2/2) + \mathbf{B}^2/8\pi$, where ϵ is the specific internal energy ($\rho\epsilon = p/(\gamma-1)$) and the adiabatic index γ is equal to 5/3 for all species. The terms $S_s^{\rho,\mathrm{m},E,p}$ represent the sources in the mass, momentum, energy, and pressure equations due to charge exchange and photoionization, for each species s. These terms are computed via two Monte Carlo kinetic solvers [12]. In the electron pressure equation, the $S_{\rm e}^p$ source term includes photoionization only. The source terms $Q_s^{\rm C}$ represent the energy transfer rate to species s due to Coulomb collisions with the other species, as described in [10].

As regards helium ions, here we introduce a modified version of the model presented in [10]. Equations (5)-(6) are now applied to two difference species in two different environments: He^{2+} ions (alpha particles) in the SW and He^{+} ions in the LISM. In fact, in the SW at 1 AU the density of He^{2+} ions is higher by approximately a factor of 1000 than that of He^{+} ions. On the other hand, the charge-exchange cross sections $\sigma(\mathrm{He} + \mathrm{He}^{2+})$ at ~ 5 keV are only ~ 3.5 times smaller than $\sigma(\mathrm{He} + \mathrm{He}^{+})$. Therefore, here we neglect He^{+} in the SW and include He^{2+} ions instead. The latter carry a significant amount of ram pressure (8% to 20% of that of SW protons) which contributes to the pressure balance at the HP. The opposite is true in the VLISM, where He^{+} ions carry about 40% of the mass of all ions, while He^{2+} ions are comparatively negligible. Therefore, in Eqs. (5)-(6), $z = 2+$ in the SW regions and $z = +$ in the LISM. This modification is the first step toward the inclusion of both He^{2+} and He^{+} everywhere in our model. Here, we have taken into account the ionization rates of He due to photoionization and charge exchange. For the latter process, for alpha particles, only the reaction $\mathrm{He}^{2+} + \mathrm{He} = \mathrm{He} + \mathrm{He}^{2+}$ is included. We fit the the helium cross-section data published by Barnett, et al. [13], obtaining the following expression

$$\sigma_{cx}(\mathrm{He} + \mathrm{He}^{2+}) = (3.142 - 0.490 \log\ \mathcal{E})^{0.967} \times 10^{-16}\ \ (\mathrm{cm}^2), \tag{9}$$

where $\mathcal{E}$ is the collision energy expressed in keV. The above expression fits the Barnett, et al. data with the mean relative error equal to 2.59% in the energy range $\mathcal{E} < 100$ keV, and the maximum relative error less than 5.5%. For hydrogen, we have been using the cross sections from Lindsay & Stebbings (LS05) [14],

$$\sigma_{cx}^{\rm LS}(\mathrm{H} + \mathrm{H}^{+}) = (4.15 - 0.531 \log\ \mathcal{E})^2 \times \left[1 - \exp\left(-\frac{67.3}{\mathcal{E}}\right)\right]^{4.5} \times 10^{-16}\ \ (\mathrm{cm}^2). \tag{10}$$

However, due to significant uncertainties in the available data, especially at low energies, no consensus has been reached so far and various formulas have been proposed (see the discussion in [15]). In this study, see Section 3.1, we provide a comparison of the results obtained using Eq. (10) with those obtained using the updated cross-sections provided by [16, 17]:

$$\sigma_{cx}(\mathrm{H} + \mathrm{H}^{+}) = (4.049 - 0.447 \log\ \mathcal{E})^2 \times \left[1 - \exp\left(-\frac{60}{\mathcal{E}}\right)\right]^{4.5} \times 10^{-16}\ \ (\mathrm{cm}^2), \tag{11}$$

The above expression is expected to provide more accurate cross-section values in the low-energy range (0.01 eV - 100 eV), while LS05 may be more accurate at high energies [17]. In the new approach, Eq. (11) is employed at $\mathcal{E} < 1$ keV while LS05 is used for $\mathcal{E} \geq 1$ keV. A different fit based on Barnett et al. data [13] was proposed in [15], where the effects on global modeling were also discussed.

Our fully 3D, hybrid-parallelized kinetic code builds on the previous hydrogen-only versions [18–21]. The Boltzmann equations for the transport of neutral atoms are solved via a Monte Carlo method [12, 19, 22] using two independent modules for H and He atoms, respectively. The solution is obtained through an iterative process consisting of two stages: the fluid stage and the kinetic stage. In the fluid stage, a fixed distribution of source terms is utilized, which is obtained from the neutral modules. On the other hand, during the kinetic stage, the distribution of plasma is fixed, and kinetic particles are run. Each stage has its own time interval that depends on whether we seek a steady-state solution or conduct a time-dependent simulation.

For a steady solution on a Cartesian grid with the base grid size of 10 AU cubed, we typically run the kinetic module every $\tau_p = 0.3$ years (about 8 time steps of the base grid for plasma) to avoid spurious instabilities or other numerical artifacts [10, 22]. To gather sufficient statistics for source terms, the kinetic modules run for $\tau_N = 700-900$ years (using $\sim 0.4\times10^9$ macroparticles per species). Both time intervals (τ_p, τ_N) must be carefully chosen to avoid numerical artifacts. These artifacts may include, among others, (i) the generation of spurious wave-like modes in the heliosheath when τ_p is too large or (ii) the enhancement of instabilities at the HP and increased stiffness in the solution when the source terms become too noisy due to low statistics, which occurs when τ_N is too small.

In a time-dependent scenario, both the time interval of the kinetic modules and the time interval of the plasma need to be significantly reduced to accurately account for the spatio-temporal scales present in both plasma and neutral atom distributions, as discussed in Section 3.3.

The kinetic modules require the VDFs of ions to be specified. In the previous studies with helium, we assumed a Maxwellian plasma [4, 10]. In Section 3.1, we discuss the effects of employing Lorentzian (kappa) VDFs for protons in the SW regions.

We use the level-set method [23, 24] to track the position of the HP. Basically, we solve the advection equation to track the position of a surface (or 'level'), $\partial_t\psi(\mathbf{x},\ t) + \mathbf{u}\nabla\psi = 0$. The SW and LISM regions are initialized at $t = 0$ with the values $\psi = -1$ and $\psi = 1$, respectively (also used as boundary conditions), so that the HP is represented by the surface $\psi = 0$ separating the two plasmas [25], see the discussion in Section 3.3. Note that this approach is not equivalent to that used, for instance in [26], where the continuity equation for the concentration of a passive scalar is used. The difference is substantial in compressible or nonuniform media.

The boundary conditions used for the simulations presented in this study are summarized in Table 1. The bottom rows of the table list the simulations, labeled from `A` to `D`. Simulation `A` corresponds to the best of models presented in [10] (`S3` in that study), where the plasma is assumed to be Maxwellian everywhere for the calculation of source terms, the LS05 cross sections are employed, and the proton density in the LISM ($n_{\rm p,LISM}$) is equal to 0.054 cm^{-3}. Simulations `B1` to `D` use a higher density of 0.068 cm^{-3}, the motivation for which is discussed in Section 3.1. Simulations `B1` and `B2` differ only in the expressions for the H charge-exchange cross sections, i.e., in `B1` we use LS05 at all energies and in `B2` we use Eq. 11 at low energies. With simulation `C`, we test the effect of using a kappa distributed protons in the SW regions. Finally, in simulation `D`, we test a new time-dependent kinetic model by simulating a nominal solar cycle (11-year periodic changes in the latitudinal extent of the fast and slow wind boundary), see details in Section 3.3.

Table 1. Boundary conditions and list of simulations presented in this study.

Parameter	Units	Value	Description
Outer boundary: LISM			
$n_{\mathrm{p,LISM}}$	cm^{-3}	0.054 and 0.068	LISM proton density
$n_{\mathrm{He^+,LISM}}$	cm^{-3}	0.00898	LISM He$^+$ density [27]
$n_{\mathrm{H,LISM}}$	cm^{-3}	0.195	LISM H density [7]
$n_{\mathrm{He,LISM}}$	cm^{-3}	0.0153	LISM He density [27]
V_{LISM}	km s^{-1}	-25.4	LISM bulk flow speed[28]
$(\lambda_{V,\mathrm{LISM}}, \beta_{V,\mathrm{LISM}})$	deg	$(255^\circ.7, 5^\circ.1)$	LISM flow direction (Ecliptic J2000) [28]
T_{LISM}	K	7,500	LISM temperature (all species) [28]
B_{LISM}	nT	0.293	ISMF strength [28]
$(\lambda_{B,\mathrm{LISM}}, \beta_{B,\mathrm{LISM}})$	deg	$(227^\circ.28, 34^\circ.62)$	ISMF direction (Ecliptic J2000) [29]
Inner boundary at 1 AU: Steady-state simulations			
$n_{\mathrm{p,1\,AU}}$	cm^{-3}	8.0	1 AU proton density
$n_{\mathrm{He^+,1\,AU}}$	cm^{-3}	10^{-4}	1 AU He$^+$ density in model A
$n_{\mathrm{He^{2+},1\,AU}}$	cm^{-3}	0.28	1 AU He^{2+} density in models B-D
$V_{\mathrm{1\,AU}}$	km s^{-1}	415	1 AU bulk flow speed
$T_{\mathrm{1\,AU}}$	K	80,000	1 AU temperature (all species)
$B_{R,\mathrm{1\,AU}}$	nT	3.75	1 AU magnetic field radial component
Inner boundary at 1 AU: Solar cycle simulation			
V_f	km s^{-1}	750	Fast SW speed (all species)
V_s	km s^{-1}	415	Slow SW speed (all species)
T_f	K	250,000	Fast SW temperature (all species)
T_s	K	80,000	Slow SW temperature (all species)
$n_{f,\mathrm{p}}$	cm^{-3}	2.2	Fast SW proton density
$n_{s,\mathrm{p}}$	cm^{-3}	7.5	Slow SW proton density
$n_{f,\mathrm{He^{2+}}}$	cm^{-3}	0.104	Fast SW He^{2+} density
$n_{s,\mathrm{He^{2+}}}$	cm^{-3}	0.26	Slow SW He^{2+} density
Simulations			
A	$n_{\mathrm{p,LISM}} = 0.054$ cm^{-3}, Maxw. plasma, LS05 cross sections (Eq. 10), S3 in [10]		
B1	$n_{\mathrm{p,LISM}} = 0.068$ cm^{-3}, Maxw. plasma, LS05 cross sections (Eq. 10)		
B2	$n_{\mathrm{p,LISM}} = 0.068$ cm^{-3}, Maxw. plasma, updated cross sections		
C	$n_{\mathrm{p,LISM}} = 0.068$ cm^{-3}, kappa SW protons, updated cross sections		
D	$n_{\mathrm{p,LISM}} = 0.068$ cm^{-3}, nominal solar-cycle with unipolar B		

3. Results and Discussion

3.1. Electron density in the VLISM and kappa distributions

One of the findings discussed in [10] is that the new models with helium underestimate the electron density observed by V1 in the VLISM. Figure 1 presents linear distributions of plasma properties obtained from steady-state simulations A - C. In Panel (a), the electron densities in the V1 data and from our models are compared. Voyager provides indirect measurements of electron density through PWS observations of plasma oscillations and quasi-thermal emission at the plasma frequency [30–33]. A commonly adopted value for the LISM proton density,

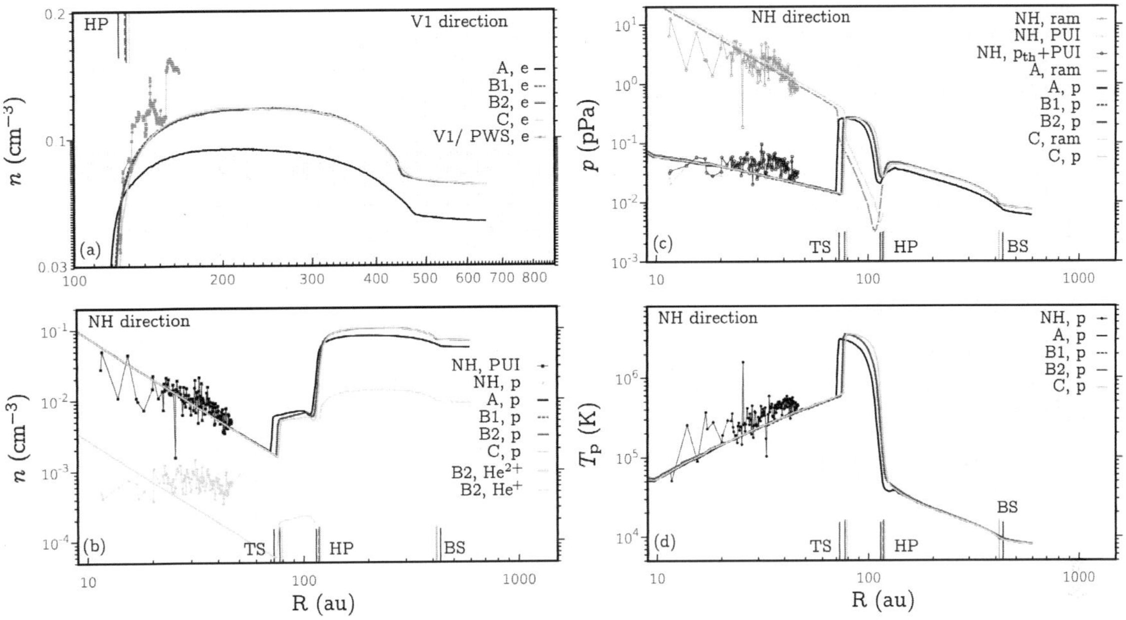

Figure 1. Comparison of plasma distributions in steady-state simulations and spacecraft data. (a) Electron density in the VLISM and comparison with V1/PWS data; (b) Proton and helium ions density and comparison with NH data; (c) Protons thermal pressure and ram pressure; (d) Proton temperature and comparison with NH data (PUIs in green and all protons in black).

$n_{\rm p,LISM} = 0.054$ cm^{-3}, was obtained with the aid of steady-state global simulations assuming, as a constraint, the fixed nucleon LISM density of 0.089 cm^{-3}, which yielded reasonable agreement between the observed heliopause (HP) position in Voyager 1 (V1) data (121.6 AU) and the one in global simulations [27]. In simulations without helium, the assumption $n_{\rm p,LISM} = 0.089$ cm^{-3} resulted in the accurate electron density reproduction (because $n_{\rm p} = n_e$), see [34]. However, if one uses $n_{\rm p,LISM} = 0.054$ cm^{-3} or lower, the V1 data are not reproduced, even when helium ions are included [4].

Recent V1 data shown in Fig. 1(a) confirm that a radial gradient exists in the electron density, with values remaining as high as 0.15 cm^{-3} in 2022 [33]. Simulation A (black curve) underestimates these values by approximately 36%. To mitigate this discrepancy in global simulations, several factors can be adjusted, including $n_{\rm p,LISM}$, $n_{\rm He^+,LISM}$, or both, as well as modifying the density of neutrals.

Based on the simulations presented in the Appendix of [10], varying the LISM H density from 0.15 cm^{-3} to 0.22 cm^{-3} does not produce significant changes in the peak of $n_{\rm e}$, which remains around 0.1 cm^{-3} in those simulations. On the other hand, increasing the He$^+$ density alone would introduce excessive ram pressure due to the mass ratio. Consequently, we have first explored the effects of increasing the LISM density of protons to ~ 0.07 cm^{-3} (degrees of ionization $\chi_H = 0.26$, $\chi_{\rm He} = 0.368$, and $(n_{\rm He}/n_{\rm H})_{\rm LISM} = 0.052$). This values are also compatible with the estimates provided in [35].

Including a realistic amount of alpha particles in the SW is sufficient to maintain the heliopause (HP) at a distance similar to previous simulations (compatible with Voyager observations). As illustrated in panel (a), the underestimation of $n_{\rm e}$ is reduced in the new simulations B1 and B2, although the values still align with the lower bound of observations, indicating the need for further improvement. Importantly, it is worth noting that the comparison with NH data remains robust in these simulations.

In all new simulations, as far as the plasma is concerned, the most evident changes as compared to case A are seen in the inner heliosheath, where the temperatures of the plasma mixture and protons (thermal + PUIs) immediately behind the TS becomes higher by $\sim$ 12.5% and $\sim$ 17.5%, respectively. Using kappa-distributed protons in the SW only (case C) does not lead to significant changes to the heliosheath thickness (compare the blue and green curves). The reason for this lies in the energy dependence of charge exchange cross sections, as explained in [36–38], and recently discussed in [39]. In fact, the energy transfer due to charge exchange, associated with the energetic tails of the Lorentzian distributions, is reduced due to smaller cross sections at high collision energy. However, notable changes in the H atoms distributions are obtained, as discussed in Section 3.2.

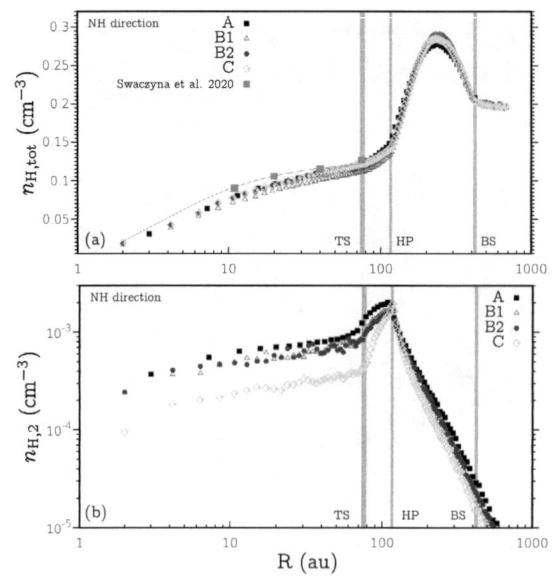

Figure 2. (a) Density distributions of H atoms along the NH direction (Ecliptic J2000 coordinates $(\lambda, \delta) = (285°, 2°)$). (b) Density of population 2 H atoms, born in the heliosheath due to charge exchange.

Using kappa-distributed protons in the SW (case C) does not produce considerable changes in the positions of the TS and HP (compare the blue and the green curves). The reason for this resides in the energy dependence of the charge-exchange cross sections [36–38]. In fact, the energy transfer due to charge exchange associated with the energetic tails of the Lorentzian distributions, is mitigated due to the smaller cross sections at high energy. In contrast, the properties of neutral H atoms vary considerably, see for instance the density distributions in Fig. 2.

3.2. Distributions of neutral hydrogen density and the effect of updated cross sections for hydrogen charge exchange

Figure 2 shows the distribution of the total density of H atoms (panel a) and the density of population 2 H atoms (those produced in the IHS, panel b), along with a comparison with the density derived in [7] based on NH data. The comparison between simulations B1 and B2 reveals that using the updated cross sections does not result in substantial differences in the plasma distributions and standoff distances of the HP and TS. However, as shown in Fig. 2, the density of H atoms becomes higher when the updated cross sections are applied. This increase varies with distance, ranging from 9.5% at 2 AU to 5.5% at the TS. Consequently, the H densities in simulations B2 and C are close to that in simulation A. The density of population 2 H atoms that reach 2 AU in simulation C decreases by 57-67%, compared to case A, however these neutrals are hotter.

3.3. Time dependent simulations with kinetic neutral hydrogen and helium atoms

Here we present the a preliminary time-dependent simulation (case D) with both kinetic neutral H and He atoms. As mentioned in Section 2, we would like to reduce the iteration time of the kinetic neutrals step to about 1 year, while keeping the same statistics, in order to resolve solar-cycle variability and large-scale fluctuations that may arise both in the plasma and in the neutral atom distributions. Because it is computationally unfeasible to increase the number of macroparticles $\sim 10^3$ times, a different strategy needs to be developed. We achieved the objective by (1) creating, for each particle (He and H atoms), many realizations of the same iteration interval, slightly randomizing the initial position and speed of the particles and eventually averaging the collected source terms. Note that the occurrence of a charge exchange event may vary in terms of both timing and spatial location, for the different realizations of one individual particle; (2) we tune the parameters of particle splitting in front of the HP to increase the actual number of macroparticles in the SW regions by a factor of ~ 3.

As far as step (1) is concerned, a similar approach was proposed in [40] and used in [41]. After collecting the source terms, only the set of newly born particles resulting from the first realization is retained, and all particles are advanced by one year. Careful consideration is given to the randomization of velocities and positions. Specifically, one must avoid (i) smoothing out gradients or inhomogeneities in the neutral atom distributions that may be present in the time-dependent scenario and (ii) excessively impacting the neutrals' VDF. To ensure a proper position randomization, we establish maximum displacements that increases linearly with the heliocentric distance, denoted as $\delta = \min(|\mathbf{x}_N|\frac{20}{850}, \delta_{\max})$ AU, with $\delta_{\max}$ equal to 20 AU in the LISM and 5 AU in the SW regions. For each stochastic realization j, we calculate the new particle position $\mathbf{x}_{N,j} = \mathbf{x}_{N,0} + a \cdot \delta$ (where $a \in [-1, 1]$ is a random number drawn from a uniform distribution). Regarding velocities, randomization only modifies the velocity vectors by mere 1%, exclusively within the LISM.

The same randomization strategy is applied in step when particles undergo the new splitting (2). The new splitting process occurs in two steps. The first step closely resembles the original version described by [42]. In this step, macroparticles are *enabled* to split when they reach predetermined specified locations in the physical domain. However, the actual splitting, i.e. the generation of a new particle with appropriately weighted mass, only takes place when a charge exchange event occurs. In the case of helium, where the mean free paths for charge exchange are large, this strategy may lead to fewer particles than desired. The second step involves a 'boost' that ensures the generation of a new physical particle, only for those particles for which splitting was enabled in the first step but in the absence of a charge exchange event. In this case, after splitting, the new position of a particle and its velocity are slightly randomized following the criteria mentioned above.

In simulation D, the domain is 1680 AU cubed, with the Sun centered at $(x, y, z) =$ (1000, 850. 850) AU. The plasma grid is Cartesian, has 5 levels of refinement, and the base grid size of $\Delta x = \Delta y = \Delta z = 10$ AU. The best resolution is ~ 0.3 AU in all directions, applied everywhere around the Sun at $R < 20$ AU, and in the upwind region ($x > 0$) extending into the LISM radially by ~ 50 AU. In the distant tail, the resolution is 1.25 AU cubed everywhere. The source terms from the kinetic modules are collected on parallelepiped-shaped Cartesian kinetic grids, as described in [20], with the finest level size of 1.25 AU.

Figure 3 presents 2D distributions of electron density, plasma pressure, and speed in the $B-V$ plane (defined by the velocity and magnetic field vectors in the unperturbed LISM, intersecting the Sun), and in the cross sections of the heliotail perpendicular to the x-axis at 900 AU from the Sun, from simulation D.

For the sake of simplicity in code testing, the heliospheric magnetic field (HMF) at the inner boundary is maintained unipolar and does not undergo polarity reversal every 11 years. A similar simulation was performed in [43] to demonstrate that even under the unipolar HMF assumption, the solar cycle influences drastically the structure of the tail because it affects the SW collimation within the two polar lobes, as compared with the steady state simulations with spherically-symmetric SW. Moreover, as discussed in [10], fluctuations and turbulence in the tail are also different from those obtained in simulations with steady and uniform b.c.'s.

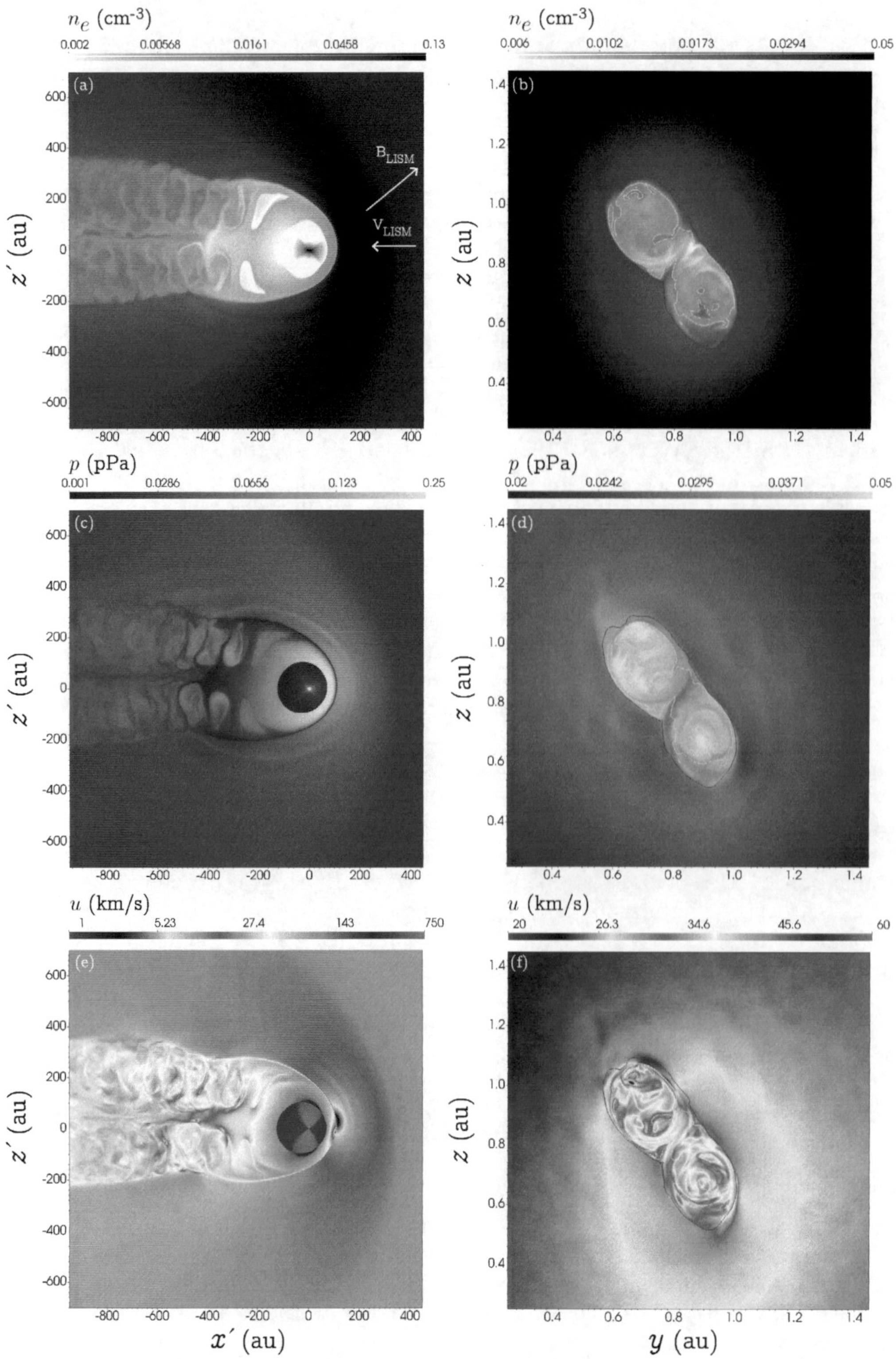

Figure 3. Two-dimensional distributions of electron density (top panels), plasma pressure (middle panels) and velocity magnitude (bottom panels) in a time-dependent simulation with kinetic H and He atoms. The left panels show quantities on the $B_\infty - V_\infty$ plane, while the right panels show cross cuts of the heliotail at 900 AU from the Sun. In the right panels, two level set contours for $\psi = -0.9$ and $\psi = +0.9$ are also shown with green and blue lines, respectively.

Even in this unrealistic scenario, the tail does not exhibit a two-fold split configuration, as shown by the level set contours in the right panels of Fig. 3. The HP position is formally represented by the contour $\psi = 0$. In this simulation, the level set is intentionally not reinitialized and is allowed to diffuse numerically to highlight the presence of a diffusion/mixing layer. In the right panels of Fig. 3, we display the contours at $\psi = -0.9$ (in green) and $\psi = +0.9$. It is important to note that (i) the presence of the mixing layer results from a combination of both numerical diffusion and actual mixing due to large-scale motions or instabilities at the HP and (ii) the level set can provide only partial information about the mixing. Moreover, it is known that in the simulations with bipolar HMF, solutions frequently exhibit small HMF values near the HP, which also favours the growth of instabilities [44]. We further clarify here that the concept of a comet-shaped heliotail does not exclude the presence of mixing of the SW and LISM plasma as discussed in [45, 46].

4. Conclusions

The most recent MHD-plasma/Kinetic-neutrals model implemented in MS-FLUKSS incorporates both hydrogen and helium atoms, treated self-consistently, and features separate fluid electrons and helium ions. This represents a state-of-the-art model for investigating global heliospheric processes and enables us to investigate the transport of neutral atoms throughout the heliosphere. Such model is critical for enhancing theoretical models and interpreting spacecraft data, and will provide predictions to assist the interpretation of future observations by the Interstellar Mapping and Acceleration Probe (IMAP) mission [47].

In this study, we presented recent updates of the model published recently in [10], that include a simplified, preliminary treatment of alpha particles; updated hydrogen cross sections from [16]; and the effects of kappa-distributed plasma. The new models suggest the need for a revision of interstellar plasma parameters, particularly the proton density, to align with Voyager PWS observations of electron density in the OHS. Furthermore, we have developed and successfully tested a time-dependent version of the kinetic model.

5. Acknowledgements

This work is supported by NASA grants 80NSSC24K0267, 80NSSC18K1212, and 80NSSC18K1649. The authors acknowledge the Texas Advanced Computing Center (TACC) at The University of Texas at Austin for providing HPC resources on Frontera supported by NSF award CISE-OAC-2031611, and also acknowledge the IBEX mission as part of NASA's Explorer Program (80NSSC18K0237). Supercomputer time allocations were also provided by NASA High-End Computing Program award SMD-17-1537. FF and NP acknowledge support from the International Space Science Institute (ISSI) in Bern, through ISSI International Team project #574 "Shocks, Waves, Turbulence, and Suprathermal Electrons in the Very Local Interstellar Medium". FF also acknowledges ISSI Team project #541 "Distribution of Interstellar Neutral Hydrogen in the Sun's Neighborhood)", for the insightful discussions on interstellar neutral hydrogen.

References

[1] Heerikhuisen J, et al. 2010 *ApJL* **708** L126–L130
[2] Reisenfeld D B, et al. 2021 *ApJS* **254** 40
[3] Wood B E, Müller H R and Möbius E 2019 *ApJ* **881** 55
[4] Fraternale F, Pogorelov N V and Heerikhuisen J 2021 *ApJL* **921** L24
[5] Swaczyna P, Rahmanifard F, Zirnstein E J, McComas D J and Heerikhuisen J 2021 *ApJL* **911** L36
[6] Swaczyna P, McComas D J, Zirnstein E J and Heerikhuisen J 2019 *ApJ* **887** 223
[7] Swaczyna P, et al. 2020 *ApJ* **903** 48
[8] Swaczyna P, et al. 2023 *ApJ* **953** 107
[9] Swaczyna P, et al. 2022 *ApJL* **937** L32

[10] Fraternale F, Pogorelov N V and Bera R K 2023 *ApJ* **946** 97
[11] Fraternale F, Pogorelov N V and Bera R K 2023 *ApJ* **953** 121
[12] Malama Y G 1991 *Ap&SS* **176** 21–46
[13] Barnett C F, Hunter H T, Fitzpatrick M I, Alvarez I, Cisneros C and Phaneuf R A 1990 *Oak Ridge National Laboratory, Report ORNL-6086* **1**
[14] Lindsay B G and Stebbings R F 2005 *JGRA* **110** A12213
[15] Bzowski M and Heerikhuisen J 2019 *ApJ* **888** 24
[16] Schultz D R, Ovchinnikov S Y, Stancil P C and Zaman T 2016 *JPhB* **49** 084004
[17] Swaczyna P, McComas D J and Schwadron N A 2019 *ApJ* **871** 254
[18] Pogorelov N V, Heerikhuisen J and Zank G P 2008 *ApJL* **675** L41
[19] Heerikhuisen J, Pogorelov N V, Florinski V, Zank G P and le Roux J A 2008 *ApJ* **682** 679–689
[20] Borovikov S N, Heerikhuisen J, Pogorelov N V, Kryukov I A and Zank G P 2008 *Numerical Modeling of Space Plasma Flows: ASTRONUM-2007* (*Astronomical Society of the Pacific Conference Series* vol 385) ed Pogorelov N V, Audit E and Zank G P p 197
[21] Borovikov S, Heerikhuisen J and Pogorelov N 2013 *Numerical Modeling of Space Plasma Flows: ASTRONUM-2012* (*Astronomical Society of the Pacific Conference Series* vol 474) ed Pogorelov N V, Audit E and Zank G P p 219
[22] Heerikhuisen J, Florinski V and Zank G P 2006 *J. Geophys. Res.* **111** A06110
[23] Osher S and Fedkiw R P 2001 *J. Comp. Phys.* **169** 463–502
[24] Osher S J and Fedkiw R P 2002 *Level Set Methods and Dynamic Implicit Surfaces* (New York: Springer-Verlag)
[25] Borovikov S N, Pogorelov N V, Burlaga L F and Richardson J D 2011 *ApJL* **728** L21
[26] Korolkov S and Izmodenov V 2021 *MNRAS* **504** 4589–4598
[27] Bzowski M, et al. 2019 *ApJ* **882** 60
[28] McComas D J, et al. 2015 *ApJS* **220** 22
[29] Zirnstein E J, Heerikhuisen J, Funsten H O, Livadiotis G, McComas D J and Pogorelov N V 2016 *ApJL* **818**
[30] Gurnett D and Kurth W 2019 *NatAs* **3** 1024–1028
[31] Ocker S, Cordes J and et al S C 2021 *Nat. Astron.* **5** 761–765
[32] Gurnett D A, Kurth W S, Burlaga L F, Berdichevsky D B, Pogorelov N V, Pulupa M and Bale S D 2021 *ApJ* **921** 62
[33] Kurth W S, Burlaga L F, Kim T, Pogorelov N V and Granroth L J 2023 *ApJ* **951** 71
[34] Pogorelov N V, Heerikhuisen J, Roytershteyn V, Burlaga L F, Gurnett D A and Kurth W S 2017 *ApJ* **845** 9
[35] Slavin J D and Frisch P C 2008 *A&A* **491** 53–68
[36] DeStefano A M and Heerikhuisen J 2017 *J. Phys. Conf. Ser.* **837** 012013
[37] Heerikhuisen J, Zirnstein E J, Pogorelov N V, Zank G P and Desai M 2019 *ApJ* **874** 76
[38] DeStefano A M and Heerikhuisen J 2020 *Phys. Plasmas* **27** 032901
[39] Bera R K, Fraternale F, Pogorelov N V, Roytershteyn V, Gedalin M, McComas D J and Zank G P 2023 *ApJ* **954** 147
[40] Heerikhuisen J, Pogorelov N V and Zank G P 2013 *Numerical Modeling of Space Plasma Flows: ASTRONUM-2012* (*Astronomical Society of the Pacific Conference Series* vol 474) ed Pogorelov N V, Audit E and Zank G P p 195
[41] Zirnstein E J, Heerikhuisen J, Pogorelov N V, McComas D J and Dayeh M A 2015 *ApJ* **804** 5
[42] Heerikhuisen J, Pogorelov N V, Florinski V, Zank G P and Kharchenko V 2009 *Numerical Modeling of Space Plasma Flows: ASTRONUM-2008* (*Astronomical Society of the Pacific Conference Series* vol 406) ed Pogorelov N V, Audit E and Zank G P p 189
[43] Pogorelov N V, et al. 2017 *SSRv* **212** 193–248
[44] Borovikov S N, Pogorelov N V, Zank G P and Kryukov I A 2008 *ApJ* **682** 1404–1415
[45] Pogorelov N V, Borovikov S N, Heerikhuisen J and Zhang M 2015 *ApJL* **812** L6
[46] Pogorelov N V and Borovikov 2015 *Numerical Modeling of Space Plasma Flows: ASTRONUM-2014* (*Astronomical Society of the Pacific Conference Series* vol 498) ed Pogorelov N V, Audit E, Colella P and Zank G P p 160
[47] McComas D J, et al. 2018 *SSRv* **214** 116

Journal of Physics: Conference Series **2742** (2024) 012012 doi:10.1088/1742-6596/2742/1/012012

CORHEL-CME: An Interactive Tool For Modeling Solar Eruptions

Jon A. Linker, Tibor Torok, Cooper Downs, Ronald M. Caplan, Viacheslav Titov, Andres Reyes, Roberto Lionello, & Pete Riley

Predictive Science Inc., 9990 Mesa Rim Road Suite 170, San Diego, CA 92121

E-mail:
linkerj@predsci.com,tibor@predsci.com,cdowns@predsci.com,caplanr@predsci.com,
titovv@predsci.com,areyes@predsci.com,lionel@predsci.com,pete@predsci.com

Abstract. Coronal Mass Ejections (CMEs) are immense eruptions of plasma and magnetic fields that are propelled outward from the Sun, sometimes with velocities greater than 2000 km/s. They are responsible for some of the most severe space weather at Earth, including geomagnetic storms and solar energetic particle (SEP) events. We have developed CORHEL-CME, an interactive tool that allows non-expert users to routinely model multiple CMEs in a realistic coronal and heliospheric environment. The tool features a web-based user interface that allows the user to select a time period of interest, and employs Regularized Biot-Savart Law (RBSL) flux ropes to create stable and unstable pre-eruptive configurations within a background global magnetic field. The properties of these configurations can first be explored in a zero-beta magnetohydrodynamic (MHD) model, followed by complete CME simulations in thermodynamic MHD, with propagation out to 1 AU. We describe design features of the interface and computations, including the innovations required to efficiently compute results on practical timescales with moderate computational resources. CORHEL-CME is now implemented at NASA's Community Coordinated Modeling Center (CCMC) using NASA Amazon Web Services (AWS). It will be available to the public in early 2024.

1. Introduction

Coronal Mass Ejections (CMEs) are magnificent displays of solar activity that propel plasma and magnetic fields outward into the solar wind. In addition to the inherent scientific interest of such phenomena, CMEs play a crucial role in space weather at Earth (e.g., [1]). When the magnetic fields associated with CMEs contain significant southward pointed fields (oppositely directed to the magnetic field at the Earth's magnetopause, i.e., negative B_z fields), they can reconnect with the Earth's magnetic field and peel away the protective magnetopause, an essential trigger to geomagnetic activity. Therefore, the magnetic structure of CMEs is of vital interest. The compression regions and shock waves created by these events also play an important role in the acceleration of solar energetic particles (SEPs) [2].

CMEs have been observed for over four decades, and while a basic picture of their magnetic structure has emerged, fundamental questions remain. One of the key reasons why these questions remain unresolved is that it is difficult to measure the magnetic field in the corona. CMEs produce ancillary signatures in a range of wavelengths that give important clues to the underlying structure of CMEs, such as EUV waves [3], EUV dimming signatures [4], and the three-part structure of CMEs observed in white light [5]. In order to reveal the physical (e.g., magnetic) structure responsible for these observed features, a model must include enough physics to reproduce observations for real events. Thermodynamic magnetohydrodynamic (MHD) models of the solar corona, which incorporate the relevant energy transport processes, have advanced to the point that they can predict EUV/X-ray emission [6, 7, 8, 9, 10, 11, 12], as well as scattering of white light observed in coronagraphs and at total solar eclipses, for direct comparison with observations.

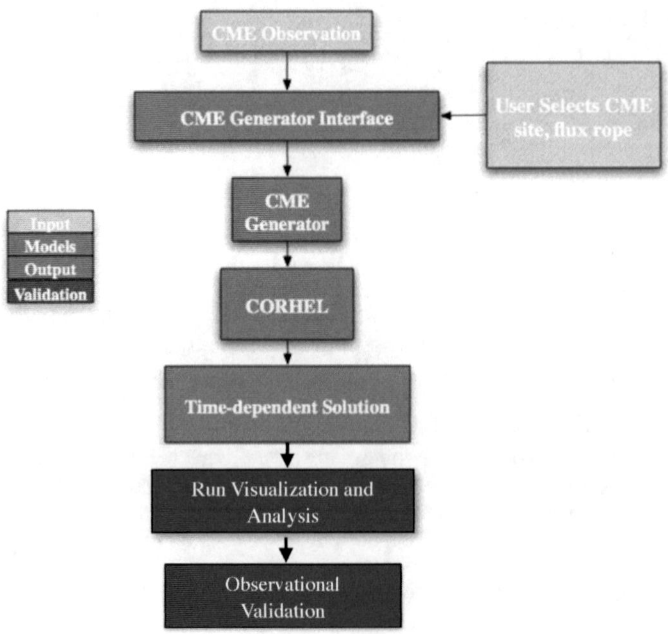

Figure 1. Framework diagram for CORHEL-CME.

Thermodynamic MHD models are now computed relatively routinely using the Magnetohydrodynamic Algorithm outside a Sphere (MAS) code within the CORona-HELiospheric (CORHEL) modeling suite. However, it is still very challenging to model CME events, especially the most violent ones. The largest and fastest CMEs usually originate in localized active regions (ARs) on the Sun. In addition to modeling the global corona, the detailed structure of AR must also be resolved. To allow non-expert users to simulate CMEs in a realistic coronal and heliospheric environment, we have developed CORHEL-CME, which provides MAS/CORHEL zero-beta and thermodynamic MHD simulations using observed photospheric magnetic fields as boundary conditions. Figure 1 shows a diagram of the main steps in CORHEL-CME. In this paper, we describe the software components of CORHEL-CME, the work flow for creating CME simulations, some of the computational shortcuts employed to make the CME simulations feasible on relatively modest computer hardware, and the standard diagnostics that are output as part of a run. CORHEL-CME is now implemented at NASA's Community Coordinated Modeling Center (CCMC) using NASA Amazon Web Services (AWS).

2. Components of CORHEL-CME

Realistically simulating CMEs for a given event requires several key elements. The background corona and solar wind must be simulated for the time period of interest; this in turn requires solar magnetic data. This data is typically in the form of full-Sun maps of the line-of-sight magnetic field measured in the photosphere, from which the radial magnetic field is inferred. Major CME events typically originate in solar active regions (ARs) where the magnetic fields are highly energized (energy above the potential state), and so a pre-eruptive configuration must be developed (also matching the photospheric field measurements). To accomplish these tasks, CORHEL-CME employs the CORHEL modeling suite (described next), and models of pre-eruptive configurations based on analytical flux rope models (described in section 2.2.1). Parameter specification for CORHEL-CME is made via a Web interface (section 3.1).

2.1. CORHEL

Advanced simulations of CMEs require a background model of the solar corona and solar wind, applicable to the time of the event that is being studied. CORHEL is a suite of coupled models and tools for describing the corona and solar wind, developed by Predictive Science Inc. (PSI). It has been delivered to the CCMC and AFRL, and solutions are available at PSI's website (`www.predsci.com`). The primary input data to CORHEL are synoptic maps of the radial photospheric magnetic field, B_r, from observatories such as the National Solar Observatories (NSO), SOLIS and GONG, and the Helioseismic and Magnetic Imager (HMI) aboard the Solar Dynamics Observatory (SDO). An essential aspect of CORHEL's modeling approach is to divide the coronal and heliospheric domains, and to use different codes and approximations in each domain. The outer boundary for the coronal solutions is placed beyond the Alfvén and sound speed critical points, at 20–30 solar radii (R_S). The coronal solutions are used to drive the heliospheric solutions (extending to $\gtrsim 230 R_S$), which can be computed in either the inertial frame or a frame co-rotating with the Sun [13]. This approach has been applied to model the propagation of observed CMEs from the low corona to 1 AU [11]. The MHD runs use MAS—a high-performance code developed over several decades for studying the structure and dynamics of the global corona and inner heliosphere [14, 15, 16, 17, 18, 19, 8]. MAS employs a semi-implicit algorithm (e.g., [20]) that allows the computational time step to greatly exceed the Courant stability limit for Alfvén and magnetoacoustic waves. This approach is necessitated by the stiffness of the MHD equations, especially when applied to strong magnetic field regions in the solar corona. The sparse matrix arising from this algorithm is inverted using iterative preconditioned Conjugate Gradient (CG) methods. MAS can be run on massively parallel CPU systems using MPI, and has been ported to GPUs using OpenACC [21] and Fortran standard parallelism [22]. In thermodynamic MHD mode, MAS incorporates a realistic energy equation; accounting for anisotropic thermal conduction, radiative losses, and coronal heating. This allows plasma densities and temperatures to be computed with sufficient accuracy to simulate extreme ultraviolet (EUV) and X-ray emission observed from space [6]. Thermal conduction is also treated implicitly using CG, or with a super time-stepping method [23]. The equations solved in MAS are shown in Appendix A of [11]. The more recent versions of MAS employ a Wave-Turbulence-Driven (WTD) approach for coronal heating and solar wind acceleration; [24, 10] describe the additional equations solved. To rapidly explore the stability and eruptive behavior of energized configurations, CORHEL-CME also employs the simpler $\beta = 0$ model (solution of the momentum equation and Faraday's law with plasma pressure = 0, e.g.,[25]), prior to employing the more computationally intensive thermodynamic MHD model.

2.2. CME Modeling

MHD simulations are a powerful tool for studying CMEs. Versions of the MAS code have been employed in state-of-the-art CME models for many years [25, 26, 27, 28, 29, 30, 13, 11, 31]. Such simulations are crucial for understanding CME initiation and evolution. Traditionally they have been time-consuming to develop and require significant expertise. The goal of CORHEL-CME is to provide routine CME models that can be used by the broader scientific community. The tool has been designed to be relatively straightforward to use, but it also incorporates enough details to produce sufficiently accurate results, including proper modeling of the pre-eruptive state and early eruption phase. The latter is particularly important if the simulations are to be used in modeling of Solar Particle Events (SPEs), since compressional fronts, formed by the expansion of the CME, can accelerate particles at heights as low as $1.3\,R_\odot$ in the corona [32]. A key component of CME simulations is specifying a pre-eruptive configuration, described in the next section. CORHEL-CME is designed to allow the user to first investigate eruptions with the $\beta = 0$ model, and, when the initial eruption properties are satisfactory, perform full Sun-to-1 AU simulations of the CME with the thermodynamic MHD model.

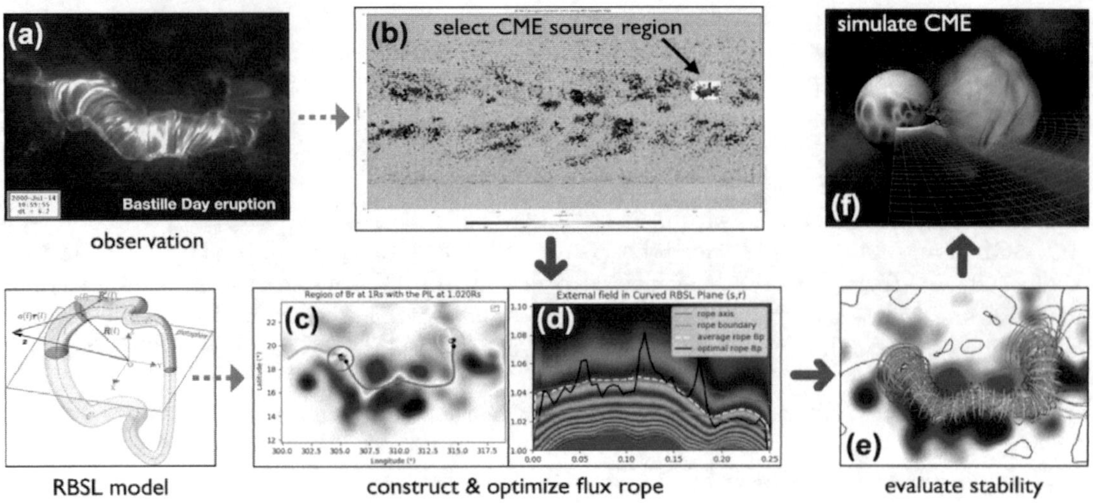

Figure 2. Main steps of modeling a CME with CORHEL-CME. See text for details.

2.2.1. Pre-Eruptive Configurations CORHEL-CME allows users to interactively perform routine simulations of CMEs based on realistic pre-eruptive conditions. Solar ARs typically exhibit strong (kilogauss) fields and a very low plasma β (ratio of plasma to magnetic pressure). Fields in equilibrium are effectively force-free. The TDM (Titov-Démoulin modified) model [33] is an extension of the TD model [34], which has been widely used to study pre-eruptive magnetic configurations (e.g., [35]) and as the initial condition for CME simulations (e.g., [36, 37, 38]). The TD model is an approximately force-free equilibrium consisting of a toroidal MFR (partly submerged below the photosphere) and a stabilizing ('strapping') potential field produced by a fictitious pair of sub-photospheric monopoles. The TDm model employs the same MFR geometry, but was designed such that the MFR can be embedded into an arbitrary, locally bipolar potential field. This facilitates the modeling of CMEs whose source regions have a relatively simple magnetic structure [31]. An earlier version of our tool, CORHEL-MAS-TDM, is available for runs on request at NASA CCMC (`https://ccmc.gsfc.nasa.gov/models/CORHEL~MAS-TDM~6.0`) and provides $\beta = 0$ modeling of CMEs in realistic solar magnetic fields. To overcome the limitation of toroidal geometry, CORHEL-CME uses the RBSL (Regularized Biot-Savart Law) model [39, 40]. This model contains an MFR of arbitrary shape, facilitating the modeling of complex CME source regions that have elongated and curved polarity inversion lines (PILs). Figure 2 shows a schematic of the model and its application to the "Bastille Day" eruption that occurred on 7/14/2000. Both CORHEL-MAS-TDM and CORHEL-CME allow the user to insert the MFRs into a given background field without altering the observed magnetogram.

3. CORHEL-CME Workflow

3.1. Web Interface

CORHEL-CME is delivered as two software packages: One package is a GUI-based web interface, the other is the CORHEL modeling suite. The web interface was developed with a client-server model using Django/Python and JavaScript. It is installed on a public-facing machine (a NASA AWS instance in the CCMC implementation) and allows users to interact with the interface via a web browser, eliminating the need for any local installation by users. The interface hides many details of CME modeling, allowing users to focus on a few free parameters that characterize pre-eruptive MFRs and their eruption. Once all parameters are selected, CORHEL-CME creates two archive files for download. One file contains all information needed to repeat or modify a

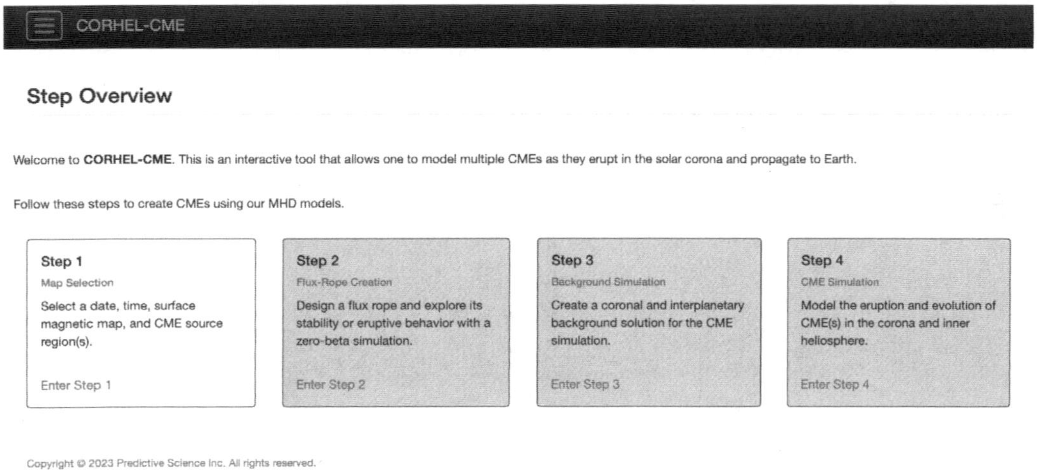

Figure 3. Navigation page of the CORHEL-CME web interface. The user is guided through the process via various steps that include magnetogram and source-region selection (Step 1), RBSL flux-rope configuration (Step 2), Set-up of background models (Step 3) and final CME simulation (Step 4).

previous interface session; the other contains the files and scripts needed to run a simulation with CORHEL/MAS. The optimum specifics of the simulation (number of cores and grid points, computational mesh, run-time of the simulation) are determined automatically.

The CORHEL modeling suite is installed on the computer where the MHD computations are to be performed. In the CCMC implementation, the archive files are automatically transferred to the computational node (at the present time, an 8-GPU NASA AWS Instance). The CORHEL software contains scripts and macros for automatic diagnostics, which are provided as a self-contained, interactive html page upon completion of the run (section 5).

Figure 2 visually illustrates the workflow for creating a CME simulation with CORHEl-CME. The user is guided through these steps via a navigation page, shown in Figure 3. First **(STEP 1)**, the user selects the date and time of the CME to be modeled (Figure 2(a)). The user is provided with choices for an appropriate surface magnetic map for the relevant Carrington rotation. The user can also select an existing coronal and heliospheric background solution at this point, if such a solution is available in the database. In that case, the surface map from that solution is loaded and used in the subsequent steps. After the data source is selected, it is then automatically downloaded from the corresponding data provider, and the user selects the location of the CME source region (typically an active region) on the map, (see Figure 2(b)).

Next **(STEP 2)**, the user designs an RBSL flux rope that is inserted into the CME source region and provides the free magnetic energy required to power the CME (see Figure 2(c)). This step includes (i) the selection of a segment of the polarity inversion line (PIL) along which the initial central section of the flux-rope axis shall be placed at an arbitrary height above the solar surface, (ii) selecting the locations of the rope footprints and its thickness, (iii) further fine-tuning of the axis geometry. The user can modify the axis path, guided by a side-view of the strapping-field contours, as shown in Figure 2(d). Once all these parameters are chosen, a downloadable tar file is produced that contains all the necessary information and commands to run a zero-beta simulation on a supercomputer. In the CCMC implementation, this file is automatically transferred to the compute server to perform the run. The user can then perform computationally inexpensive $\beta = 0$ MHD runs with varying field strengths to explore the stability of the MFR, to find the threshold for eruption, and to explore the initial CME trajectory. These runs can typically be completed in 10-30 minutes of wall clock time on multi-core or GPU systems. As interacting CMEs are an important phenomena, CORHEL-CME allows a user to implement multiple eruptions in multiple ARs, by proceeding through the steps repetitively. Performing these types of calculations can greatly increase the computational requirements, so the CCMC implementation presently limits this capability to two events in a single run.

STEP 3 provides the capability to produce a background solution for the corona and inner heliosphere (technically, these are two separate calculations). These solutions are computed for individual Carrington rotations, and are required for the Sun-to-1 AU CME simulation (Step 4). These background calculations are intentionally performed with a relatively low spatial resolution, and are later re-meshed to high resolution for CME runs (see section 4). In the present version, the user can choose between two empirical approaches for coronal heating (similar to [6]), which are briefly explained in the corresponding interface tool tips. We expect this to be updated with the WTD heating model (see section 2.1) in the near future. The heliospheric background calculation is fully automated and requires no user input. After the background run finishes, the background solution (both corona and heliospheric) is added to the database, which allows subsequent users to employ it in future calculations.

Finally (**STEP 4**), the user can perform a full Sun-to-1 AU CME run. This requires a background solution for the Carrington rotation (corresponding to the time when the CME occurred), to be present in the CORHEL-CME background-solution database. The user will then select this solution during Step 1. If no such solution is present, the user has to perform Step 3 first before he/she can proceed further. Once a background solution is selected, the user continues to Step 2 to either construct a new flux rope or to "confirm" a previously constructed one (which can be loaded into the interface by restoring the corresponding session), after which he/she can directly go to Step 4 and start the CME simulation. No further input is required in Step 4. Upon completion of each of these runs (zero-beta, background, CME), a tar file that contains an extensive run report (see Section 5) is automatically generated and can be downloaded by the user.

4. Efficiency Considerations

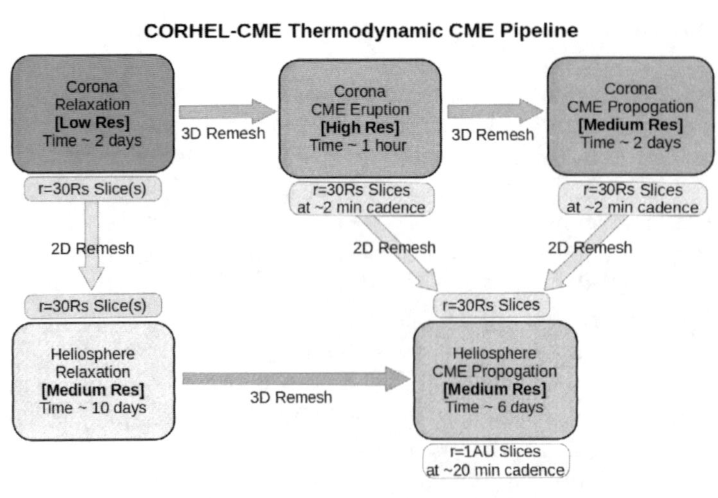

Figure 4. Flow chart showing the run pipeline for modeling a single Sun-to-1 AU CME in the corona and inner heliosphere. This corresponds to STEP 3 and STEP 4 in Figure 3.

An important goal for CORHEL-CME is to allow for real-world calculations with moderate computational resources (presently implemented on an 8-GPU card NASA AWS instance at CCMC). This requires optimizing the required steps in the calculations. The sequence of MHD calculations performed, referred to as the run pipeline, is depicted in Figure 4.

Global coronal simulations require days of physical time to approximate the corona, but only demand a few hours of computational time if ARs are resolved coarsely. On the other hand, detailed models of the pre-event energized magnetic structure in the AR require high resolution to account for specific local features, where the magnetic field is very strong. The local physical relaxation time in the AR is short (minutes to hours) because the Alfvén speed is very high in these magnetically dominated regions. These considerations have necessitated performing different steps of the CORHEl-CME pipeline with different resolutions and flexibly transitioning computed solutions between meshes of different resolution. We interpolate the simulation variables with piece-wise monotonic splines (PCHIP) and analytic integral preserving functions for the transition from one mesh to the other (a special procedure is used to obtain the vector potential **A**, see Appendix A). We refer to this as re-meshing.

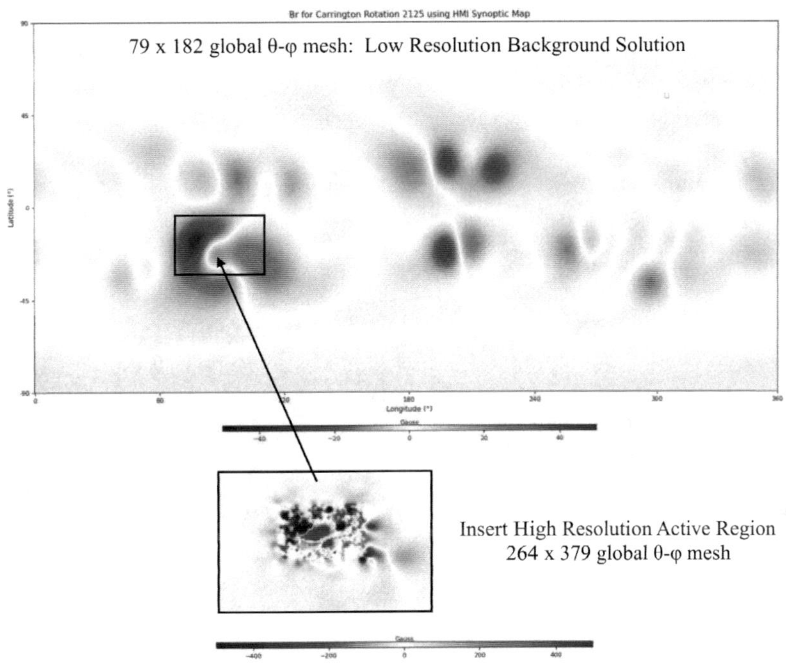

Figure 5. Latitude-longitude map of B_r boundary for low resolution background simulation of the ambient corona around the time of the July 12, 2012 CME, together with the high resolution active region that would be inserted and re-meshed for a CORHEL-CME simulation.

Re-meshing is used at several points in the run pipeline, as shown in Figure 4 (a time t = 1 corresponds to 24 minutes). For example, we compute the background simulation for the time period of interest at low resolution, to minimize run time and data storage. If we performed the background simulation at the resolution required to resolve important features of the AR magnetic field, it greatly increases the computational requirements. Instead, we interpolate our low-resolution simulation onto a higher-resolution grid, and insert the resolved active region into this background solution, prior to incorporating the RBSL flux rope (section 2.2.1). (The $\beta = 0$ flux-rope simulations are performed on the high resolution mesh.) This procedure is depicted in Figure 5. We have found that, in practice, the local perturbations introduced by this insertion resolve rapidly and do not greatly perturb the global coronal solution. There is minimal effects on the subsequent propagation of the CME, as compared to a simulation that is performed entirely at the higher resolution. We note that this re-meshing procedure differs from adaptive-mesh refinement approaches, because the *boundary condition* is modified; we are effectively solving a mathematically different problem from the one we started with. We are relying on the physical effect that this change is local to a region and does not greatly influence the global solution. We re-mesh again to an intermediate resolution after the initial eruption when the CME has strongly expanded, to compute the propagation of the CME until it fully exits the outer domain boundary of the coronal calculation at $30\,R_\odot$. The subsequent evolution is then followed in the heliospheric calculation, which is also performed more efficiently as a separate calculation.

5. Diagnostics Report

Numerical simulations of solar eruptions are typically difficult to interpret, especially if they are performed with the thermodynamic MHD model. CORHEL-CME incorporates the Psiweb Report Generator to create a self-contained interactive web report (as an HTML page), for a quick look and assessment of the MHD simulations. The report generator sets up working directories for each report, gathers report parameters from configuration files, and launches PSI's post-processing tools written in Python, Bash, and Fortran. After all analyses are performed for a given report, a self-contained web document is generated using MkDocs (see `https://www.mkdocs.org/`). Plotly's javascript library (see `https://plotly.com/javascript/`) is used for three-dimensional (3D) field line exploration, while the 2D slices and emission imagery pages are driven with custom Javascript. The report is automatically produced and tarred for download when a CORHEL-CME run has finished.

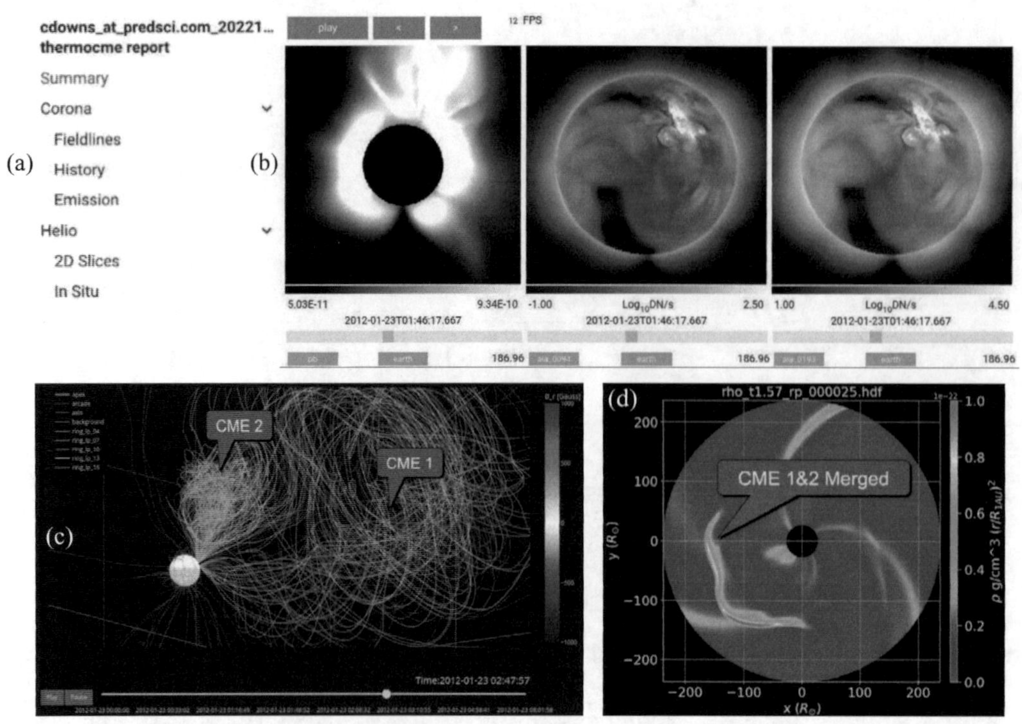

Figure 6. Field lines displayed by the CORHEL-CME interactive visualization tool, showing two flux ropes in the coronal simulation of the January 23, 2012 double-CME event. The plasma density (scaled by the distance r^2 from the Sun) in the subsequent interplanetary simulation, shown in the equatorial plane from an interactive animation in the run report.

Figure 6 shows excerpts from a report created for a simulation of the double CME event of January 23, 2012. The report provides a drop-down menu so users can choose the set of variables animated on a given page (Figure 6 (a)). Movies of EUV and X-ray emission, as well as polarization brightness (pB) as would be observed by different space-based instruments can be viewed from different vantage points in the "emission" page (Figure 6(b)). The field-line tool (Figure 6(c)) allows users to plot different sets of field lines. The user can rotate the plot, zoom into an area, and follow the evolution of the field lines via a time slider. The helio report (Figure 6(d)) provides an interactive animation of various physical quantities in the equatorial plane (30–230 $R_\odot$). Time series comparisons of the results with OMNI data can be obtained at the in situ page.

After completion of a run, the full MHD data from the simulation can also be obtained to perform more detailed analysis.

6. Summary

CORHEL-CME allows users to simulate CMEs for observed events, in coronal and heliospheric environments modeled with observed photospheric magnetic fields. A GUI interface allows users to develop pre-eruptive configurations, and investigate stable flux ropes and eruptions in $\beta = 0$ and full thermodynamic MHD. After completion of a run, a downloadable diagnostics report is created with quick-look, interactive visualizations. CORHEL-CME has been designed to operate with moderate computational resources, to allow more routine investigation of CME events. More detailed simulations can subsequently be performed, if greater computational power is available. This may be necessary for some types of scientific investigations. CORHEL-CME is currently undergoing testing at NASA CCMC, implemented on NASA AWS. It will be operational in early 2024. In the future, we will combine CORHEL-CME with the SPE Threat Assessment Tool (STAT, [41, 42]) to allow users to simulate solar particle events.

Appendix A.

The MAS MHD code solves for Faraday's law (advancement of the magnetic field), using the vector potential $\mathbf{A}$ ($\mathbf{B} = \nabla \times \mathbf{A}$). In MAS, differential operators are formed on staggered meshes to ensure that standard vector identities (e.g $\nabla \cdot \nabla \times = 0$) are preserved to round-off error. Among other advantages, this enforces $\nabla \cdot \mathbf{B} = 0$ to machine precision in the solutions. This is an important property for low-β plasma calculations, where the presence of non-divergence free fields can lead to unphysical solutions (e.g. [43]). In re-meshing our solutions, the resampling of $\mathbf{A}$ from coarse to fine meshes (or vice-versa), can produce unacceptable errors, even when higher-order, smooth methods like PCHIP are used. This occurs because the current density $\mathbf{J} = \nabla \times (\nabla \times \mathbf{A})$ (normalized units), and the errors in $\mathbf{J}$ directly impact the balance of forces. To obtain the re-meshed $\mathbf{A}^{\mathbf{RME}}$ with sufficient accuracy, we perform the following steps:

(i) Compute the 3D potential field, $\mathbf{B}^{\mathbf{pot0}}$, that corresponds to the original boundary field, B_{r0}, on the original mesh.
(ii) Obtain the nonpotential part of $\mathbf{B}$ on the original mesh: $\mathbf{B}^{\mathbf{nonpot}} = \mathbf{B} - \mathbf{B}^{\mathbf{pot0}}$.
(iii) Remesh $\mathbf{B}^{\mathbf{nonpot}}$ onto the new mesh to obtain $\mathbf{B}^{\mathbf{nonpot,RME}}$.
(iv) Obtain the re-meshed $\mathbf{J}^{\mathbf{RME}} = \nabla \times \mathbf{B}^{\mathbf{nonpot,RME}}$ by computing the curl directly on the new mesh.
(v) Prepare the new boundary field (B_{r1}) for the new mesh (e.g. insert the high resolution B_r at the boundary).
(vi) Solve for the potential field that corresponds to B_{r1} on the new mesh to obtain $\mathbf{A}^{\mathbf{pot1}}$ and $\mathbf{B}^{\mathbf{pot1}}$ (MAS employs a potential solver that finds the vector potential for a $\nabla \times \mathbf{B} = 0$ potential field).
(vii) Utilize the following gauge choice for $\mathbf{A}$:

$$\mathbf{A} = \begin{bmatrix} A_r \\ A_\theta \\ A_\phi \end{bmatrix} = \nabla \times \begin{bmatrix} \psi \\ 0 \\ 0 \end{bmatrix} + \begin{bmatrix} A_r \\ 0 \\ 0 \end{bmatrix} = \begin{bmatrix} A_r \\ \frac{1}{r}\left(\frac{1}{\sin\theta}\frac{\partial\psi}{\partial\phi}\right) \\ -\frac{1}{r}\frac{\partial\psi}{\partial\theta} \end{bmatrix}. \tag{A.1}$$

then

$$\nabla_\perp^2 A_r = -J_r^{\rm RME}; \quad \nabla_\perp^2 \psi = -B_r^{\rm nonpot,RME}. \tag{A.2}$$

Equations (A.2) are series of 2D solves at each radial slice in the coronal domain, that can be performed rapidly with preconditioned conjugate gradient methods. This determines the nonpotential portion of $\mathbf{A}$ ($\mathbf{A}^{\mathbf{nonpot}}$). The desired $\mathbf{A}^{\mathbf{RME}} = \mathbf{A}^{\mathbf{nonpot}} + \mathbf{A}^{\mathbf{pot1}}$.

Acknowledgments
This work was supported by AFOSR (contract FA9550-15-C-0001), the NASA SBIR program (contract 80NSSC19C0193), the NASA Living With a Star Strategic Capabilities program (grant 80NSSC22K0893), the NASA HTMS program (grant 80NSSC20K1274), the NSF PREEVENTS program (grant ICER1854790), the NASA HSOC program (grant 80NSSC20K1285), NASA grant 80NSSC23K0258, and the Parker Solar Probe WISPR contract NNG11EK11I to NRL (under subcontract N00173-19-C-2003 to PSI). Computational resources were provided by the NASA High-End Computing (HEC) Program through the NASA Advanced Supercomputing (NAS) Division at Ames Research Center and by the Expanse supercomputer at the San Diego Supercomputing Center through the NSF ACCESS and XSEDE programs.

References

[1] Baker D N and Lanzerotti L J 2016 *American Journal of Physics* **84** 166–180
[2] Kahler S 1992 *Annual Review of Astronomy and Astrophysics* **30** 113–141
[3] Long D M, Bloomfield D S, Chen P F, Downs C, Gallagher P T, Kwon R Y, Vanninathan K, Veronig A M, Vourlidas A, Vršnak B, Warmuth A and Žic T 2017 *Sol. Phys.* **292** 7
[4] Zhukov A N and Auchère F 2004 *Astron. Astrophys.* **427** 705–716
[5] Vourlidas A, Lynch B J, Howard R A and Li Y 2013 *Sol. Phys.* **284** 179–201
[6] Lionello R, Linker J A and Mikić Z 2009 *Astrophys. J.* **690** 902–912
[7] Downs C, Roussev I I, van der Holst B, Lugaz N, Sokolov I V and Gombosi T I 2010 *Astrophys. J.* **712** 1219–1231
[8] Downs C, Linker J A, Mikić Z, Riley P, Schrijver C J and Saint-Hilaire P 2013 *Science* **340** 1196–1199
[9] van der Holst B, Sokolov I V, Meng X, Jin M, Manchester W B I, Tóth G and Gombosi T I 2014 *Astrophys. J.* **782** 81
[10] Mikić Z, Downs C, Linker J A, Caplan R M, Mackay D H, Upton L A, Riley P, Lionello R, Török T, Titov V S, Wijaya J, Druckmüller M, Pasachoff J M and Carlos W 2018 *Nat. Astron.* **2** 913–921
[11] Török T, Downs C, Linker J A, Lionello R, Titov V S, Mikić Z, Riley P, Caplan R M and Wijaya J 2018 *Astrophys. J.* **856** 75
[12] Sachdeva N, Manchester Ward B I, Sokolov I, Huang Z, Pevtsov A, Bertello L, Pevtsov A A, Toth G, van der Holst B and Henney C J 2023 *Astrophys. J.* **952** 117
[13] Lionello R, Downs C, Linker J A, Török T, Riley P and Mikić Z 2013 *Astrophys. J.* **777** 76
[14] Mikić Z and Linker J A 1996 The large-scale structure of the solar corona and inner heliosphere *International Solar Wind 8* vol 382 ed Winterhalter D e a (AIP Conf. Proceedings) p 104
[15] Linker J A, Mikić Z, Bisecker D A, Forsyth R J, Gibson S E, Lazarus A J, Lecinski A, Riley P, Szabo A and Thompson B J 1999 *J. Geophys. Res.* **104** 9809
[16] Mikić Z, Linker J A, Schnack D D, Lionello R and Tarditi A 1999 *Phys. Plasmas* **6** 2217
[17] Riley P, Linker J A and Mikić Z 2001 *J. Geophys. Res.* **106** 15889
[18] Linker J A, Lionello R, Mikić Z, Titov V S and Antiochos S K 2011 *Astrophys. J.* **731** 110
[19] Riley P, Linker J A, Lionello R and Mikic Z 2012 *J. Atmos. Solar-Terr. Phys.* **83** 1–10
[20] Lionello R, Mikić Z and Linker J A 1999 *J. Comp. Phys.* **152** 346–358

Journal of Physics: Conference Series **2742** (2024) 012012 doi:10.1088/1742-6596/2742/1/012012

[21] Caplan R M, Linker J A, Mikić Z, Downs C, Török T and Titov V S 2019 GPU Acceleration of an Established Solar MHD Code using OpenACC *ASTRONUM* (*Journal of Physics Conference Series* vol 1225) p 012012

[22] Caplan R M, Stulajter M M and Linker J A 2023 Acceleration of a production solar mhd code with fortran standard parallelism: From openacc to 'do concurrent' *2023 IEEE International Parallel and Distributed Processing Symposium Workshops (IPDPSW)* pp 582–590

[23] Caplan R M, Mikić Z, Linker J A and Lionello R 2017 Advancing parabolic operators in thermodynamic MHD models: Explicit super time-stepping versus implicit schemes with Krylov solvers *ASTRONUM* (*Journal of Physics Conference Series* vol 837) p 012016

[24] Downs C, Lionello R, Mikić Z, Linker J A and Velli M 2016 *Astrophys. J.* **832** 180

[25] Mikić Z and Linker J A 1994 *Astrophys. J.* **430** 898

[26] Linker J A, Lionello R, Mikić Z and Amari T 2001 *J. Geophys. Res.* **106** 25165

[27] Linker J A, Mikić Z, Lionello R, Riley P, Amari T and Odstrcil D 2003 *Phys. Plasmas* **10** 1971

[28] Riley P, Linker J A, Mikić Z, Odstrcil D, Zurbuchen T H, Lario D and Lepping R P 2003 *J. Geophys. Res.* **108** 2

[29] Riley P, Lionello R, Mikić Z, Linker J, Clark E, Lin J and Ko Y K 2007 *Astrophys. J.* **655** 591–597

[30] Riley P, Lionello R, Mikić Z and Linker J 2008 *Astrophys. J.* **672** 1221–1227

[31] Downs C, Warmuth A, Long D M, Bloomfield D S, Kwon R Y, Veronig A M, Vourlidas A and Vršnak B 2021 *Astrophys. J.* **911** 118

[32] Kozarev K A and Schwadron N A 2016 *Astrophys. J.* **831** 120

[33] Titov V S, Török T, Mikic Z and Linker J A 2014 *Astrophys. J.* **790** 163

[34] Titov V S and Démoulin P 1999 *Astron. Astrophys.* **351** 707–720

[35] Kliem B, Titov V S and Török T 2004 *Astron. Astrophys.* **413** L23–L26

[36] Török T and Kliem B 2005 *Astrophys. J. Lett.* **630** L97–L100

[37] Manchester IV W B, Vourlidas A, Tóth G, Lugaz N, Roussev I I, Sokolov I V, Gombosi T I, De Zeeuw D L and Opher M 2008 *Astrophys. J.* **684** 1448–1460

[38] Lugaz N, Roussev I I and Sokolov I V 2009 The August 24, 2002 coronal mass ejection: when a western limb event connects to earth *IAU Symposium* (*IAU Symposium* vol 257) ed N Gopalswamy & D F Webb pp 391–398

[39] Titov V S, Downs C, Mikić Z, Török T, Linker J A and Caplan R M 2018 *Astrophys. J. Lett.* **852** L21

[40] Titov V S, Downs C, Török T, Linker J A, Caplan R M and Lionello R 2021 *Astrophys. J. Suppl.* **255** 9

[41] Linker J A, Caplan R M, Schwadron N, Gorby M, Downs C, Torok T, Lionello R and Wijaya J 2019 Coupled MHD-Focused Transport Simulations for Modeling Solar Particle Events *ASTRONUM* (*Journal of Physics Conference Series* vol 1225) p 012007

[42] Young M A, Schwadron N A, Gorby M, Linker J, Caplan R M, Downs C, Török T, Riley P, Lionello R, Titov V, Mewaldt R A and Cohen C M S 2021 *Astrophys. J.* **909** 160

[43] Brackbill J U and Barnes D C 1980 *Journal of Computational Physics* **35** 426–430

Journal of Physics: Conference Series **2742** (2024) 012013 doi:10.1088/1742-6596/2742/1/012013

Space Weather with Quantified Uncertainties: Improving Space Weather Predictions with Data-Driven Models of the Solar Atmosphere and Inner Heliosphere

Nikolai V. Pogorelov[1], Charles N. Arge[2], Ronald M. Caplan[3], Phillip Colella[4], Jon A. Linker[3], Talwinder Singh[1], Brian Van Straalen[4], Lisa Upton[5], Cooper Downs[3], Christopher Gebhart[4], Dinesha V. Hegde[1], Carl Henney[6], Shaela Jones[2], Craig Johnston[2], Tae K. Kim[1], Andrew Marble[7], Syed Raza[1], Miko M. Stulajter[3], and James Turtle[3]

[1] Department of Space Science and Center for Space Plasma and Aeronomic Research, The University of Alabama in Huntsville, Huntsville, AL 35899, USA
[2] NASA Goddard Space Flight Center, 8800 Greenbelt Road, Greenbelt, MD 20771, USA
[3] Predictive Science Inc., 9990 Mesa Rim Road Suite 170, San Diego, CA 92121. USA
[4] Lawrence Berkeley National Laboratory, 1 Cyclotron Rd., Berkeley, CA 94720, USA
[5] Southwest Research Institute, 6220 Culebra Road, San Antonio, TX 78238, USA
[6] Air Force Research Laboratory, Space Vehicles Directorate, Kirtland AFB, NM 87117, USA
[7] Cooperative Institute for Research in Environmental Sciences, University of Colorado, Boulder, CO 80305, USA

E-mail: nikolai.pogorelov@uah.edu

Abstract. To address Objective II of the National Space Weather Strategy and Action Plan "Develop and Disseminate Accurate and Timely Space Weather Characterization and Forecasts" and US Congress PROSWIFT Act 116–181, our team is developing a new set of open-source software that would ensure substantial improvements of Space Weather (SWx) predictions. On the one hand, our focus is on the development of data-driven solar wind models. On the other hand, each individual component of our software is designed to have accuracy higher than any existing SWx prediction tools with a dramatically improved performance. This is done by the application of new computational technologies and enhanced data sources. The development of such software paves way for improved SWx predictions accompanied with an appropriate uncertainty quantification. This makes it possible to forecast hazardous SWx effects on the space-borne and ground-based technological systems, and on human health. Our models include (1) a new, open-source solar magnetic flux model (OFT), which evolves information to the back side of the Sun and its poles, and updates the model flux with new observations using data assimilation methods; (2) a new potential field solver (POT3D) associated with the Wang–Sheeley–Arge coronal model, and (3) a new adaptive, 4-th order of accuracy solver (HelioCubed) for the Reynolds-averaged MHD equations implemented on mapped multiblock grids (cubed spheres). We describe the software and results obtained with it, including the application of machine learning to modeling coronal mass ejections, which makes it possible to improve SWx predictions by decreasing the time-of-arrival mismatch. The tests show that our software is formally more accurate and performs much faster than its predecessors used for SWx predictions.

1. Introduction

The solar wind (SW) emerging from the Sun is the main driving mechanism of solar events which may lead to geomagnetic storms that are the primary causes of space weather (SWx) disturbances that affect the magnetic environment of Earth and may have hazardous effects on the space-borne and ground-based technological systems, as well as human health. For this reason, accurate modeling of the background SW is a necessary part of SWx forecasting. Geomagnetic storms are caused both by SW stream interactions and by the largest solar coronal disturbances called coronal mass ejections (CMEs) and their interplanetary counterparts, interplanetary coronal mass ejections (ICMEs). The connection of the ambient interplanetary magnetic field to CME-related shocks and impulsive solar flares determines where solar energetic particles propagate. Therefore, modeling stream interaction background and ICMEs propagating through it on the basis of observational data is a key area of research in solar and heliospheric physics. Such modeling should necessarily be built on mathematically- and physically-consistent connections between eruptive events, magnetic phenomena on the Sun, and SW structures in the solar atmosphere and inner heliosphere (IHS). Substantial success has been achieved in global numerical modeling of the IHS. However, the prediction capability remains far from being satisfactory [1, 2]. There are several reasons for that. Firstly, since global numerical modeling of inner heliosphere (IHS)is a time-dependent problem, developing time-dependent boundary conditions (b.c.'s) that would incorporate remote and *in situ* observations of the Sun in a self-consistent way remains an important research challenge. Secondly, numerical approaches used to attack SWx problems are rather outdated and may be missing the important features of discontinuity formation, interaction, and propagation from the Sun towards Earth.

We are developing a new data-driven, time-dependent model of the solar corona and IHS to predict the solar wind properties at the Earth's orbit. This model is being implemented in a publicly available software. In addition to simulation of the background SW, we are particularly investigating the physics of the evolution, eruption, and propagation of CMEs between the Sun and Earth in the IHS. For this purpose, we are currently applying our data-constrained, flux-rope-based, constant-turn CME model [3]. The latter is based not only on the traditionally used STEREO Sun-Earth Connection Coronal and Heliospheric Investigation (SECCHI) and Solar and Heliospheric Observatory's Large Angle and Spectrometric Coronograph (SOHO/LASCO) coronagraphs, but also on photospheric observations providing us with the estimates of the toroidal and poloidal magnetic fluxes, and the sign of helicity. The enhancement of the observational data base that constrain simulation results is one of the strategic approaches to improve SWx forecasting. It also makes it possible to improve our understanding of basic physical processes occurring in the IHS.

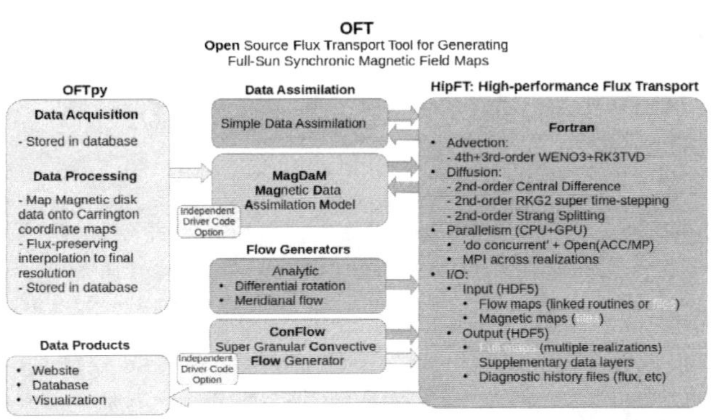

Figure 1: *Summary chart of the Open Flux Transport software suite.*

Journal of Physics: Conference Series **2742** (2024) 012013 doi:10.1088/1742-6596/2742/1/012013

2. The Open Flux Transport Model (OFT)

The Open-source Flux Transport (OFT) software suite is a complete system for generating full-Sun magnetograms through acquiring and processing observational data, generating realistic convective flows, and running a flux transport model. It is summarized in Figure 1.

2.1. Data Acquisition and Processing: OFTpy

Assimilation of magnetograms from solar observatories is the key input to flux transport models. We have developed OFTpy, a python codebase that facilitates acquisition and processing of data products to a computation-ready format for our flux transport model HipFT described in Sec. 2.3. The current code supports full processing of Joint Science Operations Center (JSOC) HMI_M720s line-of-sight (LOS) magnetograms to a radial-flux Carrington Rotation (CR) Map. This extensible software also serves as a prototype/example for future users to incorporate magnetograms from alternative observatories. In future iterations, we plan to expand capabilities to include NSO GONG LOS magnetograms, and HMI vector magnetograms. For data acquisition of HMI LOS magnetograms, OFTpy acts as a Python wrapper for product query and download from Stanford University's JSOC. Figure 2 shows an example. OFTpy allows the user to find and download the required data product over a specified time range and cadence. Processing steps include: high resolution interpolation to a CR map, downsampling by integration, and conversion from line-of-sight to radial. Additional pre-processing calculates the assimilation weights for use in data assimilation. The default assimilation functionality is to set weights in the pre-processing steps, however the python map-object class has been designed with a flexible set of layers that will allow more complex user-defined assimilation routines.

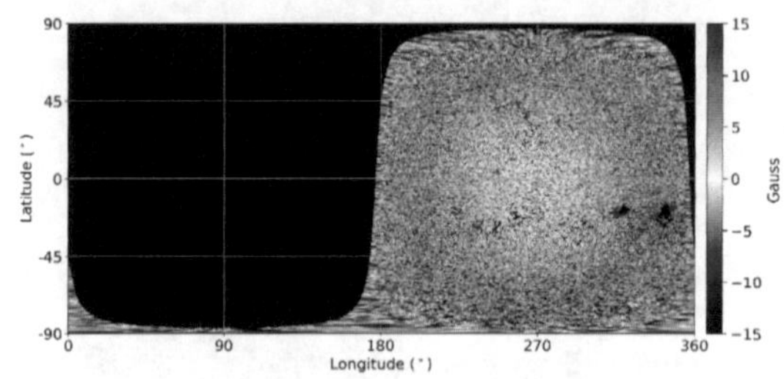

Figure 2: *An HMI 720s magnetogram, observed on 2021/01/01 23:59:52, converted to radial flux, projected into Carrington Coordinates, and re-binned to 1024x512 resolution with OFTpy*

2.2. Convective Flow Generator: ConFlow

Supergranules are convective structures in the Sun associated with the magnetic network and most readily observed in Doppler velocity images. These structures have flow velocities of $\sim$ 500 m/s, diameters of about 30,000 km, and lifetimes of about a day. Typically, surface flux transport (SFT) models use a diffusion coefficient to account for the mixing that occurs from the turbulent convective motions. This diffusion causes the magnetic flux in active regions to weaken and spread out uniformly. An alternative approach is to explicitly model the turbulent surface flows produced by convection, as is done in the AFT SFT model [4]. The convective flows are created by using vector spherical harmonics to create an evolving convective velocity field that captures details of the turbulent motion, including the power spectra (up to l of $\sim$ 350) and the finite lifetimes of the convective cells. On its own, the simulations of these velocities are visually indistinguishable from Doppler observations of the Sun. When incorporated into an SFT model, the use of realistic photospheric supergranular velocity fields drives the evolution of the flux. This produces an evolving magnetic network pattern similar to the quiet-sun network observed on the Sun, far surpassing the realism of SFT models that use only a diffusion coefficient alone.

Courtesy of Dr. David Hathaway, we have been allowed to implemented his convective flow models into a stand-alone code called ConFlow that will be made available as open-source (the ConFlow module in Figure 1). ConFlow provides the poloidal and toroidal components of the convective flows using an improved specification of the input spectral coefficients of the vector spherical harmonics. In the version of the flow model used in [4], these spectral coefficients were defined as discrete values determined over a trial-and-error iterative process to minimize the differences between the output Doppler velocity spectrum and the one measured by SDO/HMI. For ConFlow, this has been updated with an improved specification for the poloidal and toroidal input spectral coefficients that are now defined analytically for each spherical harmonic of degree l. In addition, to avoid aliasing artifacts due to the sampling limits of the default resolution of our flux transport model (1024x512), the input spectra are cut off at $l = 340$, which acts as a low-pass filter that keeps most of the supergranular convection cells, but rejects higher-frequency convective cells like granulation that cannot be resolved.

2.3. High-performance Flux Transport (HipFT)
HipFT is the computational core of OFT. It implements advection, diffusion, and data assimilation for the solar surface on a logically rectangular non-uniform spherical grid. It is parallelized for use with multi-core CPUs and GPUs (both NVIDIA and Intel) using a combination of Fortran's standard parallel 'do concurrent', OpenMP Target data directives, and MPI. It is designed to be modular, incorporating various differential rotation, meridional flow, supergranular convective flow, and data assimilation models. It can also compute multiple realizations in a single run, spanning multiple choices of parameters, resulting in an ensemble of maps.

HipFT uses a combination of numerical methods focused on accuracy and performance. To avoid the excessive diffusivity of upwind schemes for advection, we have implemented a flux preserving WENO3 [5] scheme that is up to fourth-order accurate in space in the regions of smooth solutions and first-order accurate near high gradients. It is combined with the strong stability preserving SSPRK(4,3) [6] time stepping scheme, which doubles the stable time step limit. For diffusion, we use the second-order explicit Gegenbauer extended stability Runge-Kutta scheme [7] (RKG2). We also iterate the diffusion using a novel practical time step limit (PTL) [8] that ensures the diffusion is accurate at any size of the overall time step. The advection and diffusion operators are second-order Strang split.

In Fig. 3, we show an example production run over 11-years from 2011 to 2022 using convective flows from ConFlow and data assimilation with OFTpy HMI data, along with $175\,\mathrm{km}^2$ diffusion. The 11 year run took ~ 13 hours to compute on a single RTX 3090Ti NVIDIA GPU.

The HipFT code is publicly available for use on GitHub[1]. Full details of the HipFT code, including its capabilities, algorithms, performance, and validation tests are described in an upcoming publication.

3. WSA solar wind model using a high-performance PFSS+CS model
The simplest description of the coronal magnetic field, based on a photospheric boundary map of B_r, is a potential (current-free) field model, i.e., a solution of Laplace's equation, $\nabla^2\Phi = 0$. Potential field source-surface (PFSS) and potential field current-sheet (PFCS) models have been computed for many years, typically using algorithms based on a spherical harmonic expansion. In the PFSS model, the field at the top boundary (the source-surface, typically at 2-2.5$R_\odot$) is assumed to be radial. The PFCS model uses a Laplace equation solve to extend the field from the source surface to 20-30$R_\odot$ and reproduce the latitudinally independent behavior of $|B_r|$ observed by Ulysses. We have developed a publicly available code POT3D[2] [9], that solves Laplace's equation using finite differences on a non-uniform logically-rectangular spherical grid. POT3D uses MPI to exploit massively parallel computations with 3D data decomposition. It allows user-specified meshes that can be made highly nonuniform. POT3D computes solutions on CPUs and GPUs far more rapidly than traditional spherical harmonic solvers, allowing for efficient high resolution calculations. The

[1] `https://github.com/predsci/HipFT`
[2] `https://github.com/predsci/POT3D`

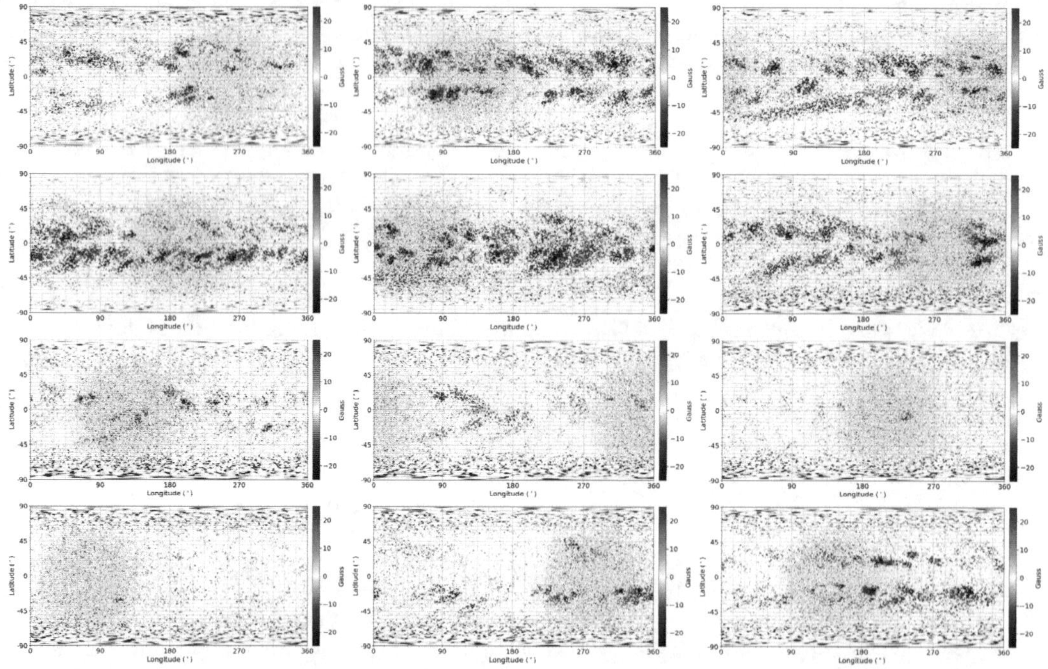

Figure 3: *Example of output B_r maps from an 11-year OFT/HipFT run. The run used ConFlow super granular flows with a spectral cutoff of $l = 340$ and a diffusion coefficient of $175\,km^2/s$. Left to right and top to bottom: maps on January 1st from 2011 to 2022.*

GPU acceleration is accomplished using Fortran's standard parallelism ('do concurrent') along with OpenACC directives for data management. In [9], we analyzed the various qualitative and quantitative differences that naturally arise by using different maps as input, and illustrated how coronal morphology and open flux depend most strongly on the outer boundary condition. We also have shown how large-scale morphologies and the open magnetic flux are remarkably insensitive to model resolution, while the surface mapping and embedded magnetic complexity vary considerably. This establishes important context for past, current, and future applications of the PF for coronal and solar wind modeling.

A key requirement for having POT3D serve as the coronal model in our SWQU effort is for it to provide coronal magnetic field solutions that agree well with those from the spherical harmonic potential field solver previously used in the WSA model [10, 11] (we denote this overall model as SH-WSA, and that using POT3D as POT3D-WSA). In Fig. 4 we show side-by-side comparisons between the SH-WSA modelresults with thsoe using POT3D-WSA. We compared the asymptotic predicted solar wind speed and interplanetary magnetic field (IMF) polarity at the outer boundary of the models positioned at 21.5 $R_\odot$ and analyzed sources of discrepancies between the two. While we found that slight differences between the field line tracing methods resulted in a handful of field lines near the magnetic neutral line connecting to dramatically different locations on opposite sides of the magnetic neutral line, they were not the dominant source of discrepancy. The primary sources were due to either low-lying closed flux arcades in polar coronal holes that showed up in one model and not the other, or slight differences in magnetic connectivity in regions where the solar wind velocity equation is particularly sensitive. Inter-model discrepancies were compared to intra-model discrepancies in the presence of small perturbations of model computational settings and the

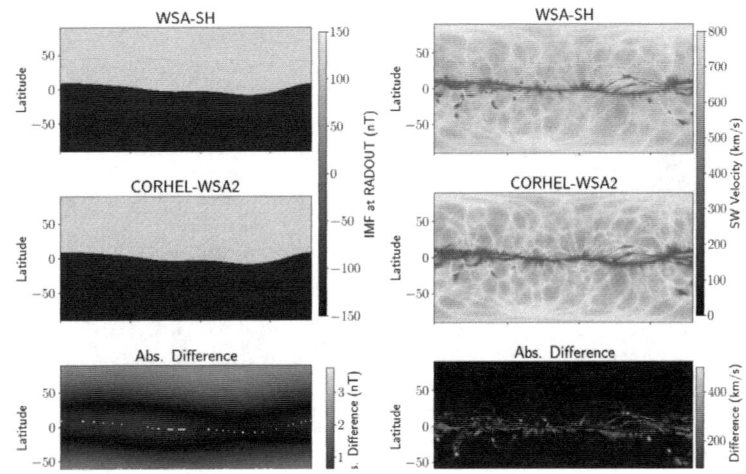

Figure 4: *Comparison between PFSS+PFCS+WSA model outer boundaries located at 21.5Rs for 2020 Feb 02 20:00. The left column shows the magnetic field and the right shows the model-derived terminal solar wind speed. The top and middle panels are calculated using the WSA-SH and WSA-POT3D models, respectively. The bottom panel shows the absolute difference between the results of the two codes.*

two results were found to be comparable. We also compared inter-model discrepancies to variation within model ensembles created using ADAPT photospheric magnetic flux maps and found that the ensemble variation of the two model was significantly higher than their computational uncertainties. A paper reporting the results of this study has been submitted for publication.

4. Modeling the inner heliosphere

4.1. Development of new capabilities in Chombo.

One of the goal of this research is to develop a new code for SWx MHD applications, called HelioCubed, having the following major features.

- The underlying discretization of space is a locally-structured mapped multiblock grid using equiangular cubed-sphere mappings. The use of such a spatial discretization does not have pole singularities, and is well-suited for the remaining choices we have made about the discretization method.
- The discretion of the MHD equations uses a fourth-order-accurate finite-volume discretization in space, combined with a method of lines discretization in time based on a fourth-order-accurate Runge–Kutta method. Discrete spherical symmetry is preserved under time evolution. HelioCubed also supports block-structured adaptive mesh refinement (AMR).
- HelioCubed runs efficiently on high-performance computing platforms using MPI + domain decomposition for distributed-memory parallelism, as well as supporting the use of GPUs for the single MPI rank calculations. It is implemented using Proto, a high-performance, high-productivity framework for implementing locally-structured grid methods for partial differential equations being developed at LBNL. Our main reason for using Proto is that it provides a mechanism for supporting the use of GPU-based systems that is not available in the older softer frameworks such as Chombo [12] that are based on call-backs to Fortran, while providing support for obtaining high performance on such systems for high-order methods.

Journal of Physics: Conference Series **2742** (2024) 012013 doi:10.1088/1742-6596/2742/1/012013

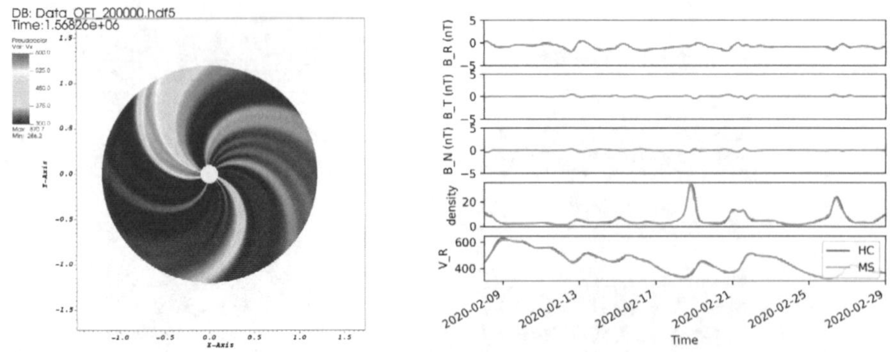

Figure 5: *Left panel: Solar wind solution calculated with HelioCubed using time-independent boundary conditions at UT 2020-02-02T19:59:53, created using OFT-POT3D-WSA-HelioCubed chain of models. Right panel: The comparison of SW solution at Earth using same boundary conditions with MS-FLUKSS (orange) and HelioCubed (blue).*

Both CPU and GPU execution of Proto have been deployed using CUDA on NVIDIA GPU architectures and using HIP on AMD GPU architectures. The CUDA version has been optimized for use with NVIDIA data center GPUs such as the V100, A100, and H100. The initial implementation of the new testing infrastructure, including continuous integration testing, has been performed.

The development of HelioCubed is a complex undertaking, with multiple major components. For this reason, we have factored the development process into three components that can be developed concurrently, with a relatively small effort needed to integrate them together at the end.

4.2. HelioCubed: A code for solving MHD equations

We have extended the algorithm for systems of hyperbolic conservation laws described in [13, 14] to the case of MHD. This method uses a method-of-lines approach, with a 4th-order accurate Runge–Kutta method (RK4) for time discretization. The right-hand sides to the conervation laws are computing using fluxes computed at each time using a 4th-order finite-volume method. This is combined with dissipation methods for steep gradients and discontinuities, such as a nonlinear limiter, and high-order linear and non-linear artificial viscosities, that preserve 4th-order in smooth parts of the solution. Divergence cleaning has been conducted using the Powell method [15]. Finally, we use a specialized version of the finite-volume discretization for abstract spherical coordinate systems (for which equiangular cubed-sphere and classical spherical coordinates are special cases), which, when applied to discrete spherical solutions, preserves spherical symmetry to to machine precision.

To carry out the development of the MHD method, we implemented a stand-alone test bed for the case of classical spherical coordinates. We translate boundary conditions generated by the POT3D coronal model [9] into a format that is compatible with both MS-FLUKSS [16] and HelioCubed, using the Proto interface to the HDF5 library to intake boundary conditions, output data, and save checkpoint files. The software is capable of executing solar wind simulations using both time-independent and time-dependent boundary conditions. The functionality to extract data along spacecraft trajectories has been incorporated into HelioCubed. Furthermore, the Constant Turn Flux Rope CME model [3, 17] has been assimilated into HelioCubed. Left panel of figure 5 shows our preliminary results for data-driven solar wind simulations performed using HelioCubed.

We have also validated our results using direct comparison of the solution with the solution created with MS-FLUKSS using the same boundary conditions, as shown in the right panel of Figure 5.

4.3. Block-structured AMR in Cubed-Sphere Coordinates

The spatial discretization for HelioCubed is based on a mapped-multiblock discretization [18] of a spherically-symmetric domain. In this approach, the unit sphere is discretized as the union of six smooth logically-rectangular blocks that align with each other at patch boundaries. The coordinatization of a 3D spherical domain is ideologically identical to that of a mapped-multiblock domain, which is de facto a tensor product of a spherical domain with a radial coordinate. Thus, we are using the methods developed in [13, 14] for a single rectangular domain to represent the coupling across the blocks.

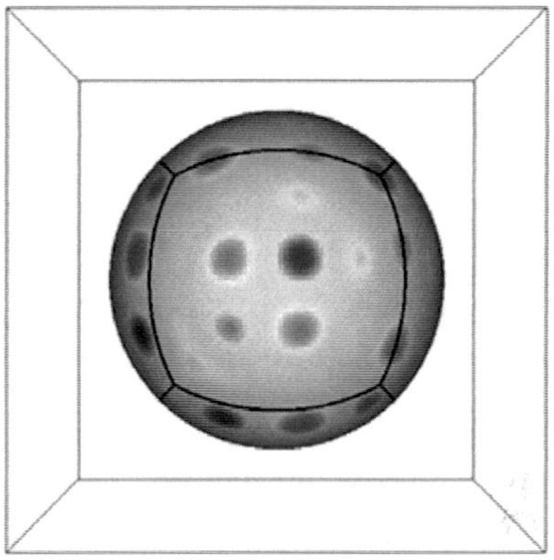

Figure 6: *Sample output from applying a 4th order Laplace operator to a mapped multiblock dataset with "cubed sphere shell" geometry. Each of the 6 blocks is oriented with its center normal to the positive and negative Cartesian coordinate axes and is composed of four* $32 \times 32 \times 1$ *patches. When applying the Laplace operator, the 24 patches are distributed across 4 MPI processes and load-balanced to place contiguous patches in the same block on the same CPU. The results vary smoothly at block boundaries, so that no numerical artifacts due to interpolated block boundary conditions are introduced.*

The Mapped-Multiblock Framework. We have implemented the algorithms [18] specialized to the case of abstract spherical coordinates. This includes the methods for representing multiblock coordinate systems and specifying data on the unions of rectangles defined for such coordinate systems. The 4th-order accurate methods are used for computing ghost cells at block boundaries.

AMR on Mapped Grids. We have developed the AMR capabilities required by HelioCubed for MHD. These include basic representations of data on unions of rectangles, including data objects for data on AMR hierarchies, a version of the refinement-in-time Runge–Kutta time discretization for AMR grids described in [14], and 4th-order accurate interpolation in space and time at refinement boundaries.

Single-Patch Tools for Mapped Grids. We have implemented a number of tools that are needed for mapped grids, primarily for transforming between logical coordinate space, or in physical space. These include the calculations of metric terms for the mappings, product and quotient operators, and stencils for transforming between various centerings. These are complicated tasks to be implemented with the 4th-order accuracy, and especially for vector- and tensor-valued patch data. We have also developed specialized versions of these and other operators that are required to implement the versions of our MHD algorithm that preserves spherically symmetric solutions.

These tools are factored into mapped-multiblock, AMR, and single-patch tools, but in fact many of them are used in two or three contexts. For example, the single-patch tools can be called for a single patch or across multiple patches in AMR or mapped-multiblock union of rectangles. A second example is the Proto interface for HDF5 files, which can be written for all three contexts and displayed by the VisIt visualization tool [19] in the abstract-coordinate and physical spaces. In Figure 6, we show an example of a flux divergence calculation on mapped-multiblock coordinates obtained using our HDF5 capability and VisIt.

Journal of Physics: Conference Series **2742** (2024) 012013 doi:10.1088/1742-6596/2742/1/012013

4.4. Ensemble modeling of CMEs.

Flux-rope-based magnetohydrodynamic modeling of coronal mass ejections (CMEs) is a promising tool for prediction of the CME arrival time and magnetic field at Earth. In [3], we introduced a constant-turn flux rope model and used it to simulate the 12-July-2012 16:48 CME in the IHS. We constrained the initial parameters of this CME using the graduated cylindrical shell (GCS) model and the reconnected flux in post-eruption arcades. We correctly reproduced all the magnetic field components of the CME at Earth, with an arrival time error of approximately 1 hour. The error that small is not just a particular advantage of the proposed method. In reality, the chosen CME had properties easily identifiable from the observational data. We further estimated the average subjective uncertainties in the GCS fittings, by comparing the GCS parameters of 56 CMEs reported in multiple studies and catalogs. We determined that the GCS estimates of the CME latitude, longitude, tilt, and speed have average uncertainties of 5.74°, 11.23°, 24.71°, and 11.4% respectively. Using these average uncertainties, we have created 77 ensemble members for the 12-July-2012 CME. We found that 55% of our ensemble members correctly reproduce the sign of the magnetic field components at Earth. We also determined that the uncertainties in GCS fitting can widen the CME arrival time prediction window to about 12 hours for the 12-July-2012 CME. On investigating the forecast accuracy introduced by the uncertainties in individual GCS parameters, we conclude that the half-angle and aspect ratio have little impact on the predicted magnetic field of the 12-July-2012 CME, whereas the uncertainties in longitude and tilt can introduce relatively large spread in the magnetic field predicted at Earth.

In [17], we developed an innovative technique for predicting CME arrival times, which utilizes magnetohydrodynamic simulations with data-informed flux-rope-based CMEs introduced into a data-driven solar wind background. For the six CMEs we analyzed, the Mean Absolute Error (MAE) in arrival time was roughly 8 hours. We further refined our arrival time predictions by implementing ensemble modeling and comparing the collective solutions with data from STEREO-A and STEREO-B heliospheric imagers. This involved using our simulations to generate synthetic J-maps. The machine-learning (ML) technique known as lasso regression was employed for this comparison. By adopting this strategy, we were able to cut down the MAE to approximately 4 hours. Additionally, another ML method built on neural networks (NNs) allowed us to lower the MAE to about 5 hours when HI data from both STEREO-A and STEREO-B were accessible. NNs proved capable of delivering comparable MAE even when only the STEREO-A data were utilized. Moreover, our approaches resulted in quite promising standard deviation values for arrival time.

5. Conclusions: SWx community engagement

We are developing publicly available software for SWx forecasts with the aim to make it formally more accurate and order of magneitude faster than any existing codes. We work with a clear understanding of the particular requirements an open-source software should satisfy as a potential prototype of the new operational tool. This includes active engagement with the operational end users at NASA/JSC Space Radiation Analysis Group (SRAG) and NOAA/NWS Space Weather Prediction Center (SWPC) to better understand their operational needs and constraints. In consultation with end users, we are adding new features to our software to help simplify the SWx community efforts to create new operational tools and products toward improving SWx analyses and predictions. These efforts are accomplished by a close coordination with the transition centers, such as Community Coordinated Modeling Center (CCMC), Short-term Prediction Research and Transition Center (SPoRT), SWPC operations and SWPC's Space Weather Prediction Testbed (SWPT).

References

[1] Riley P, Linker J A, Mikic Z, Caplan R M, Downs C and Thumm J L 2019 *Astrophys. J.* **884** 18

[2] Temmer M, Scolini C, Richardson I G, Heinemann S G, Paouris E, et al. 2023 *arXiv e-prints* arXiv:2308.04851 (*Preprint* 2308.04851)

[3] Singh T, Kim T K, Pogorelov N V and Arge C N 2022 *The Astrophysical Journal* **933** 123 URL https://dx.doi.org/10.3847/1538-4357/ac73f3

[4] Hathaway D H, Williams P E, Dela Rosa K and Cuntz M 2010 *Astrophys. J.* **725** 1082–1090 (*Preprint* 1008.4385)

[5] Li S 2006 Weno schemes for cylindrical and spherical grid. *CSC* (Citeseer) pp 177–183

[6] Gottlieb S, Ketcheson D and Shu C W 2011 *Strong stability preserving Runge-Kutta and multistep time discretizations* (World Scientific)

[7] O'Sullivan S 2019 *Journal of Computational Physics* **388** 209–223

[8] Caplan R M, Johnston C D, Daldoff L K S and Linker J A 2024 *Sumbitted to the Journal of Physics: Conference Series, ASTRONUM 2023*

[9] Caplan R M, Downs C, Linker J A and Mikic Z 2021 *The Astrophysical Journal* **915** 44 URL https://dx.doi.org/10.3847/1538-4357/abfd2f

[10] Arge C N, Odstrcil D, Pizzo V J and Mayer L R 2003 Improved Method for Specifying Solar Wind Speed Near the Sun *AIP Conf. Proc. 679: Solar Wind Ten* pp 190–193

[11] Arge C N, Henney C J, Hernandez I G, Toussaint W A, Koller J and Godinez H C 2013 Modeling the corona and solar wind using ADAPT maps that include far-side observations *Solar Wind 13* (*American Institute of Physics Conference Series* vol 1539) ed Zank G P, et al., pp 11–14

[12] Adams M, Colella P, Graves D T, Johnson J, Keen N, et al. URL https://commons.lbl.gov/display/chombo/Chombo+-+Software+for+Adaptive+\Solutions+of+Partial+Differential+Equations

[13] Colella P, Dorr M, Hittinger J and Martin D 2011 *Journal of Computational Physics* **230** 2952 – 2976 ISSN 0021-9991

[14] McCorquodale P and Colella P 2011 *Commun. Appl. Math. Comput. Sci.* **6** 1–25 URL https://doi.org/10.2140/camcos.2011.6.1

[15] Powell K G, Roe P L, Linde T J, Gombosi T I and Zeeuw D L D 1999 *Journal of Computational Physics* **154** 284 – 309 ISSN 0021-9991 URL http://www.sciencedirect.com/science/article/pii/S002199919996299X

[16] Pogorelov N, Borovikov S, Heerikhuisen J, Kim T, Kryukov I and Zank G 2014 Ms-flukss and its application to modeling flows of partially ionized plasma in the heliosphere *Proceedings of the 2014 Annual Conference on Extreme Science and Engineering Discovery Environment* XSEDE '14 (New York, NY, USA: ACM) pp 22:1–22:8 ISBN 978-1-4503-2893-7 URL http://doi.acm.org/10.1145/2616498.2616499

[17] Singh T, Benson B, Raza S A Z, Kim T K, Pogorelov N V, et al. 2023 *The Astrophysical Journal* **948** 78 URL https://dx.doi.org/10.3847/1538-4357/acc10a

[18] McCorquodale P, Dorr M, Hittinger J and Colella P 2015 *Journal of Computational Physics* **288** 181–195 ISSN 0021-9991

[19] Childs H, Brugger E, Whitlock B, Meredith J, Ahern S, et al. 2012 Visit: An end-user tool for visualizing and analyzing very large data *High Performance Visualization–Enabling Extreme-Scale Scientific Insight* pp 357–372

Journal of Physics: Conference Series **2742** (2024) 012014 doi:10.1088/1742-6596/2742/1/012014

Modeling of TeV Galactic Cosmic-ray Anisotropy based on Intensity Mapping in an MHD Model Heliosphere

M. Amenomori[1], Y. W. Bao[2], X. J. Bi[3], D. Chen[4], T. L. Chen[5], W. Y. Chen[3], Xu Chen[4], Y. Chen[2], Cirennima[5], S. W. Cui[6], Danzengluobu[5], L. K. Ding[3], J. H. Fang[3,7], K. Fang[3], C. F. Feng[8], Zhaoyang Feng[3], Z. Y. Feng[9], Qi Gao[5], Q. B. Gou[3], Y. Q. Guo[3], Y. Y. Guo[3], Y. Hayashi[10], H. H. He[3], Z. T. He[6], K. Hibino[11], N. Hotta[12], Haibing Hu[5], H. B. Hu[3], K. Y. Hu[3,7], J. Huang[3], H. Y. Jia[9], L. Jiang[3], P. Jiang[4], H. B. Jin[4], K. Kasahara[13], Y. Katayose[14], C. Kato[10], S. Kato[15], I. Kawahara[14], T. Kawashima[15], K. Kawata[15], M. Kozai[16], Labaciren[5], G. M. Le[17], A. F. Li[3,9,18], H. J. Li[5], W. J. Li[3,9], Y. Li[4], Y. H. Lin[3,7], B. Liu[19], C. Liu[3], J. S. Liu[3], L. Y. Liu[4], M. Y. Liu[5], W. Liu[3], H. Lu[3], T. Makishima[14], Y. Masuda[10], S. Matsuhashi[14], M. Matsumoto[10], X. R. Meng[5], Y. Meng[3,7], A. Mizuno[15], K. Munakata[10], Y. Nakamura[15], H. Nanjo[1], C. C. Ning[5], M. Nishizawa[20], R. Noguchi[14], M. Ohnishi[15], S. Okukawa[14], S. Ozawa[21], X. Qian[4], X. L. Qian[22], X. B. Qu[23], T. Saito[24], M. Sakata[25], T. Sako[15], T. K. Sako[15], T. Sasaki[11], J. Shao[3,9], T. Shibasaki[26], M. Shibata[14], A. Shiomi[26], H. Sugimoto[27], W. Takano[11], M. Takita[15], Y. H. Tan[3], N. Tateyama[11], S. Torii[28], H. Tsuchiya[29], S. Udo[11], R. Usui[14], H. Wang[3], S. F. Wang[5], Y. P. Wang[5], Wangdui[5], H. R. Wu[3], Q. Wu[5], J. L. Xu[4], L. Xue[8], Z. Yang[3], Y. Q. Yao[4], J. Yin[4], Y. Yokoe[15], Y. L. Yu[3,7], A. F. Yuan[5], L. M. Zhai[4], H. M. Zhang[3], J. L. Zhang[3], X. Zhang[2], X. Y. Zhang[8], Y. Zhang[3], Yi Zhang[30], Ying Zhang[3], S. P. Zhao[3], Zȟaxisangzhu[5], X. X. Zhou[9], Y. H. Zou[3,7], and N. V. Pogorelov[31]

[1]Department of Physics, Hirosaki University, Hirosaki 036-8561, Japan.
[2]School of Astronomy and Space Science, Nanjing University, Nanjing 210093, China.
[3]Key Laboratory of Particle Astrophysics, Institute of High Energy Physics, Chinese Academy of Sciences, Beijing 100049, China.
[4]National Astronomical Observatories, Chinese Academy of Sciences, Beijing 100101, China.
[5]Department of Mathematics and Physics, Tibet University, Lhasa 850000, China.
[6]Department of Physics, Hebei Normal University, Shijiazhuang 050016, China.
[7]University of Chinese Academy of Sciences, Beijing 100049, China.
[8]Institute of Frontier and Interdisciplinary Science and Key Laboratory of Particle Physics and Particle Irradiation (MOE), Shandong University, Qingdao 266237, China.
[9]Institute of Modern Physics, SouthWest Jiaotong University, Chengdu 610031, China.
[10]Department of Physics, Shinshu University, Matsumoto 390-8621, Japan.
[11]Faculty of Engineering, Kanagawa University, Yokohama 221-8686, Japan.
[12]Faculty of Education, Utsunomiya University, Utsunomiya 321-8505, Japan.
[13]Faculty of Systems Engineering, Shibaura Institute of Technology, Omiya 330-8570, Japan.

[14]Faculty of Engineering, Yokohama National University, Yokohama 240-8501, Japan.
[15]Institute for Cosmic Ray Research, University of Tokyo, Kashiwa 277-8582, Japan.
[16]Polar Environment Data Science Center, Joint Support-Center for Data Science Research, Research Organization of Information and Systems, Tachikawa 190-0014, Japan.
[17]National Center for Space Weather, China Meteorological Administration, Beijing 100081, China.
[18]School of Information Science and Engineering, Shandong Agriculture University, Taian 271018, China.
[19]Department of Astronomy, School of Physical Sciences, University of Science and Technology of China, Hefei 230026, China.
[20]National Institute of Informatics, Tokyo 101-8430, Japan.
[21]National Institute of Information and Communications Technology, Tokyo 184-8795, Japan.
[22]Department of Mechanical and Electrical Engineering, Shangdong Management University, Jinan 250357, China.
[23]College of Science, China University of Petroleum, Qingdao 266555, China.
[24]Tokyo Metropolitan College of Industrial Technology, Tokyo 116-8523, Japan.
[25]Department of Physics, Konan University, Kobe 658-8501, Japan.
[26]College of Industrial Technology, Nihon University, Narashino 275-8575, Japan.
[27]Shonan Institute of Technology, Fujisawa 251-8511, Japan.
[28]Research Institute for Science and Engineering, Waseda University, Tokyo 162-0044, Japan.
[29]Japan Atomic Energy Agency, Tokai-mura 319-1195, Japan.
[30]Key Laboratory of Dark Matter and Space Astronomy, Purple Mountain Observatory, Chinese Academy of Sciences, Nanjing 210034, China.
[31]Department of Space Science and Center for Space Plasma and Aeronomic Research, University of Alabama in Huntsville, 320 Sparkman Drive, Huntsville, AL 35899, USA.

Abstract. Cosmic rays do not arrive at the Earth uniformly. Recent experiments have consistently observed small anisotropies with amplitudes of $\sim$0.2% in the arrival directions of cosmic rays at TeV energies. We perform the modeling of the cosmic-ray anisotropy at TeV energies using the intensity-mapping method based on Liouville's theorem. This work indicates unrealistically small-scale ansiotropies with $<\sim 10^\circ$ angular scales in the distribution of the cosmic-ray relative intensity at the outer boundary of the heliosphere. We would possibly need to resolve an issue that the experimental data covers ten years while the Magnetohydrodynamics (MHD) model heliosphere used in this work is only a single snapshot at a certain moment. Performing the intensity-mapping for multiple snapshots of the MHD model heliosphere and taking the average of the results would improve the results of this work.

1. Introduction

Galactic cosmic rays (CRs) do not arrive at the Earth uniformly. Anisotropies with amplitudes of roughly 0.2% have been observed at tera-electronvolt (TeV) energies by recent cosmic-ray experiments (e.g. [1, 2, 3, 4, 5, 6]), such as a large-scale deficit region named 'Loss-Cone' and a large-scale excess region named 'Tail-In'. The origins of these anisotropic features have yet to be revealed, although the anisotropy reflects the propagation of CRs in the magnetic fields of the heliosphere and the surrounding interstellar medium.

Previous papers [7, 8] utilize the 'intensity-mapping' method, in which a model heliosphere is reconstructed by Magnetohydrodynamics (MHD) simulations, CR trajectories in the magnetic fields of the MHD model heliosphere are calculated, and Liouville's theorem is employed to map the distribution of the CR intensity at the Earth to the outer boundary of the heliosphere. Then, the obtained anisotropies of the CR intensity at the outer boundary is interpreted in physical terms by modeling the intensity distribution based on a superposition of CR flows. It was indicated in the previous paper [7] that, dominant in the interstellar medium outside the heliosphere is a dipole flow of CRs along the interstellar magnetic field ($B_{\rm ISM}$), and that a density gradient of CRs also exists in the direction of the Vela supernova remnant.

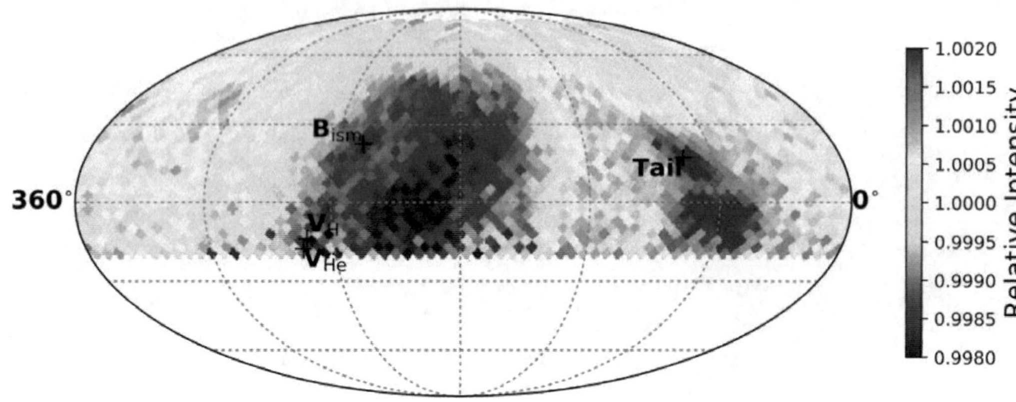

Figure 1. CR relative intensity ditribution in equatorial cooridnates obtained from the data of the Tibet ASγ experiment taken from November 1999 to May 2010.

In order to make the modeling conclusive, however, two technical issues must be resolved. Firstly, the previous work [7] employed protons with monochromatic energy of 4 TeV for the intensity-mapping, although the energy change of those protons due to the acceleration and deceleration in the MHD model heliosphere were taken into account. Since CRs arriving at the Earth are composed of different atomic nuclei with different energies, the energy spectrum and the composition of observed CRs must be taken into account in the intensity-mapping process. Secondly, the χ^2/d.o.f. of the fitting between the experimental data and the best-fit model anisotropy was 4.5 in the previous paper, which was not good enough. The χ^2/d.o.f. value must be reduced to ~ 1 by improving the modeling of the distribution of the CR intensity at the heliospheric outer boundary.

2. Intensity Mapping

In this work, we use the data obtained from November 1999 to May 2010 by the Tibet ASγ experiment. Figure 1 shows the obtained CR relative intensity distribution in equatorial coordinates. We pixelize the sky in our field of view ($-20° <$ decl. $< 80°$) in 2056 pixels, using the HEALPix algorithm [9] with $N_{\rm side} = 16$. Each pixel has an approximate size of $3.7° \times 3.7°$. We estimate the rigidity distribution of observed CRs using detailed Monte Carlo (MC) simulations of air-shower generation and detector response. We generate air showers using CORSIKA v7.4000 [11] with EPOS LHC [12] for the high-energy hadronic interaction model and FLUKA v2011.2b [13, 14] for the low-energy hadronic interaction model, in the energy range from 0.3 TeV to 10 PeV based on a model of the CR energy spectrum and chemical composition derived from direct measurements [10]. We feed the generated air showers into the detector response simulation developed by GEANT v4.10.00 [15] and analyze them in the same way as the experimental data. The obtained rigidity distribution of CRs is shown in Figure 2 for three declination bands.

We use the fourth-order Runge-Kutta method to calculate the trajectories of CRs in the MHD model heliosphere. To smooth out possible seasonal effects, we place the Earth at four positions on the ecliptic plane at a distance of 1 AU from the Sun. Using the HEALPix algorithm [9] we pixelize the sky with $N_{\rm side} = 32$ in the declination range from $-20°$ to $80°$, and shoot CR anti-partiles from each pixel center into the MHD model heliosphere, sampling their rigidity

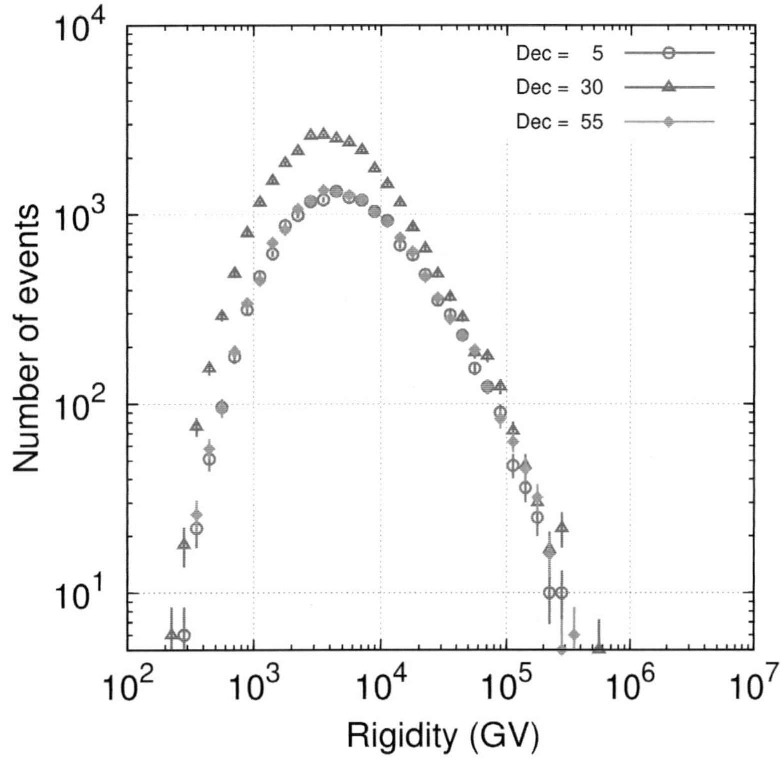

Figure 2. Distribution of CR rigidity as seen by the Tibet ASγ experiment for three declination bands, reproduced with detailed MC simulations.

from Figure 2. We use the same MHD model heliosphere as the one employed in [7]. We track the CRs in the MHD model heliosphere and record their momentum directions at the outer boundary, and then the CR intensity at the outer boundary in the CR momentum direction ($I_{\rm ISM}$) is regarded as equal to that at the Earth ($I_{\rm E}$). We evaluate the distribution of the relative intensity of CRs at three boundaries at distances of 630 AU, 1580 AU and 3980 AU from the Sun, with the nose direction truncated at the surface where the direction (strength) of the magnetic field becomes different from that outside the heliosphere by $< 0.1^\circ$ (0.1%).

The 'declination bias' exists in the experimental data — the average of the CR intensity has been normalized to unity in each declination band since the detection efficiency of the experiment has not been absolutely calibrated along declination. The following steps, therefore, need to be taken to estimate the $I_{\rm ISM}$ distribution at the outer boundary: 1) construct $I_{\rm ISM}$, a model of the CR intensity distribution at the outer boundary, 2) map $I_{\rm ISM}$ to that at the Earth $I_{\rm E}$ based on the CR trajectory calculation in the MHD model heliosphere, 3) normalize the average of $I_{\rm E}$ to unity in each declination band, and 4) obtain χ^2 between the experimental data and the normalized $I_{\rm E}$. We repeat these four steps until the χ^2 is minimized, and obtain the best-fit model of $I_{\rm ISM}$. We express the distribution of the CR relative intensity at the outer boundary in spherical harmonics Y_{lm} at step 1):

$$I_{\rm ISM} = 1 + \sum_{l=1}^{l_{\rm max}} \sum_{m=-l}^{l} f_{lm} Y_{lm} + I_{\rm CG}, \tag{1}$$

where f_{lm}'s are free fitting parameters. We increase the maximum order $l_{\rm max}$ of the spherical harmonics until we get the χ^2/d.o.f. close to unity.

3. Results and Discussions

The results are summarized in Figure 3. Panels (a), (b) and (c) are the best-fit model distributions of the CR relative intensity at the outer boundaries at distances of 630 AU, 1580 AU and 3980 AU from the Sun, respectively. The maximum order $l_{\rm max}$ of the spherical harmonics Y_{lm} in Eq.(1) is $l_{\rm max} = 4$ for Panel (a), $l_{\rm max} = 8$ for Panel (b), and $l_{\rm max} = 20$ for Panel (c), and the obtained χ^2/d.o.f. values are 0.96 (probability 89%), 0.98 (71%) and 0.94 (95%), respectively. From Figure 3, one can notice that $l_{\rm max}$ decreases as the distance of the boundary from the Sun decreases. In terms of physics, it should be a natural assumption that only large-scale anisotropies exist in the distribution of the CR intensity at the helisopheric outer boundary, and that small-scale anisotropies are added to the intensity distribution by the modulation of CR trajectories in the heliospheric magnetic field. The tendency seen in panels (a) – (c) in Figure 3, however, disagrees with this assumption. Considering that the MHD model heliosphere employed in this work has been refined already to give reasonable agreement with experimental results (for details see [7]), we consider that our intensity mapping needs to be improved further. One possible issue would be that, the CR scattering with magnetic irregularities in the heliosphere is not taken into account in the intensity mapping process when we calculate CR trajectories in a single snapshot of the magnetic field structure of the MHD model heliosphere, while the experimental data used in this work, covering a period of ~ten years, contains this CR scattering effect. This discrepancy would be resolved by performing the intensity-mapping for multiple snapshots of the MHD model heliosphere and taking the average of the results, which would lead to remove the unnatural tendency in Figure 3 and enable us to derive CR flows in the interstellar medium outside the heiosphere.

Acknowledgments

The collaborative experiment of the Tibet Air Shower Arrays has been conducted under the auspices of the Ministry of Science and Technology of China and the Ministry of Foreign Affairs

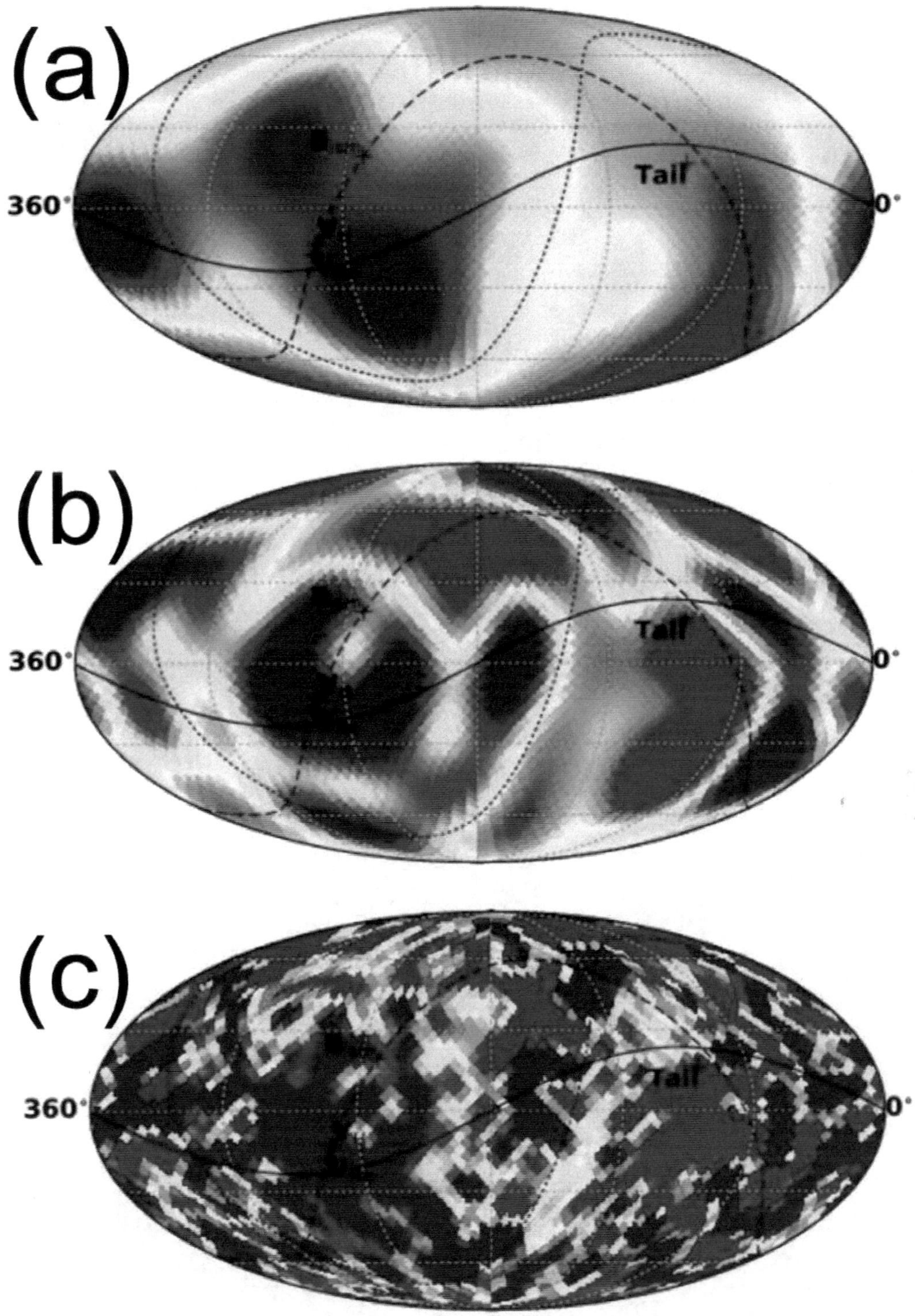

Figure 3. Best-fit model distributions of the CR relative intensity at the heliospheric outer boundaries at distances of 630 AU (a), 1580 AU (b) and 3980 AU (c) from the Sun, respectively.

of Japan. This work was supported in part by a Grant-in-Aid for Scientific Research on Priority Areas from the Ministry of Education, Culture, Sports, Science and Technology, and by Grants-in-Aid for Science Research from the Japan Society for the Promotion of Science in Japan. This work is supported by the National Natural Science Foundation of China under Grants No. 12227804, No. 12275282 and No. 12073050, and the Key Laboratory of Particle Astrophysics, Institute of High Energy Physics, CAS. This work is also supported by the joint research program of the Institute for Cosmic Ray Research (ICRR), the University of Tokyo.

References

[1] A. U. Abeysekara *et al.*, Astrophys. J., **871**, 96 (2019)
[2] A. U. Abeysekara *et al.*, Astrophys. J., **865**, 57 (2018)
[3] M. G. Aartsen *et al.*, Astrophys. J., **826**, 220 (2016)
[4] B. Bartoli *et al.*, Astrophys. J., **809**, 90 (2015)
[5] M. Amenomori *et al.*, Astrophys. Space Sci. Trans., **6**, 49 (2010)
[6] M. Amenomori *et al.*, Science, **314**, 439 (2006)
[7] M. Zhang *et al.*, Astrophys. J., **889**, 97 (2020)
[8] M. Zhang *et al.*, Astrophys. J., **790**, 5 (2014)
[9] K. M. Górski *et al.*, Astrophys. J., **622**, 759 (2005)
[10] M. Shibata *et al.*, Astrophys. J., **716**, 1076 (2010)
[11] D. Heck *et al.*, CORSIKA: a Monte Carlo Code to Simulate Extensive Air Showers Report FZKA-6019 (Forschungszentrum Karlsruhe, 1998)
[12] T. Pierog *et al.*, Phys. Rev. C, **92**, 034906 (2015)
[13] A. Ferrari *et al.*, Report CERN-2005-10, INFN/TC_05/11, SLAC-R-773 (CERN European Organization for Nuclear Research, 2005)
[14] T. T. Böhlen *et al.*, Nucl. Data Sheets, **120**, 211 (2014)
[15] S. Agostinelli *et al.*, Nucl. Instrum. Meth. A, **506**, 250 (2003)

Journal of Physics: Conference Series **2742** (2024) 012015 doi:10.1088/1742-6596/2742/1/012015

Modeling electron acceleration during the contraction of a magnetic island

Atit Deuja and Haihong Che

Center for Space Plasma and Aeronomic Research (CSPAR), University of Alabama in Huntsville, Huntsville, AL 35899, USA

Department of Space Science, University of Alabama in Huntsville, Huntsville, AL 35899, USA

E-mail: ad0159@uah.edu

Abstract. Magnetic reconnection releases the magnetic energy through the contraction of multi-magnetic island leading to the electron acceleration as proposed by Drake et. al in 2006. However, how the released magnetic energy is converted into electron's kinetic energy is still theoretically not well understood. We model in particular the kinetic process assuming the adiabatic contraction of magnetic island that induces electric field which is proportional to the vector potential of the magnetic island and approximate the magnetic island with an ellipse. Under this model, we show that the energy gain is achieved through the work of inductive electric field. We further show that the curvature drift which is along the inductive electric field dominates the energy gain. We compared our model with the magnetic island formed by tearing instability in a 2.5D particle-in-cell simulation of magnetic reconnection and found the results from the model consistent with that of the simulation.

1. Introduction

Particle acceleration is an essential phenomenon observed in various explosive events: for example, solar flares, magnetospheric substorms, gamma ray brusts, etc. in space [1, 2, 3], astrophysical [4, 5, 6] and also in laboratory [7, 8, 9] plasma. The magnetic reconnection (MR) is thought to be a major driver for these explosive events [10, 11, 12]. The MR is a fundamental plasma process that converts the magnetic energy into electromagnetic energy, heat and particle's kinetic energy via topological rearrangement of magnetic field lines.

In recent decades, extensive studies [8, 13, 14, 15, 16, 17, 18, 19, 20, 21, 22, 23, 24] have been carried out to understand the role of MR in particle acceleration within space and astrophysical plasmas. Specifically, the focus has been in understanding the processes involved in mildly relativistic particle acceleration during the events such as solar flares and magnetospheric substorms. But these studies posited an inherent challenge known as the "injection problem" [16], as the cosmic ray acceleration mechanism proposed by Fermi in 1949 necessitates pre-acceleration of particles to achieve a mildly relativistic speed for Fermi acceleration to occur [25]. The contraction of multi-magnetic island acceleration mechanism proposed by Drake [15] in 2006 is one of the commonly invoked model in solar flares. This mechanism suggests that during the contraction of multi-magnetic islands which are produced by tearing instability, electrons are accelerated during the bounce motion within the converging island and from the repetitive interaction among the randomly distributed magnetic islands. However, recent studies show that the contraction of magnetic island is an adiabatic invariant process with particle exhibiting adiabatic motion [21, 26, 27]. The contraction timescale for a particle to exhibit such adiabatic motion requires it to be long and hence the acceleration process due to the contraction of magnetic island is a slow and inefficient process [19] in contrast to the Fermi acceleration which is a stochastic process.

It is yet unclear exactly what essential physical process governs the electron's motion in the contracting magnetic island and specifically whether the electron can achieve stochastic motion by the bouncing among the multiple magnetic islands. This requires the information of the time duration of electron confinement within a magnetic island, which is governed by the electron's interaction with the island. Recent studies [17, 21] based on fluid model and test particles showed that the curvature drift is a dominant source of electron heating but the underlying physics for the energy gain is still in question. In this paper, we aim to answer this question by presenting a model to grasp the essential physical process of the contracting magnetic island and then use test particles to study electron interaction with the islands. We compare the results from the model with the particle-in-cell (PIC) simulations to evaluate our model. The elongated magnetic island in the model is initially approximated by an ellipse, and assuming adiabatic contraction, the contracting magnetic island induces an electric field such that the curl of induced electric field is proportional to the magnetic field of the island. Under this model, the particle moving along the contracting magnetic island near the Alfven speed [15, 21] gets a kick intermediated by the induced electric field every time it passes through the ends of the magnetic island. We use test particle simulations to study particle motion, energy gain and drift velocities. The energy gain process in this scenario is dominated by curvature drift term where the electron gains energy through the work done by the induced electric field.

2. Modeling the contracting magnetic island

When the width of current sheets nears the ion inertial scale, the magnetic shear in current sheets leads to the onset of tearing instability [16] triggering the formation of multiple magnetic islands. The contraction of magnetic island is pictured by the gradual change in its geometry from elongated to a round due to the internal magnetic tension to relax the stored magnetic energy. The contraction of magnetic island is an adiabatic invariant process with the growth rate γ of the tearing instability being significantly smaller than the electron gyro-frequency Ω_e and the time scale Δt of one-cycle of electron motion along the magnetic island [28].

Figure 1 shows the schematic diagram of the contraction model of a two dimensional (2D) magnetic island. In the model, the elongated magnetic island is initially approximated by an ellipse in xy-plane which later contracts to form a circle releasing the magnetic energy where both the magnetic flux Φ and volume V remains approximately conserved for an in-compressible plasma with the radius of circle $r = \sqrt{ab}$ where a and b are the length of the semi-major and semi-minor axis of the ellipse.

The corresponding vector potential $\mathbf{A}_z$ of the elongated 2D magnetic island approximated by an ellipse is given by,

$$\mathbf{A}_z = -d_i B_0 \left[\frac{(x - x_0)^2}{a^2} + \frac{(y - y_0)^2}{b^2} \right] \hat{\mathbf{z}} \tag{1}$$

where d_i is the ion inertial length, B_0 is the asymptotic reconnection magnetic field, (x_0, y_0) is the center of the island, and a and b are respectively the length of semi-major and semi-minor axis. The magnetic field $\mathbf{B}$ is given by,

$$\mathbf{B} = \nabla \times \mathbf{A}_z = d_i B_0 \left[\frac{-2(y - y_0)}{b^2} \hat{\mathbf{x}} + \frac{2(x - x_0)}{a^2} \hat{\mathbf{y}} \right] \tag{2}$$

Journal of Physics: Conference Series **2742** (2024) 012015 doi:10.1088/1742-6596/2742/1/012015

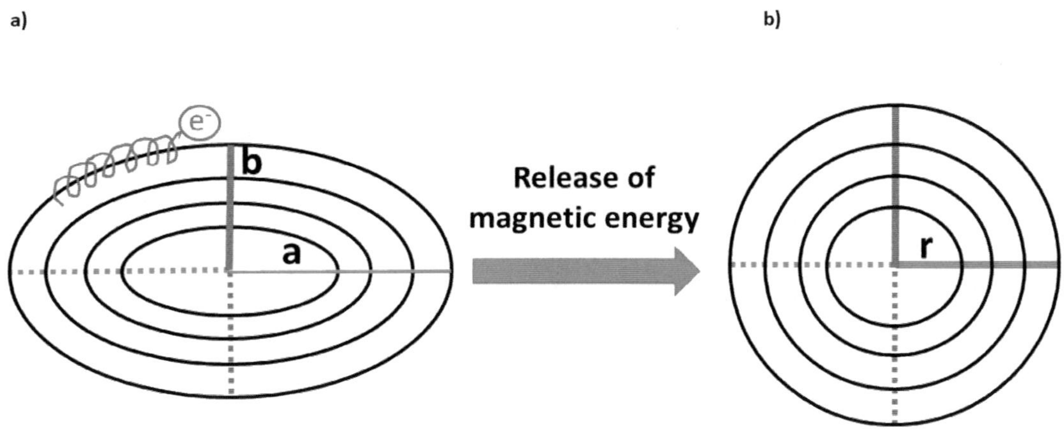

Figure 1. A schematic diagram of contraction model of a 2D elongated magnetic island described by equation (1). a) An elongated magnetic island contracts releasing magnetic energy and becomes b) a rounded magnetic island.

The current density $\mathbf{J}$ can then be calculated as,

$$\mathbf{J} = \frac{c}{4\pi}\nabla \times \mathbf{B} = constant \tag{3}$$

where c is the speed of light. The slowly varying magnetic field induces electric field via Faraday's induction law i.e.

$$\nabla \times \mathbf{E}_{ind} = -\frac{1}{c}\frac{\partial \mathbf{B}}{\partial t} \tag{4}$$

Using equation 2, the induced electric field can be written as,

$$\mathbf{E}_{ind} = -\frac{1}{c}\frac{\partial \mathbf{A}_z}{\partial t} \tag{5}$$

We assume that the induced electric field is proportional to the vector potential,

$$\mathbf{E}_{ind} = C_R \frac{d_i B_0}{c}\left[\frac{(x-x_0)^2}{a^2} + \frac{(y-y_0)^2}{b^2}\right]\hat{\mathbf{z}} \tag{6}$$

where C_R is a coefficient and the $\mathbf{E}_{ind}$ is out-of-plane along z-axis. The current density calculated from the model is constant, which is not the case for a realistic magnetic island and hence it is a model's shortcoming. The model primarily emphasizes on the process of magnetic energy release through the contraction of magnetic island and acceleration of electrons via the inductive electric field, rather than exact MR process, which is better captured by the PIC simulation. Therefore, we extract the fundamental aspects of the contraction process and illustrate that the model effectively demonstrate the underlying physics, as we confirmed this through the comparison with the results from the PIC simulation.

3. Simulations

To evaluate our model, we use the electric field **E** and magnetic field **B** data calculated from 2.5D PIC magnetic reconnection simulation using the p3d code [29]. The PIC simulation initializes from the doubly periodic two Harris current sheets with reconnection magnetic field configuration given by [15],

$$B_x = B_0 \left[tanh\left(\frac{y - L_y/4}{w_0}\right) - tanh\left(\frac{y - 3L_y/4}{w_0}\right) - 1\right] \tag{7}$$

where w_0 represents the half-thickness of the current sheets and L_y represents simulation domain length in y-directions. The simulation is carried out in the presence of a uniform guide field $B_g = 1.0B_0\hat{\mathbf{z}}$ and the particle density is given by [21],

$$n = n_0 \left[sech^2\left(\frac{y - L_y/4}{w_0}\right) + sech^2\left(\frac{y - 3L_y/4}{w_0}\right)\right] \tag{8}$$

where n_0 is a peak density at the center of the current sheet superimposed over a uniform ambient density $0.2n_0$. The PIC code solves the relativistic Newton-Lorentz equations and Maxwell equations to trace particles' positions and trajectories in velocity space. In the simulation, the mass ratio used is $m_e/m_i = 0.01$, the speed of light is $c = 20v_A$ and the initial temperatures are set to $T_e/m_i v_A^2 = 1/12$ for electrons and $T_i/m_i v_A^2 = 5/12$ for ions, where m_e represents electron mass, m_i represents ion mass and $v_A = B_0/(4\pi n_0 m_i)^{1/2}$ denotes the Alfven speed. The simulation spatial dimensions are $L_x \times L_y = 64d_i \times 32d_i$, where $d_i = c/(4\pi n_0 e^2/m_i)^{1/2}$ is the ion-inertial length with ion charge e, and consists of 4096×2048 cells. The boundary conditions used in the PIC simulation are periodic in both x and y-direction and the total simulation time used is $t = 64.0\Omega_{ci}^{-1}$ with $\Omega_{ci} = eB_0/m_i c$.

We carried out 2.5D test particle simulations to investigate particle motion, energy gain and drift velocities in the magnetic islands, comparing results from both the model and PIC simulation. We solved the Newton-Lorentz equation to advance test particle and calculate its trajectory using Boris method [30]. We used an electron as a test particle with the same mass ratio $m_e/m_i = 0.01$ as in PIC simulation, and the charge $q = -e$. The electron injected into the simulation domain has the dimensions of $L_x \times L_y = 4d_i \times 2d_i$ that consists of 256×128 cells which is sufficient for the study of one single magnetic island. The spatial and temporal resolution in the test particle simulations are same to that of PIC simulation. The electron is treated under guiding center limit with a velocity $\mathbf{v}$ given by;

$$\mathbf{v} = \mathbf{v}_{||} + \mathbf{v}_L + \mathbf{v}_E + \mathbf{v}_{cur} + \mathbf{v}_{\nabla B} \tag{9}$$

where the parallel velocity to the magnetic field $\mathbf{v}_{||} = \mathbf{B(v.B)}/B^2$, the Larmor velocity is $\mathbf{v}_L$, the $\mathbf{E} \times \mathbf{B}$ drift is,

$$\mathbf{v}_E = \frac{c\mathbf{E} \times \mathbf{B}}{B^2} \tag{10}$$

the gradient drift is,

$$\mathbf{v}_{\nabla B} = \frac{m_e v_\perp^2}{2q} \frac{c\mathbf{B} \times \nabla B}{B^3} \tag{11}$$

and the curvature drift is,

$$\mathbf{v}_{cur} = \frac{m_e v_{||}^2}{q} \frac{c\mathbf{B} \times (\hat{\mathbf{b}}.\nabla)\hat{\mathbf{b}}}{B^2} \tag{12}$$

respectively with the perpendicular velocity to the magnetic field $\mathbf{v}_\perp = \mathbf{v} - \mathbf{B(v.B)}/B^2$. The particle drift $\mathbf{v}_E$ acts along the xy-plane whereas the $\mathbf{v}_{\nabla B}$ and $\mathbf{v}_{cur}$ drift is out-of-plane along z-axis.

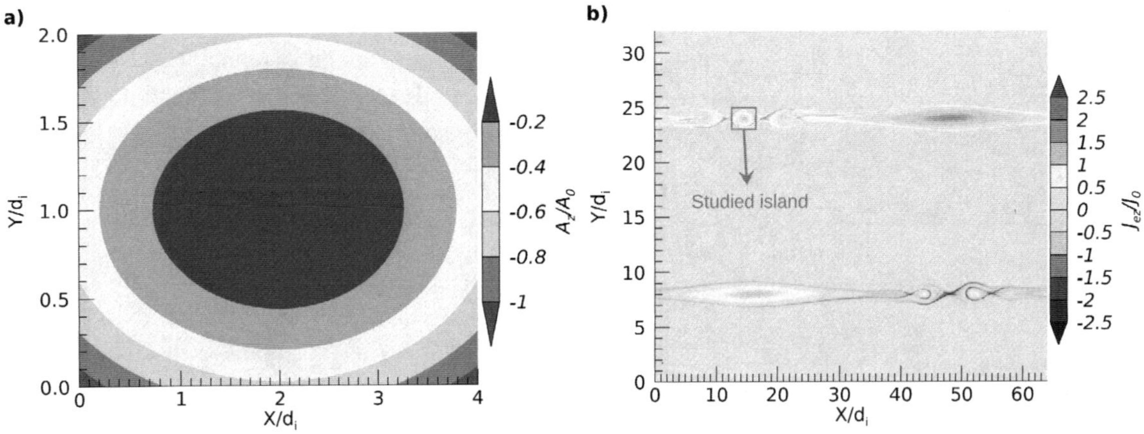

Figure 2. Contour of out-of-plane component of a) vector potential $\mathbf{A}_z$ from the model with $a = 2.83$ and $b = 1.26$ and b) electron current density $\mathbf{J}_{ez}$ from the PIC simulation at time $\Omega_{ci}t = 19.8$ in the presence of the guide field $B_g = 1.0B_0$. The highlighted square region (red) in b) denotes the studied magnetic island.

Table 1. Field description of test particle simulations

Name	Description
`Model`	$\mathbf{B} = (-2(y-y_0)/b^2, 2(x-x_0)/a^2, 0)$; $\mathbf{E} = (0, 0, E_{ind})$
`S1`	$\mathbf{B} = (B_x, B_y, 0)$; $\mathbf{E} = (0, 0, E_z)$

To initialize the test particle simulation, length of semi-major axis a and semi-minor axis b of the ellipse approximating the magnetic island (as shown in figure 1) are chosen in such a way that the specification of the magnetic island in particular the magnetic field of the model closely matches to that of the studied magnetic island from the PIC simulation. Figure 2 shows the magnetic islands from (a) model and (b) PIC simulation. The magnetic island in the model is represented by an ellipse with dimension $a = 2.83$ and $b = 1.26$, centered at $(x_0, y_0) = (2, 1)$. This specific choice of a and b generates magnetic field that scales close to the magnetic field configuration of the island marked by the solid red square in figure 2 (b). We present the contour plot of the out-of-plane vector potential $\mathbf{A}_z$ normalized by the magnitude $A_0 = d_iB_0$ to analyze the modeled island in figure 2 (a) whose magnitude increases from center towards the edge of the island. Figure 2 (b) shows the contour of out-of-plane electron current density $\mathbf{J}_{ez}$ obtained from the PIC simulation which dominates the current on kinetic scale. The studied magnetic island highlighted by the solid red square has domain size $L_x \times L_y = 4d_i \times 2d_i$ equal to that of modeled island which is also described in the xy-plane and consists 256 × 128 cells. The magnitude of the electron current density J_{ez} is expressed in the unit of $J_0 = n_0ev_A$.

Table 1 shows the field description of the test particle simulations carried out in this study. The `Model` (in Table 1) describes the test particle simulation performed in magnetic island from the model whose field is given by equation 2 and equation 6. To evaluate our model, we performed a test particle simulation `S1` under similar field conditions i.e. we take into account only x and y-components magnetic field data and out-of-plane electric field data from the PIC simulation.

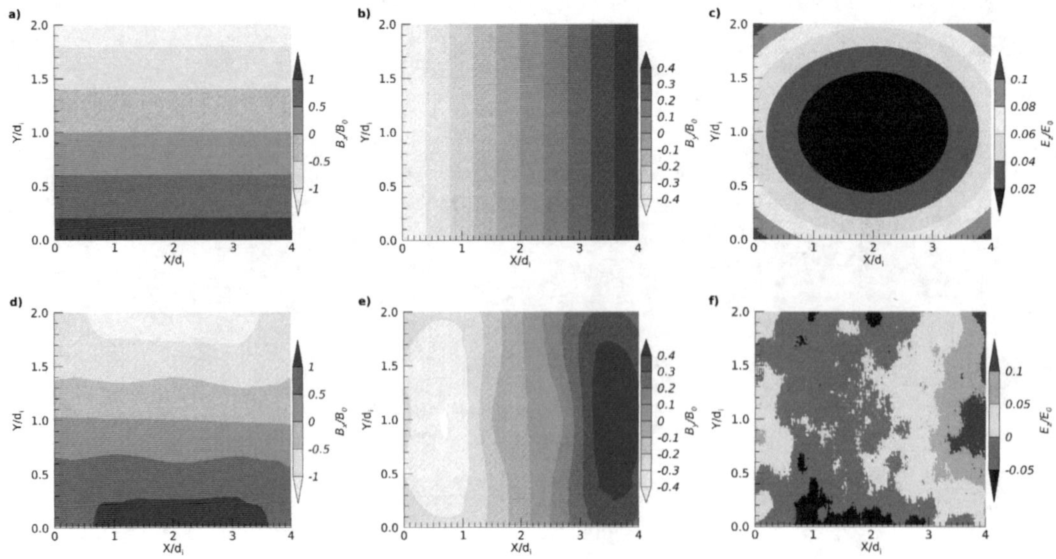

Figure 3. Top panel: a) x-component of the magnetic field $\mathbf{B}_x$, b) y-component of the magnetic field $\mathbf{B}_y$ and c) out-of-plane electric field $\mathbf{E}_z = \mathbf{E}_{ind}$ with $C_R = 0.1$ from the model. Bottom panel: d) x-component of the magnetic field $\mathbf{B}_x$, e) y-component of the magnetic field $\mathbf{B}_y$ and f) out-of-plane electric field $\mathbf{E}_z$ from the PIC simulation.

4. Comparison of results from model and PIC simulation

Figure 3 shows the comparison of fields' components of elongated magnetic island from the model (top panel) with that of the studied island from the PIC simulation (bottom panel). The contours of magnetic field from the modeled magnetic island closely matches to that of the studied magnetic island from the PIC simulation. The x-component of the magnetic field $\mathbf{B}_x$ for both model and PIC simulation magnetic island (see figure 3 (a) and (d)), increases along the y-direction from the $y = y_0$ line on both sides and remains constant along the x-direction, whereas the y-component of the magnetic field $\mathbf{B}_y$ (figure 3 (b) and (e)), increases along the x-direction from the $x = x_0$ line on both sides and remains constant along the y-direction. The near matching profiles of both the $\mathbf{B}_x$ and $\mathbf{B}_y$-components of the modeled magnetic island compared to that of simulation island confirms the elliptic configuration of the magnetic island. Nevertheless, near at the edge of the simulation island, the magnetic field components experience distortion as a result of their interaction with its neighbouring magnetic islands. The induced electric field $\mathbf{E}_{ind}$ (figure 3 (c)) from the model with a coefficient $C_R = 0.1$ is compared with the z-component of electric field $\mathbf{E}_z$ from the PIC simulation. The magnitude of electric field is expressed in the unit of $E_0 = v_A B_0/c$. The model $\mathbf{E}_{ind}$ is out-of-plane along positive z-direction whose magnitude increases from center to the edge of the magnetic island whereas the simulation $\mathbf{E}_z$ is more complex with its field components align in both positive and negative direction. The color contour in figure 3 (f) shows that the positive field components dominates the $\mathbf{E}_z$, where the work is done by the mean electric field on electron velocity $\mathbf{v}$ and hence the electron gains energy for both the model and PIC simulation fields.

Figure 4 shows the comparison of electron trajectory, energy gain and the drift velocities calculated from the test particle simulation `Model` (top panel) with that of test particle simulation `S1` (bottom panel). The electron's trajectory (red solid line in figure 4 (a)) is plotted in the background of induced electric field $\mathbf{E}_{ind}$ (color contour) whereas the electron trajectory in

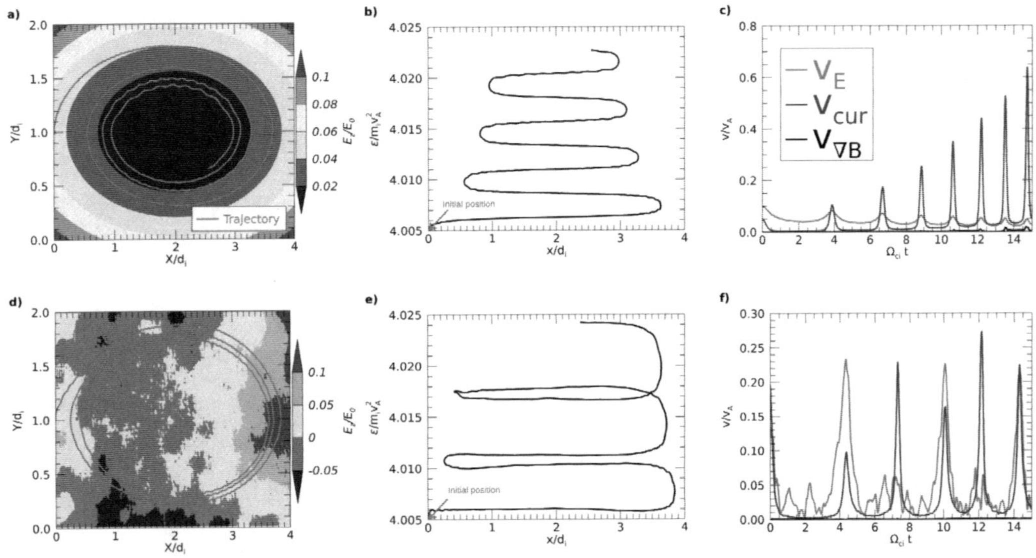

Figure 4. Top panel: a) Electron trajectory, b) energy gain and c) drift velocities calculated from the test particle simulation `Model`. Bottom panel: d) Electron trajectory, e) energy gain and f) drift velocities calculated from the test particle simulation `S1`.

figure 4 (d)) is plotted in the background of electric field $\mathbf{E}_z$ (color contour). For a positive out-of-plane electric field, the electron moves along the magnetic field lines and slowly drifts inward into the magnetic island where it gains energy every time it bounces from the edge of the contracting magnetic island. Since the mean electric field $\mathbf{E}_z$ in figure 4 (d) is positive, the electron trajectory for `Model` is consistent with that of `S1`. Electron's energy as a function of x position is shown in figure 4 (b) and figure 4(e). The electron's energy in figure 4 (b) remains nearly constant throughout the orbit but except at the edge where the curvature of the magnetic field is maximum, the electron gains energy through the work by the inductive electric field from the contracting magnetic island which is validated by figure 4 (e) obtained from the test particle simulation `S1`. Electron drift velocities evolution profile as a function of time is shown in figure 4 (c) and figure 4 (f). The magnitude of $\mathbf{E} \times \mathbf{B}$ drift v_E, curvature drift v_{cur} and gradient drift $v_{\nabla B}$ are respectively denoted by red, blue and black solid line. The curvature drift increases as the electron drift towards the center of the island and becomes a dominant since the magnitude of parallel velocity increases due to the acceleration of electron at the ends of the island by the induced electric field. On the other hand the v_E drift decreases as the magnitude of field components decreases along its trajectory. The gradient drift which depends upon the perpendicular energy of the electron (see figure 5) is very small compared to other drifts as the contraction of the magnetic island mostly affects parallel velocity of the electron.

Figure 5 shows that the magnetic moment μ and kinetic energy $k.\varepsilon.$ evolution profile from the test particle simulations (a) `Model` and (b) `S1`. The μ is expressed in the unit of $\mu_0 = m_i v_A^2/B_0$ and the $k.\varepsilon.$ is expressed in the unit of $m_i v_A^2$. For both `Model` and `S1`, the increase in total kinetic energy is mostly contributed by the increase in $k.\varepsilon.$ due to parallel velocity. The magnitude of parallel velocity increases as the particle slowly drifts inward into the contracting magnetic island. The increase of parallel velocity is sharp at the ends of the magnetic island where the curvature of magnetic field strength is maximum and therefore the kinetic energy due to this parallel velocity increases sharply in this region. The magnetic moment μ for both `Model` and `S1` remains nearly constant throughout the trajectory except at the ends where the Larmor velocity

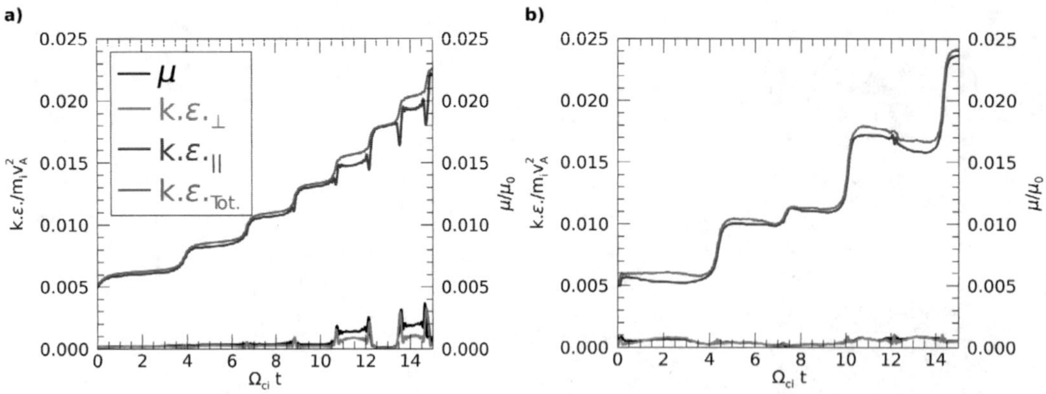

Figure 5. Magnetic moment and kinetic energy evolution profile from the test particle simulations a) `Model` and b) `S1`. The magnetic moment μ (black), kinetic energy due to perpendicular velocity $k.\varepsilon._{\perp}$ (red), kinetic energy due to parallel velocity $k.\varepsilon._{||}$ (blue) and the total kinetic energy $k.\varepsilon._{Tot.}$ are shown respectively in figures above.

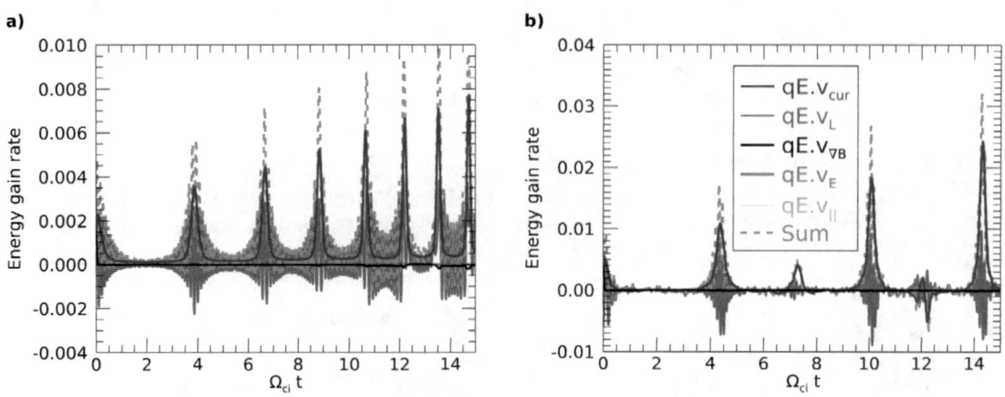

Figure 6. Electron's energy gain rate corresponding to the different velocity terms described by equation (13) from the test particle simulations a) `Model` and b) `S1`. The contributions of various velocity terms to energy gain, in descending order of their magnitude, are as follows: curvature drift (blue), Larmor velocity (green) and gradient drift (black). The $\mathbf{E}\times\mathbf{B}$ drift (red) and parallel velocity (orange) have zero magnitude and sum of all these terms is represented by Sum (dashed magenta).

varies sharply.

Under the guiding center limit, the electron moving with velocity $\mathbf{v}$ gains energy through the work done by the electric field $\mathbf{E}$. The rate of energy gain is given by

$$\frac{dW}{dt} = q\mathbf{E}.(\mathbf{v}_{||} + \mathbf{v}_L + \mathbf{v}_E + \mathbf{v}_{cur} + \mathbf{v}_{\nabla B}) \tag{13}$$

Figure 6 shows the rate of energy gain of an electron obtained from the test particle simulations a) `Model` and b) `S1`. The energy gain rate is expressed in the unit of $ev_A^2 B_0/c$. The electric field for both `Model` and `S1` is out-of-plane along z-direction. Hence, the $q\mathbf{E}.\mathbf{v}_{||}$ and $q\mathbf{E}.\mathbf{v}_E$ term vanishes (magnitude is zero in figure 6) as both $\mathbf{v}_{||}$ and $\mathbf{v}_E$ are perpendicular to $\mathbf{E}$. The curvature

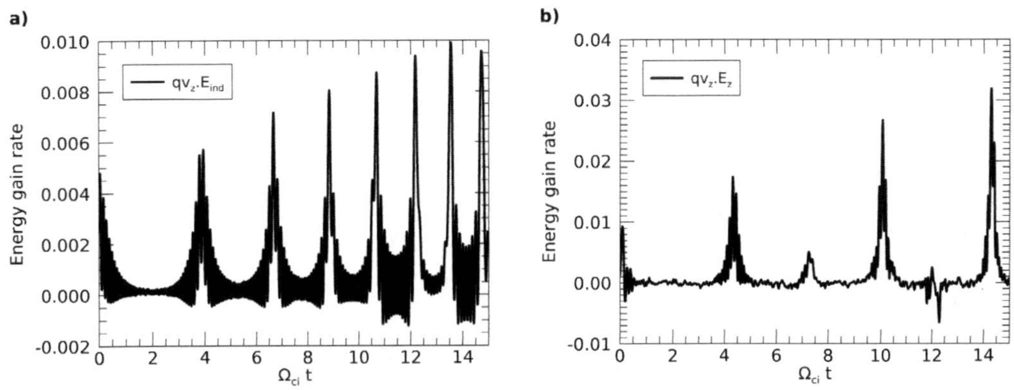

Figure 7. Electron's energy gain rate as a function of time from the test particle simulations a) `Model` and b) `S1`.

drift term dominates the electron's energy gain. Although gradient drift term is negligible, it decreases the electron's energy. The Larmor velocity also contributes in net increase of electron's energy. Therefore, their sum results in total energy gain which can also be seen from figure 7. The energy gain rate by the electron moving with velocity $\mathbf{v}$ as the function of time is shown in figure 7. The rate of energy gain can be rewritten in the form

$$\frac{dW}{dt} = q\mathbf{E}.\mathbf{v} = qE_x v_x + qE_y v_y + qE_z v_z \quad (14)$$

Since the electric field is out-of-plane along z-direction, only the last term i.e. $qE_z v_z$ in equation (14) contributes the electron's energy gain. The resemblance of energy gain rate due to various velocity terms between the `Model` and `S1` in figure 6 and figure 7 suggests that the out-of-plane electric field $\mathbf{E}_z$ is dominated by the inductive electric field generated by the contraction of the magnetic island and the energy gain is achieved through the work done by out-of-plane induced electric field.

5. Conclusion

The contraction of the magnetic island in the magnetic reconnection is modeled approximating an elliptical magnetic island together with an out-of-plane induced electric field. In the model, the particle's motion is nearly adiabatic and the released magnetic energy is converted into the kinetic energy of particles via the work done by the inductive electric field. The model predicts that the energy gain process is dominated by the curvature drift term and the prediction is verified by the test particle simulations on the fields generated by 2.5D particle-in-cell simulation showing electron acceleration via the contraction of a magnetic island.

6. Acknowledgments

The authors acknowledge partial support by NSF CAREER award 2144324 and NASA Heliophysics ECIP funding No. 80NSSC19K1106. The simulations were supported by the NASA High-End Computing (HEC) Program through the NASA Advanced Supercomputing (NAS) Division at Ames Research Center.

7. References

[1] Kopp R A and Pneuman G W 1976 *Sol. Phys.* **50** 85
[2] Phan T D et al. 2000 *Nature* **404** 848
[3] Lin R P et al. 2003 *ApJ* **595** L69
[4] Giannios D, Uzdensky, D A, and Begelman M C 2009 *MNRAS* **395** L29
[5] Hoshino M and Lyubarsky Y 2012 *Space Sci. Rev.* **173** 521
[6] Zhang H, Li X, Guo F and Giannios D 2018 *ApJ* **862** L25
[7] Ji H, Yamada M, Hsu S and Kulsrud R 1998 *Phys. Rev. Lett.* **80** 3256
[8] Chien A et al. 2023 *Nat. Phys.* **19** 254
[9] Stenzel R L and Gekelman W 1984 *Adv. Space Res.* **4** 459
[10] Øieroset M, Lin R P, Phan T D, Larson D E and Bale S D 2002 *Phys. Rev. Lett.* **89** 195001
[11] Krucker S, Hudson H S, Glesener L, White S M, Masuda S, Wuelser J P and Lin R P 2010 *ApJ* **714** 1108
[12] Oka M et al. 2018 *Space Sci. Rev.* **214** 82
[13] Egedal J, Daughton W and Le A 2012 *Nat. Phys.* **8** 321
[14] Uzdensky D A, Cerutti B and Begelman M C 2011 *ApJ* **737** L40
[15] Drake J F, Swisdak M, Che H and Shay M A 2006 *Nature* **443** 7111
[16] Che H and Zank G P 2019 *J. Phys.: Conf. Series* **1332** 012003
[17] Majeski S and Hantao J 2023 *Phys. Plasmas* **30** 042106
[18] Che H and Zank G P 2020 *ApJ* **889** 11
[19] Che H, Zank G P and Benz A O 2021 *ApJ* **921** 135
[20] Zank G P, le Roux J A, Webb G M, Dosch A and Khabarova O 2014 *ApJ* **797** 28
[21] Dahlin J T, Drake J F and Swisdak M 2014 *Phys. Plasmas* **21** 092304
[22] Li X, Guo F, Li Hui and Li G 2017 *ApJ* **843** 21
[23] Li X, Guo F, and Li Hui 2019 *ApJ* **879** 5
[24] Guo F et. al 2019 *ApJL* **879** L23
[25] Fermi E 1949 *Phys. Rev.* **75** 1169
[26] Guidoni S E, DeVore C R, Karpen J T and Lynch B J 2016 *ApJ* **820** 60
[27] Dahlin J T 2020 *Phy. Plasmas* **27** 100601
[28] Daughton W, Roytershteyn V, Karimabadi H, Yin L, Albright B J, Bergen B and Bowers k J 2011 *Nat. Phys.* **7** 539
[29] Zeller A, Biskamp D, Drake J F, Rogers B N, Shay M A and Scholer M 2002 *J. Geophys. Res.: Space Phys.* **107** 1230
[30] Boris J P 1970 Relativistic plasma simulation-optimization of a hybrid code *Proceedings of the 4th Conference on Numerical Simulation of Plasmas* ed Boris J P and Shanny R A (Washington D.C.: Naval Research Laboratory) pp 3-67

Journal of Physics: Conference Series **2742** (2024) 012016 doi:10.1088/1742-6596/2742/1/012016

Peculiarities of charged particle kinetics in spherical plasma

V I Kolobov[1,2], R R Arslanbekov[2] and H Liang[1]

[1]The University of Alabama in Huntsville, Huntsville, AL 35899, USA
[2]CFD Research Corporation, Huntsville, AL 35806, USA

vladimir.kolobov@uah.edu

Abstract. We describe transient kinetic simulations of partially ionized weakly-collisional plasma around spherical bodies absorbing or emitting charged particles. Numerical solutions of kinetic equations for electrons and ions in 1D2V phase space are coupled to an electrostatic solver using the Poisson equation or quasineutrality condition for small Debye lengths. The formation of particle groups and their contributions to electric current flow and screening of charged bodies by plasma are discussed for applications to Langmuir probes and solar wind.

1. Introduction

Problems associated with screening spherical bodies in plasmas (Langmuir probes, dust particles, satellites in the ionosphere, solar wind, etc.) are remarkably similar. In all these problems, one type of charged particle gets trapped near an electrically charged body by the electric fields, which are induced by the body and self-generated by the surrounding plasma. Due to plasma screening, the trapped particles modify the electric field profile around the body compared to the Coulomb profile (in a vacuum). Dust particles in plasma are charged negatively, and the trapped particles are positive ions [1]. Sun is charged positively, and the trapped particles are electrons [2]. In the case of Langmuir probes, both situations are possible [3]. The floating potential of the probe corresponds to a negatively charged probe, ensuring net zero current to the surface and ion trapping around the probe.

Thermo- and photo-electron emission often occurs from probes and dust particles. Sun emits electrons and ions, and their fluxes must be balanced in the dynamic quasi-steady state of ambient solar wind. Kinetic equations for collisionless non-magnetized plasma have striking similarities with those for magnetized solar wind plasma, especially for the magnetic field of a magnetic monopole. However, we note an essential difference between unmagnetized and magnetized spherical plasma. In the first case, trapped particles circle the sphere on closed orbits around the force center. Trapped particles bounce radially along the magnetic field lines in the second case.

The kinetic theory of plasma around satellites [3] and the kinetic theory of ambient solar wind [4] have been developed. However, their implementation for practical simulations of probes under different conditions is still missing, and relations between the kinetic and commonly used MHD models for the solar wind are unclear. Effects of collisions on particle trapping and screening of charged objects in plasma remain a subject of active research [5,6]. Several recent papers have emphasized the importance of the electric fields in the solar wind for explaining the measured electron velocity distribution functions [6.7], which usually contain three electron groups: core, strahl (beam), and halo. The latest paper [8] shows that the halo could form from the strahl without scattering due to electron reflection by the ambipolar electric field.

Journal of Physics: Conference Series **2742** (2024) 012016 doi:10.1088/1742-6596/2742/1/012016

This paper describes a kinetic model of spherical plasma coupling grid-based kinetic solvers for electrons and ions in phase space with electrostatic solvers for the electric field. We use a Poisson and an iterative solver for the electric field based on the quasineutrality condition. The latter is crucial for the solar wind due to the tiny value of the Debye length close to the Sun compared to the spatial scale of interest (about one AU). We illustrate the current capabilities of our model for applications to Langmuir probe and solar wind. We also introduce a quasi-diffusion kinetic model for trapped electrons, which could be a part of a future hybrid code for more efficient treatment of different electron groups.

2. Kinetic model of spherical plasmas

Assuming spherical symmetry, we reduce the general 3D3V phase space to a 1D2V phase space, which consists of the radial distance r, the magnitude of velocity v, and the cosine of the angle between the velocity and the radial direction μ. The kinetic equation in the conservative form has the form [9]:

$$\frac{\partial Y}{\partial t}+\frac{\partial}{\partial r}(\mu v Y)+\frac{\partial}{\partial \mu}\left\{\left[\frac{v(1-\mu^2)}{r}+\frac{qE}{m}\frac{1-\mu^2}{v}\right]Y\right\}+\frac{\partial}{\partial v}\left[\left(\frac{qE}{m}\mu\right)Y\right]=S(r,v,\mu), \quad (1)$$

where $Y=r^2v^2f$, $f=f(r,v,\mu)$ is the phase space density, q and m are the charge and mass of the given species, respectively, E is the electric field, and $S(r,v,\mu)$ is a collision operator describing different types of collisions and wave-particle interactions [10]. The electrostatic potential ϕ or the normalized potential energy $U=\frac{e\phi}{k_BT_e}$ is usually found from the Poisson equation:

$$\nabla^2\phi=\frac{e}{\varepsilon_0}(n_e-n_i) \Longrightarrow \tilde{\nabla}^2 U=\frac{r_0^2}{\lambda_{De}^2}(\tilde{n}_e-\tilde{n}_i) \quad (2)$$

where $\tilde{n}_e$ and $\tilde{n}_i$ are the normalized electron and ion densities. We normalize the kinetic equation with a typical length, r_0, typical velocity v_0 (the thermal speed v_{th}), and typical time scale t_0, and use the Poisson equation for the normalized electric field $E_0=4\pi e n_0 r_0$, where n_0 is a typical plasma density.

There are two challenges to the numerical solution of this system. First, solving the Poisson equation in the limit of small Debye length (at $r_0 \gg \lambda_D$) becomes a bottleneck. The formation of a thin sheath with a strong electric field near the surface makes using fine physical mesh and short time steps necessary due to the CFL limits in phase space. These factors lead to prohibitive memory and runtime requirements for code stability and convergence, requiring another method for calculating electric fields. Second, the electron and ion mass disparity makes our explicit solvers very expensive during the transient process (when $\tilde{n}_e \neq \tilde{n}_i$) due to CFL limitations. This challenge can be resolved using an artificially small mass ratio, which does not substantially affect the steady solution [11]. The method is based on the fact that in steady-state collisionless electrostatic plasma, the solution with a reduced mass ratio can be scaled to the one for the actual mass ratio. It has been successfully used for speed-limited particle-in-cell (SLPIC) simulations [11].

We solve the kinetic equations for electrons and ions using the same mesh in phase space, which can be dynamically adapted as described in [9]. The Poisson solver uses separate mesh and the Thomas algorithm to calculate the self-consistent electric field based on the number density of electrons and ions obtained from the phase space density. Below, we describe an iterative method of calculating electric fields for problems with small Debye lengths. This coupled kinetic solver for charged particles and the electric field applies to many spherical plasma problems. Below, we demonstrate its current capabilities for simulations of Langmuir probes and solar wind. We also introduce reduced kinetic models for trapped particles.

3. Langmuir Probes

Simulations of spherical probes were conducted at $\frac{T_i}{T_e} = 1$ for different $m_r = \frac{m_i}{m_e}$ and λ_{De}. Figure 1 shows the results for $m_r = 4$ and $\tilde{\lambda}_{De} = 0.02, 0.05, 0.1, \text{and } 0.2$ (from left to right). The electric potential $\tilde{\phi}(\tilde{r})$ is calculated self-consistently by the Poisson solver. The flux is reasonably consistent with the analytical predictions. The size of the space charge sheath is about $\tilde{\lambda}_{De}$. At $\tilde{\lambda}_{De} = 0.02$, the potential has oscillations near the left boundary, which is unphysical and indicates problems with the Poisson solver.

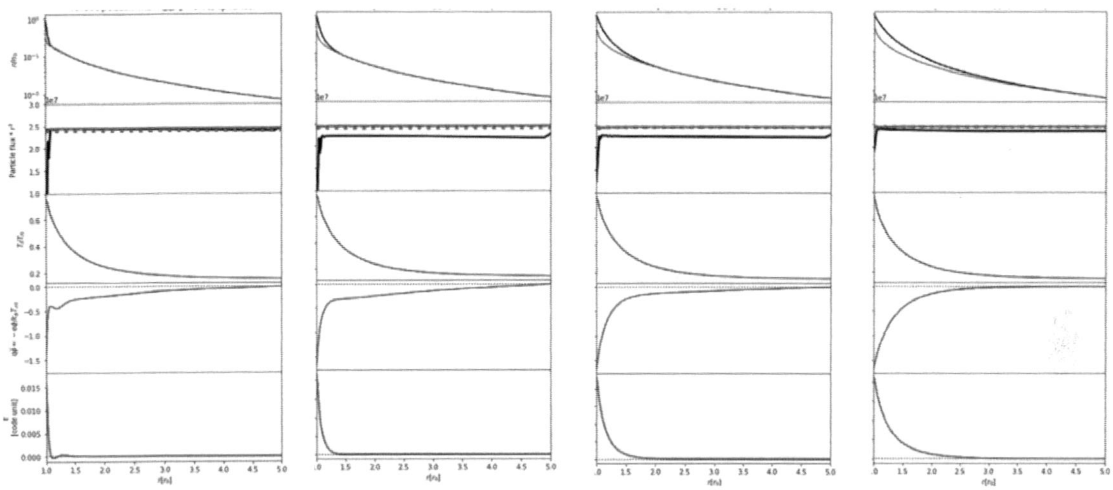

Figure 1 The profiles of density, the particle flux multiplied by r^2, temperature, electron potential energy, $-\frac{\tilde{\phi}(\tilde{r})}{\tilde{\lambda}_{De}^2}$, and electric field for $m_r = 4$ and different $\tilde{\lambda}_{De}$. The black (blue) solid line is for electrons (ions). For the particle flux, the green (red) dashed line shows the analytical solution for electrons.

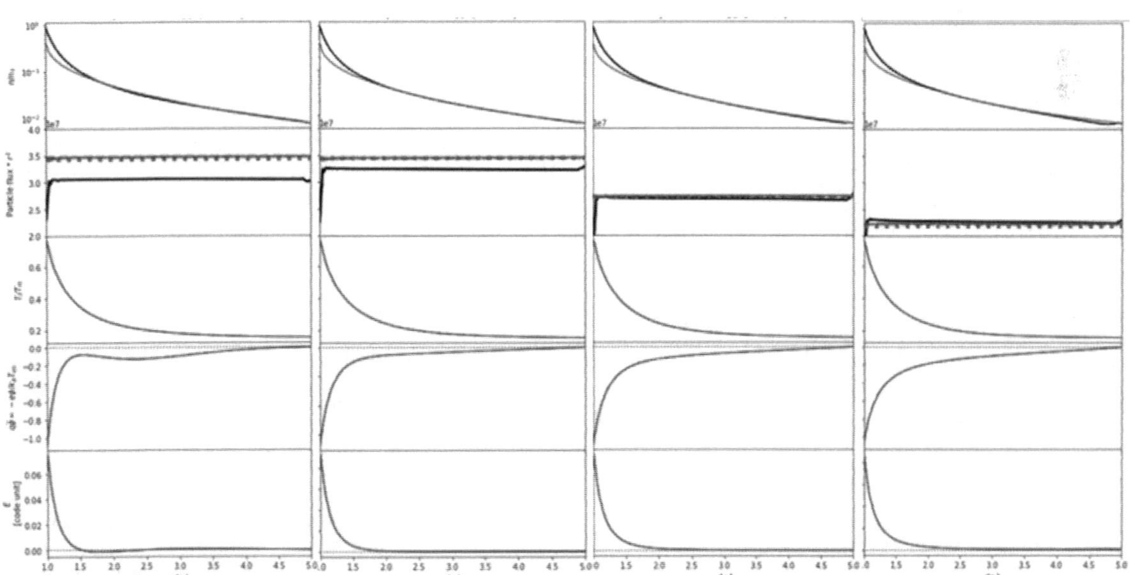

Figure 2 The profiles of density, the particle flux multiplied by r^2, temperature, electron potential energy, $-\frac{\tilde{\phi}(\tilde{r})}{\tilde{\lambda}_{De}^2}$, and the electric field at $\tilde{\lambda}_{De} = 0.1$ and different $m_r = 2, 4, 8$ and 16. The black (blue) solid line is for electrons (ions). For the particle flux, the green (red) dashed line shows the analytical solution for electrons (ions).

Journal of Physics: Conference Series **2742** (2024) 012016 doi:10.1088/1742-6596/2742/1/012016

Figure 2 shows the results for $\tilde{\lambda}_{De} = 0.1$ and $m_r = 2, 4, 8$ and 16 (from left to right). The calculated $\frac{\tilde{\phi}_L}{\tilde{\lambda}_{De}^2}$ is different for each case. The larger m_r, the easier the system reaches equilibrium. The flux is reasonably consistent with the analytical prediction for $m_r = 4, 8, 16$. The location $\tilde{r}_c$ where the potential nearly reaches its asymptotic behavior increases with increasing m_r. This is consistent with the kinetic theory of solar wind, which predicts $\tilde{r}_c \approx (m_i/m_e)^{1/4}$ [12]. Most of the electrons at $r < \tilde{r}_c$ are returning to Sun, while the electrons at $r > \tilde{r}_c$ include escaping electrons and those trapped on both sides due to the electric and magnetic mirror forces [11]. The trapped electron trajectories can only be filled up via collisions in steady simulations. In our transient simulations, trapped electrons appeared without collisions due to transient effects [13].

4. Kinetic Model of Solar Wind

The kinetic equation for charged particles in magnetized plasma coincides with Eq. (1) for a magnetic monopole [14]. In collisionless (exospheric) models, analytical solutions of the kinetic equations for electrons and ions can be obtained. For a Maxwellian distribution of injected particles at the boundary, the analytical solution gives for the electron and ion fluxes [2,12]. In our code, the potential on the right boundary ϕ_R was set to zero, and the potential on the left boundary ϕ_L, was determined to equalize the ion and electron fluxes in a steady state. The calculated ϕ_L was used as a boundary condition for the Poisson solver to find $\phi(r)$ at different λ_{De}.

This method of calculating the potential drop, $\Delta\varphi^{nocol}$, ensuring equal fluxes of electrons and ions assumes a collisionless plasma and does not consider numerical diffusion, transients, and other factors. Therefore, our code has implemented a more general method based on an interactive potential adjustment to equalize the electron and ion particle fluxes. The scheme uses an adjustable relaxation parameter α (which is proportional to the sphere capacitance), and the analytical value of $\Delta\varphi^{nocol}$ as an initial condition:

$$\Delta\varphi^{n+1} = \Delta\varphi^{n} + \alpha\Delta\varphi^{nocol}\frac{j_e^n - j_i^n}{|j_e^n + j_i^n|} \quad (3)$$

with $\Delta\varphi^{n=0} = \Delta\varphi^{nocol}$. This scheme allowed converging towards a solution with equal electron and ion particle fluxes.

We first obtained 1D2V solutions assuming constant ion velocity ($v_i = const$) and the ion density profile $n_i(r) = n_0(R/r)^2$. Boldyrev et al. [12] mentioned that the ion velocity does not change considerably the electric potential profile in the solar wind. Figure 3 illustrates the electron-trapping potential and the EVDF obtained in our simulations. The results show that the EVDF for most electrons is close to isotropic, which justifies using a simpler model for these electrons.

We have implemented a model based on the two-term spherical harmonic expansion (SHE) in velocity space, which reduces the hyperbolic kinetic equation in 1D2V phase space to a parabolic diffusion-type equation for an isotropic part of the EVDF in the 1D1U phase space (U is kinetic energy) [15]. This model, below P1 for brevity, was coupled to an iterative solver for the self-consistent electric field. Figure 4 shows the radial profiles of solution variables and the corresponding Electron Energy Distribution Functions (EEDF's) (the isotropic part of EVDFs) at two locations obtained with the P1 model.

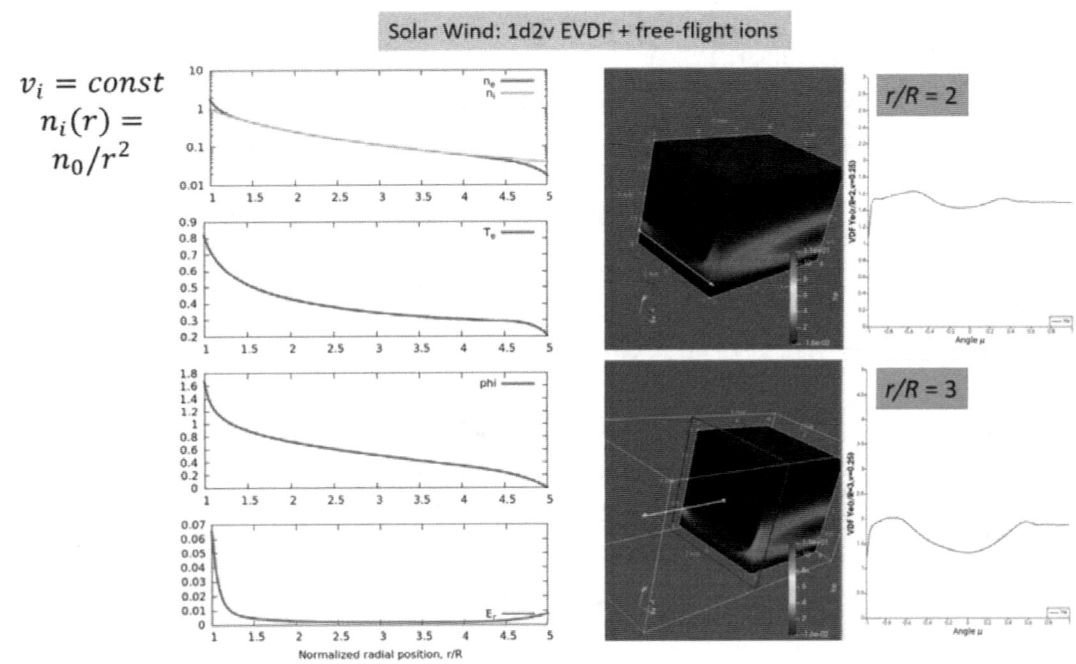

Figure 3. 1D2V solutions for Debye length-to-radius ratio $\lambda_{De}/R = 0.1$. Particle densities, electron temperature, electrostatic potential, and electric field (left). EVDFs at two locations (right).

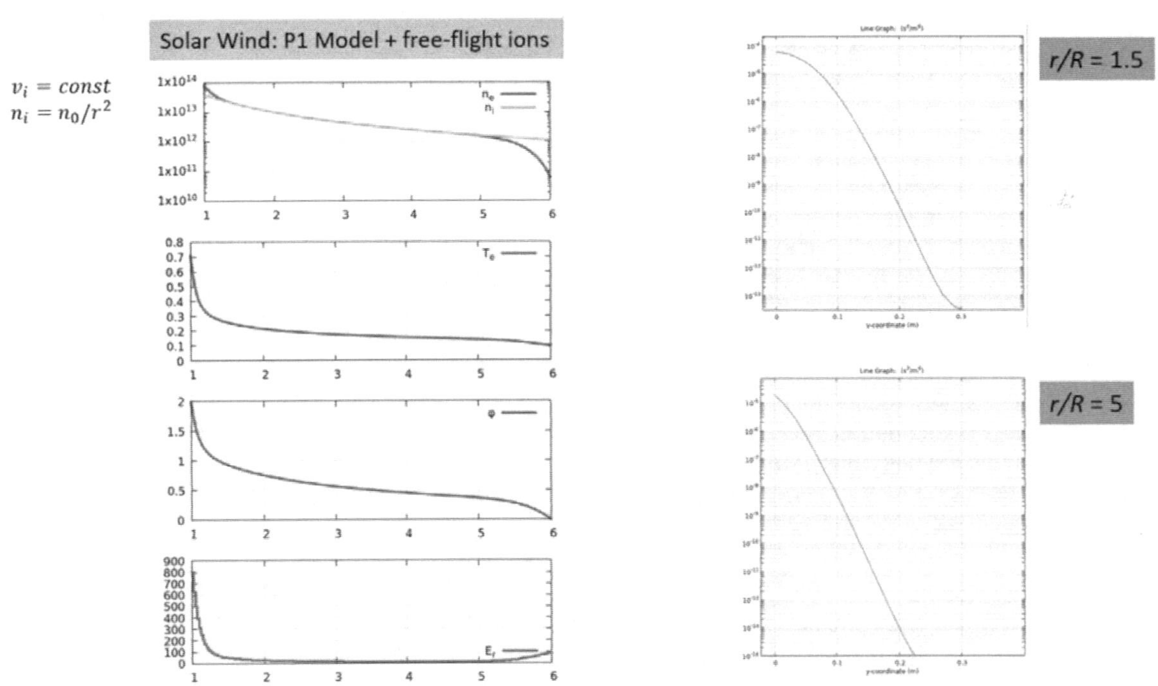

Figure 4. The P1 solutions for Debye length-to-radius ratio $\lambda_{De}/R = 0.1$. Particle densities, electron temperature, electrostatic potential, and electric field (left). EEDFs at two locations (right).

Figure 5 compares the results of 1D2V and P1 models. The trapping potential for electrons forms in both cases. The decrease of electron temperature from the Sun's surface (predicted in [12]) is well reproduced by the P1-model code, with very similar (normalized) electron temperature magnitudes. The electrostatic potential and electric field predicted with the 1D2V and P1 models agree quantitatively.

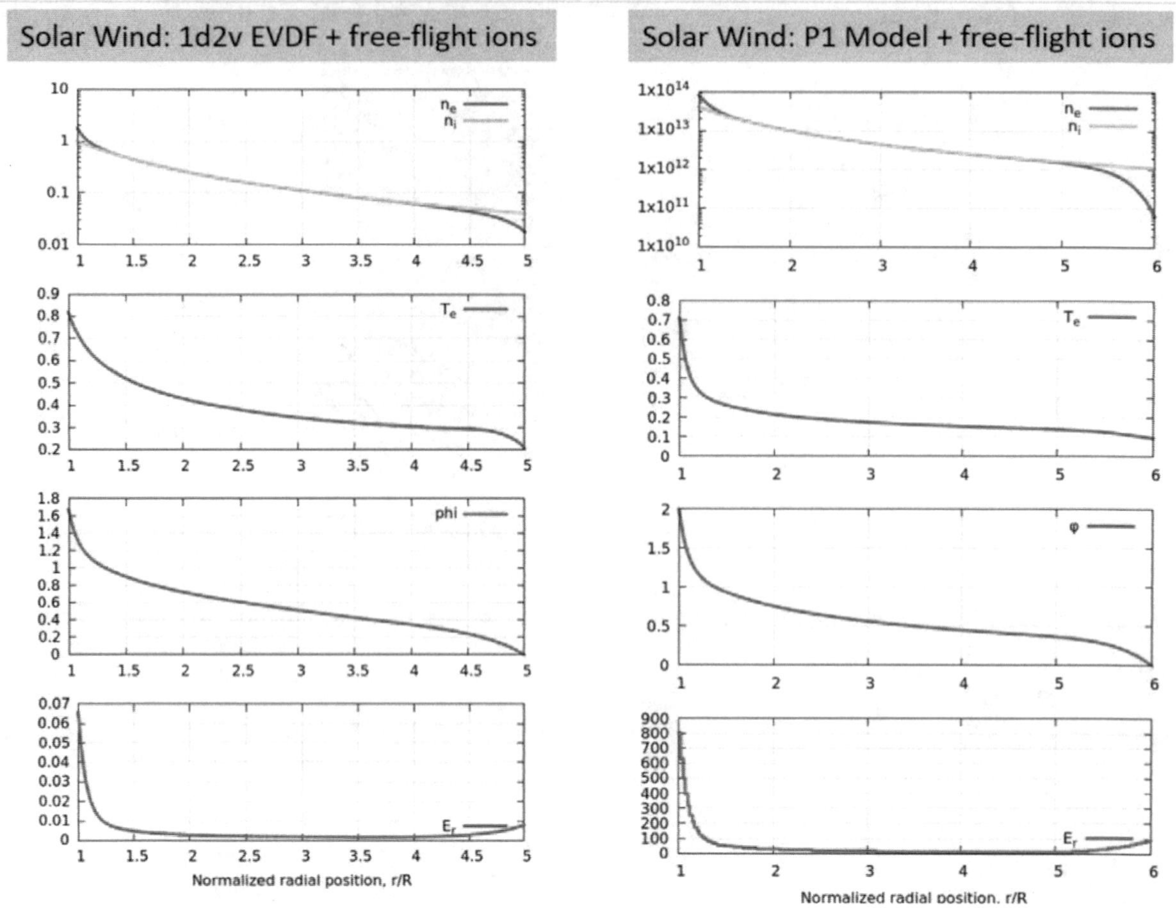

Figure 5. The 1D2V (left) and P1-model (right) solutions for Debye length-to-radius ratio $\lambda_{De}/R = 0.1$.

The overall agreement between the results of the 1D2V and P1 models is encouraging. The P1 model simulates returning electrons at a significantly reduced cost compared to the 1D2V model. The discrepancies in magnitudes of plasma density (top panels) and electric field (bottom panels) appear because our 1D2V code operates in normalized units, and the P1 model currently uses SI units. The magnitudes are relatively close in the same units. More detailed comparisons (e.g., for the particle and energy fluxes controlled by runaway anisotropic electrons) are required to validate the P1 model further for a broader range of conditions.

To enable calculations for smaller Debye lengths, we implemented an iterative scheme for computations of (steady-state) electric potential based on the paper [16]. We used two schemes:

$$\varphi^{n+1}(r) = \varphi^n(r) + \alpha\varphi^{nocol}(r)\frac{n_e^n(r)-n_i^n(r)}{|j_e^n(r)+j_i^n(r)|} \tag{4}$$

or

$$\varphi^{n+1}(r) = \varphi^n(r) + \alpha\varphi^{nocol}(r)ln[n_e^n(r)/n_i^n(r)] \tag{5}$$

Approximate analytic solutions for the plasma potential $\varphi^{nocol}(r)$ were applied as initial conditions to speed up convergence. The relaxation parameter α could be a function of r.

This Quasi-Neutral Field Solver (QNFS) allowed the formation of well-converged quasi-neutral solutions for low-resolution meshes in phase space without resolving space charge sheath. We have used a radially varying weight function for the relaxation parameter and verified that solutions are Debye-length independent. Figure 6 shows an example of calculations with a coarse static grid.

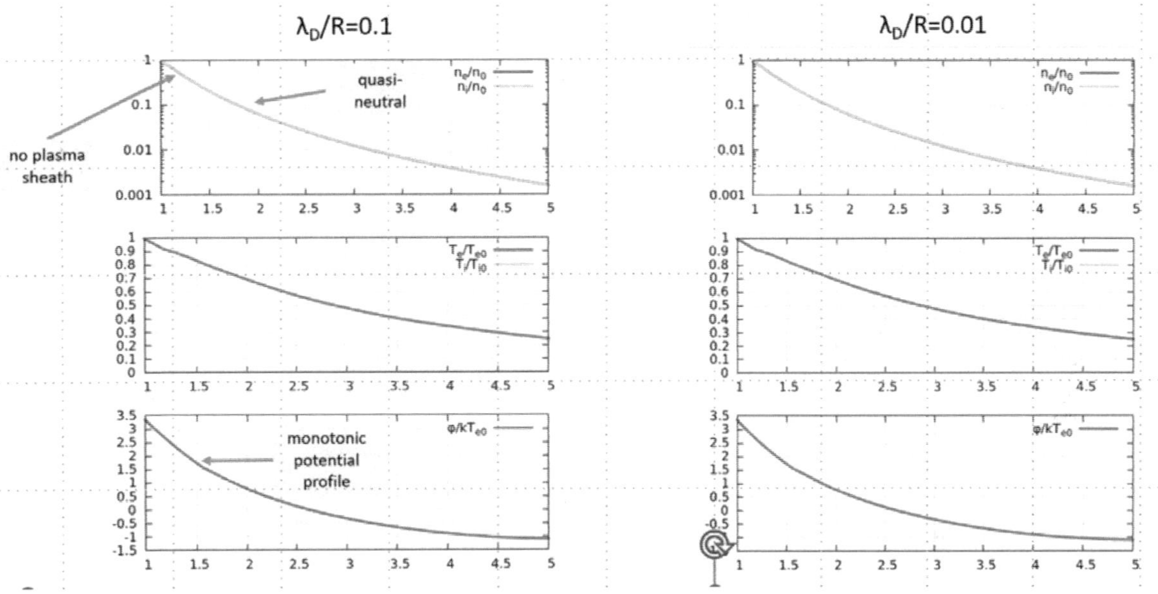

Figure 6 Converged spatial distributions for different plasma densities (Debye lengths).

We have enhanced QNFS for dynamically varying phase space mesh. Figure 7 shows a well-converged solution with Adaptive Mesh Refinement (AMR) in phase space (baseline grid + 1-level-up refinement). The dynamic AMR allows convergence on a coarse grid and then applies the iterative QNFS scheme. In these simulations, phase-space AMR was activated after some iteration steps (after detecting prior convergence), and this was the reason for the observed jumps on the left part of Figure 7. These jumps did not affect the steady-state solutions. With this method, we first observed somewhat noisy solutions for the electric field and improved them using a noise-filtering technique. The QNFS module enables efficient computations under high plasma density conditions (extremely low Debye lengths), and we plan to expand the QNFS module further for simulations with ion transport and collisions.

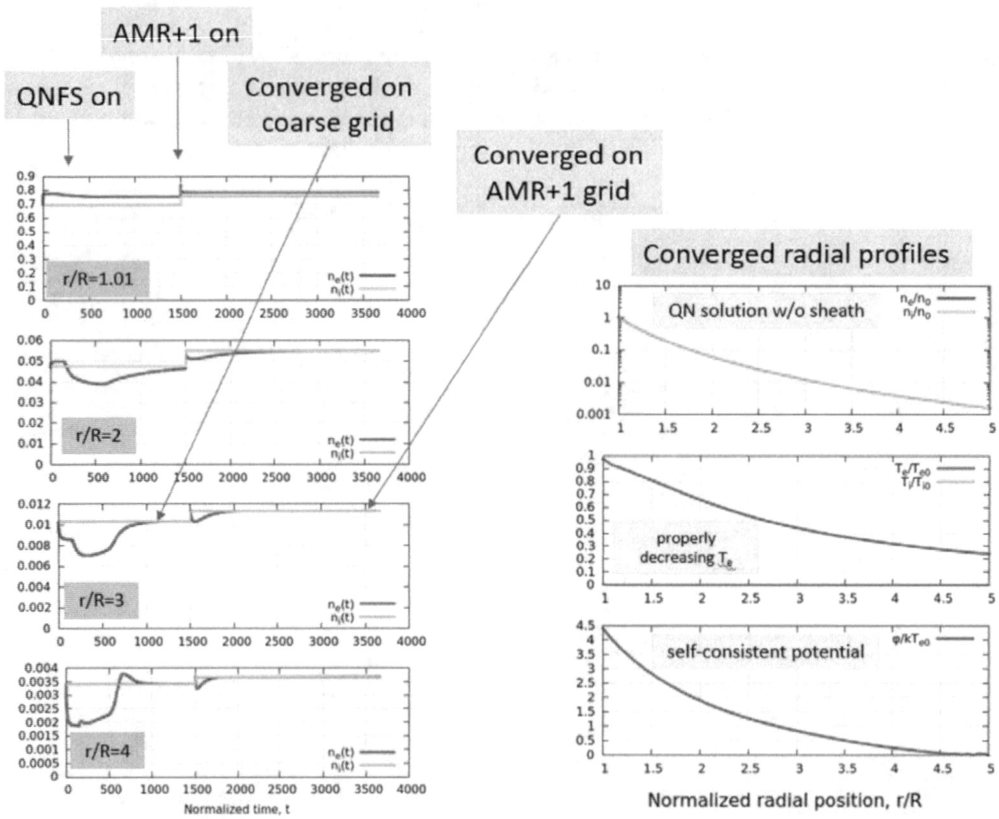

Figure 7. Electron density convergence history with QNFS at different distances from the surface (left) and converged radial plasma profiles (right).

5. A quasi-diffusion model for electrons

Quasi-diffusion models have been used for neutron, photon, and electron kinetics [17,18]. They have a broader applicability range than the P1 model described above. To evaluate the quasi-diffusion models proposed in [18], we have implemented a hybrid model of partially-ionized spherical plasma in COMSOL. The model uses a Fokker-Planck kinetic solver for the EEDF, a free-flight fluid model for ions, and the Poisson equation for the electric field. Electrons and ions are injected from an internal sphere and exist through the external sphere. Elastic collisions of electrons with neutral gas atoms and Coulomb interactions among electrons are considered. Self-consistent distributions of the electric potential were found for different values of the potential drop between the spheres by solving a transient problem until a steady state. We observed that steady-state electron flux was reduced by varying the potential drop from 1.5 V to 3 V.

Below, we illustrate the model results for spheres with 1 and 30 cm radius in Argon gas at a pressure of 1 mTorr. The electrons and ions are injected with densities 10^{18} cm^{-3} and equal temperatures of 1eV, corresponding to the Debye length at the sphere location of 0.007 mm. Figure 8 shows the spatial distributions of the electron density and temperature and the EEDF at a specific spot for different values of the Coulomb collision frequencies at the assumed potential drop of 1.5 V between the spheres.

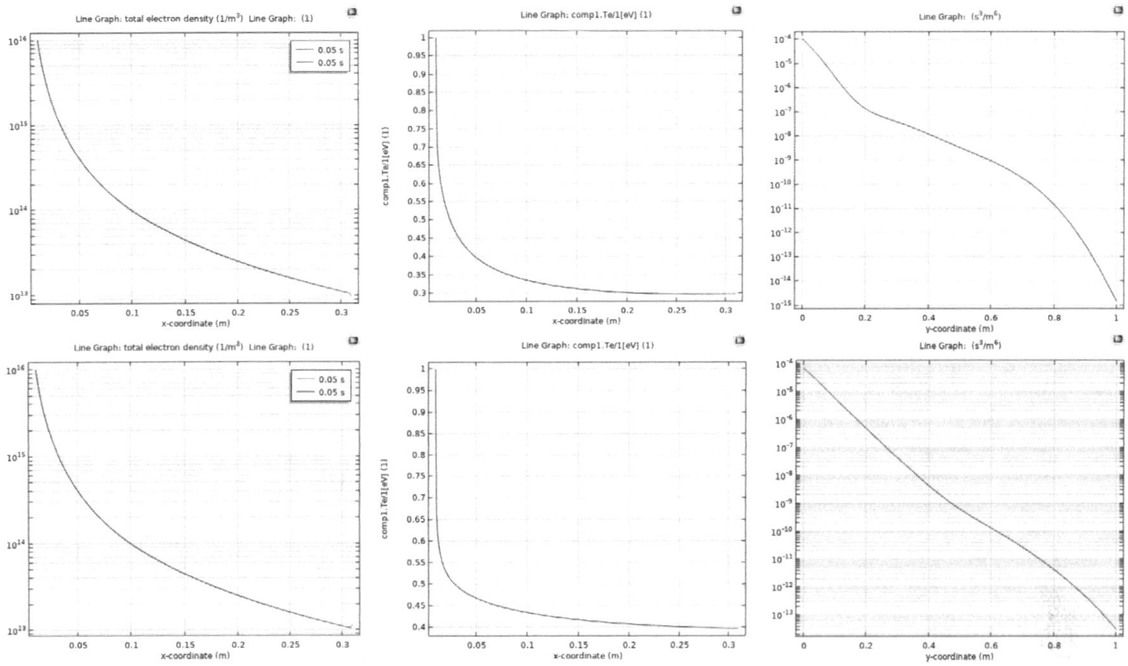

Figure 8 The spatial distributions of the electron density and temperature and the EEDF at a specific location for different values of the Coulomb collision frequencies: $\nu_{ee} = 0$ *(top) and* $\nu_{ee} \neq 0$ *(bottom).*

Figure 9 illustrates the effects of the Coulomb collisions on the spatial distributions of the electric potential. Minimal changes occur when Coulomb collisions are taken into account.

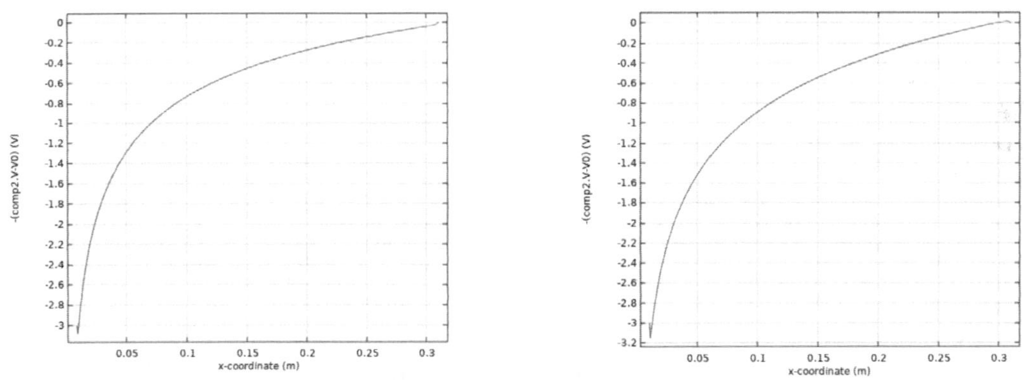

Figure 9 The spatial distribution of the potential without (left) and with Coulomb collisions (right).

To summarize, partially-ionized spherical plasma with densities up to 10^{18} m^{-3} could be modeled with COMSOL using a Poisson solver. For the considered test of the quasi-diffusion model, a potential well trapping most electrons formed and $T_e(r)$ dropped in a very similar fashion to full 1D2V results (from 1 to ~0.25). The EEDF of trapped electrons remained close to a Maxwellian (following the injected EEDF) even without Coulomb collisions—however, a significant "hot" tail with a temperature higher than the injected temperature formed. Coulomb collisions resulted in the transition to a Maxwellian EEDF with the same "temperature" for all electron energies, i.e., heating the cold, trapped electrons by the hot tail electrons. The spatial distributions of electron density $n(r)$ and $T_e(r)$ remained practically unchanged. This behavior indicates a proper numerical scheme, as Coulomb collisions among electrons should not affect the EEDF moments. A simple version of the quasi-diffusion model used in these simulations appears promising.

Journal of Physics: Conference Series **2742** (2024) 012016 doi:10.1088/1742-6596/2742/1/012016

6. Conclusions

We have implemented a kinetic model of weakly collisional plasma around spherical bodies absorbing or emitting charged particles. Grid-based numerical solutions of kinetic equations for electrons and ions in 1D2V phase space have been coupled to a Poisson solver or an iterative solver enforcing the quasineutrality condition for small Debye lengths. The formation of particle groups and their contributions to electric current flow and screening of charged bodies by plasma have been studied for applications to Langmuir probes and solar wind. We plan to implement a hybrid model for electrons in future work, integrating a 1D2V solver with a 1D1U quasi-diffusion solver for more efficient simulations of different electron groups.

Acknowledgments

This work was supported by the NSF EPSCoR project OIA-2148653 and NASA SBIR project 80NSSC-21-C-0527.

References

[1] Kiselyov A A, Dolgonosov M S and V. L. Krasovsky V L 2015 Formation of trapped-ion population in the process of charging of an absorbing sphere in a collisionless plasma, *Europhys. Lett.* **111** 15001
[2] Meyer-Vernet N 2007 *Basics of the Solar Wind*, Cambridge University Press
[3] Al'pert Ya L, Gurevich A V and Pitaevskii L P 1965 *Space Physics with Artificial Satellites* (Plenum, New York)
[4] Lemaire J and Scherer M 1973 Kinetic Models of the Solar and Polar Winds, *Reviews of Geophysics* **11** 427
[5] Filippov A V 2021 Distribution of Electrons and Ions Near an Absorbing Spherical Body in a Nonequilibrium Plasma, *Journal of Experimental and Theoretical Physics* **132** 148
[6] Berčič L, Landi S, and Maksimović M 2021 The Interplay Between Ambipolar Electric Field and Coulomb Collisions in the Solar Wind Acceleration Region, *Journal of Geophysical Research: Space Physics* **126** e2020JA028864
[7] Halekas J S, et al. 2022 The Radial Evolution of the Solar Wind as Organized by Electron Distribution Parameters, *The Astrophysical Journal* **936** 53
[8] Horaites K and Boldyrev S 2022, The heliospheric ambipolar potential inferred from sunward-propagating halo electrons, *MNRAS* **515** 5135
[9] Kolobov V I, Arslanbekov R R and Levko D 2029 Boltzmann-Fokker-Planck kinetic solver with adaptive mesh in phase space, *AIP Conference Proceedings* **2132** 060011
[10] Tang B, Zank G P and Kolobov V I 2022 Numerical Modeling of Suprathermal Electron Transport in the Solar Wind: Effects of Whistler Turbulence with a Full Diffusion Tensor, *ApJ* **924** 113
[11] Werner G R, Robertson S, Jenkins T G, Chap A M and Cary J R 2022 Accelerated steady-state electrostatic particle-in-cell simulation of Langmuir probes, *Physics of Plasmas* **29** 013502
[12] Boldyrev S, Forest C and Egedal J 2020 Electron temperature of the solar wind, *Proceedings of the National Academy of Sciences* **117** 9232
[13] Sánchez-Arriaga G 2013 A direct Vlasov code to study the non-stationary current collection by a cylindrical Langmuir probe, *Phys. Plasmas* **20** 013504
[14] Kolobov V I, Arslanbekov R R and Levko D 2020 Electron Groups in Solar Wind and Gas Discharge Plasmas, *J. Phys.: Conf. Ser.* **1623** 012006
[15] Humphrey N A and Kolobov V I 2023, Electron kinetics in a positive column of AC discharges in a dynamic regime, *Plasma Sources Sci. Technol.* **32** 085017
[16] Meyer-Vernet N and Issautier K 1998, Electron temperature in the solar wind: Generic radial variation from kinetic collisionless models, *JGR Space Physics* **103** 29705
[17] Gol'din V Ya 1964 A quasi-diffusion method of solving the kinetic equation, *USSR Computational Mathematics and Mathematical Physics* **4** 136
[18] Koterov V N and. Shcheprov A V 1991 Calculation of spatially nonequilibrium transport of electrons in a gas-discharge plasma, *U.S.S.R. Comput. Math. Math. Phys.* **31** 56

Journal of Physics: Conference Series **2742** (2024) 012017 doi:10.1088/1742-6596/2742/1/012017

Numerical modeling of non-equilibrium neutrino radiation field by solving quantum kinetic equation

Hiroki Nagakura
Division of Science, National Astronomical Observatory of Japan, 2-21-1 Osawa, Mitaka, Tokyo 181-8588, Japan

E-mail: hiroki.nagakura@nao.ac.jp

Abstract. Numerical modeling of neutrino quantum kinetics is a new frontier in the study of core-collapse supernova (CCSN) and binary neutron star merger (BNSM). The noticeable feature in the quantum kinetics is neutrino flavor conversion driven by neutrino self-interactions. Although there remain many unresolved issues in their non-linear properties of flavor conversions, rapid progress has been made in the last few years. In this article, we introduce the current status for the study of collective neutrino oscillations, which are representative characteristics in self-induced flavor conversions, and then we show our recent results: global quantum kinetic simulations in CCSN and BNSM environments.

1. Introduction

Neutrinos are the most mysterious elementary particles among those known to exist in our universe. There are comelling experimental evidences that the intriguing particle can undergo flavor conversions or oscillations, representating one of the quantum kinetic features beyond the standard model of particle physics. The flavor conversion implies that the eigenstates between flavors and masses are different from each other. Determining neutrino masses, mixing angles, and nature (Dirac or Majorana) are important but remaining fundamental issues in not only neutrino physics but also cosmology and even astrophysics. In fact, the neutrino oscillation was a key process to solve the solar neutrino problem, and nowadays it is highly likely that they also play pivotal roles in high-energy astrophysical phenomena such as core-collapse supernova (CCSN) and binary neutron star merger (BNSM). Neutrino quantum kinetics in these explosive phenomena are the subject in this article.

During formations of CCSN and BNSM, matter density can exceed nuclear saturation density ($\rho \sim 10^{14}\mathrm{g/cm}^3$, where ρ denotes the mass density of baryons) and its temperature can also reach $> 10^{11}$K. In such extremely hot and dense environments, neutrinos can be produced abundantly through various weak interactions. Since neutrinos are the least interacting particles, they are the dominant lepton- and energy carriers in the vicinity of proto-neutron star (NS) or black hole (BH). This exhibits that different fluid elements can exchange their energy and leptons via neutrinos, which has an influence on fluid dynamics and nucleosynthesis. It should also be stressed that neutrinos are observable particles, in which key information on microphysical properties in high density environments (equation-of-state, neutrino-matter interactions, and nuclear reactions) are imprinted. Developing accurate models of neutrino kinetics is, hence, mandatory for the study of CCSN and BNSM.

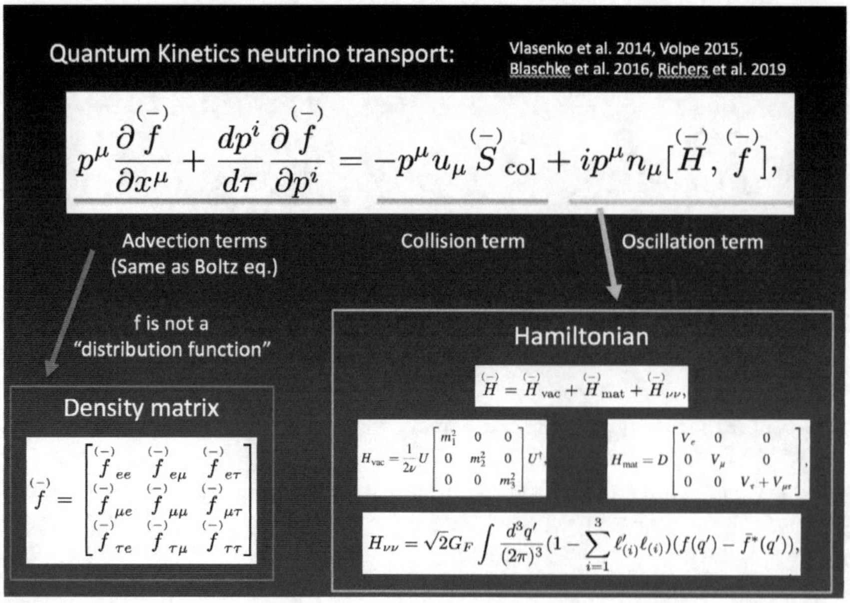

Figure 1. Quantum kinetic equation for neutrinos.

During the last decades, neutrino-radiation hydrodynamic simulations have been a frontier in computational astrophysics (see [1] as a recent review), which is due to the complex, multi-physics, and multi-dimensional natures of CCSN and BNSM physics. It has been expected that neutrino kinetics in these phenomena can be fully determined by solving Boltzmann equation. In this theory, the distribution function of neutrinos provides all fundamental quantities to describe neutrino kinetics including neutrino-matter interactions. Although their direct numerical simulations are computationally expensive, the numerical method and implementations on massive parallel supercomputer facilities have been already established (see, e.g., [2, 3]). It is worthy to note that the first axisymmetric CCSN simulations with full Boltzmann neutrino transport in our group was carried out in 2018 [4]. Nowadays, we performed the Boltzmann CCSN simulations in 3D [5] and with general relativistic treatment [6]. With improving other input physics (equation-of-state and neutrino-matter interactions; see, e.g., our recent review of [10]), the community, no doubt, keeps progressing towards first-principles CCSN models.

One thing we do notice here is, however, that neutrino flavor conversions induced by neutrino self-interactions have been neglected in CCSN and BNSM models. It is important to note that linear stability analyses of flavor conversions based on recent CCSN models have commonly suggested that neutrino-flavor conversions occur in post-shock regions, driven by fast mode [8, 7, 9] and collisional instability [11]. In BNSM environments, both instabilities would also be ubiquiouts [12, 13]. These facts suggest that neutrino kinetics should be modeled with taking into account flavor conversions, which can not be handled by Boltzmann equation. The most straightforward and simplest extension to include effects of flavor conversions is the mean-field quantum kinetic equation, or QKE (see, e.g., [14, 15]). The QKE has abilities to capture essential features of various flavor instabilities, including fast mode and collisional instability. This has motived us to develop QKE solvers for modeling of neutrino radiation fields in CCSNe and BNSMs.

Although there are many previous works solving QKE, they mostly focused on local properties of flavor conversions. This is due to the fact that there is a large disparity of scales between

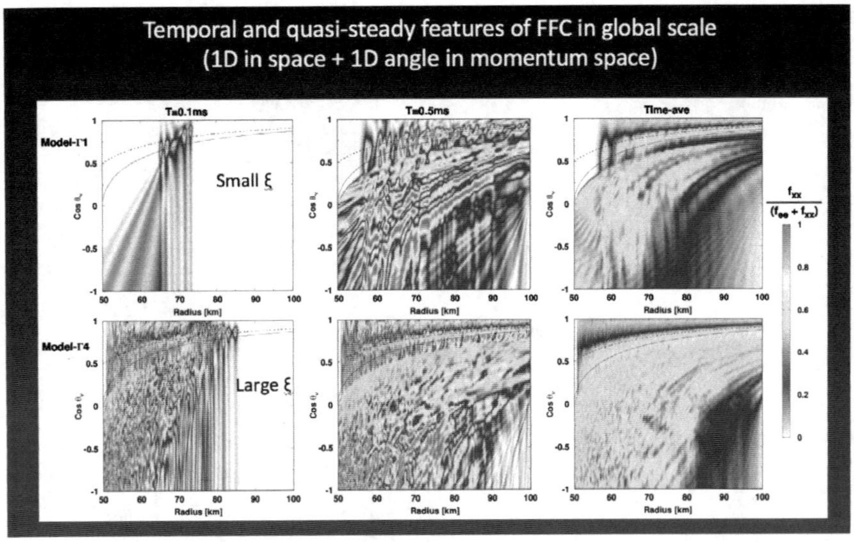

Figure 2. Spherically symmetric FFC simulations. The top and bottom panels show the results of the same simulations but with different attenuation parameters. The bottom one represents a more realistic result than that in the top panels. See the text for more details.

neutrino flavor conversions and astrophysical phenomena, if neutrino flavor conversions are driven by neutrino self-interactions. This is the subject of this article. Although these studies focusing on local properties are very important, radiation fields in non-linear phases would be qualitatively different from those in CCSN and BNSM. This is because the global advection of neutrinos characterizes the neutrino angular distributions and energy spectram. This motivated us to develop a new code: GRQKNT (General-Relativistic Quantum-Kinetics Neutrino Transport) [16], in which we solve general relativistic QKE with multi-energy, multi-angle, three-flavor frameworks, and essential neutrino-matter interactions (or collision terms). We also applied the code to understand global properties of neutrino radiation fields accompanied by fast neutrino flavor conversion (FFC). Below, let us at first describe the essence of GRQKNT code, and then we will present some essential findings in our recent studies.

2. GRQKNT

A new general-relativistic quantum-kinetics neutrino transport code, GRQKNT, was developed, aiming at studying global neutrino radiation field with neutrino oscillations in CCSN and BNSM environments. The details of input physics, methodologies, and basic tests in GRQKNT are presented in [16]. Here, we describe the design briefly.

We solve QKE (see Fig. 1) with a conservative form. The formalism is essentially the same as that used in our Boltzmann code. Although many modules in GRQKNT are transplanted from our Boltzmann code, there are mainly four different features. First, we solve density matrices of neutrinos and antineutrinos with a flavor-basis, in which the off-diagonal components represent the coherency of different flavor states. This indicates that the number of dimension in flavor space needs to be increased from 6 (for Boltzmann) to 18 (for QKE). Next, GRQKNT includes neutrino oscillation Hamiltonian, which has vacuum, matter, and neutrino self-interaction potentials. Third, the emission, absorption, and scattering processes are incorporated in GRQKNT code. Although they are a minimum set of neutrino-matter interactions for the study of CCSN and BNSM, its formulations and implementations are consistent with quantum

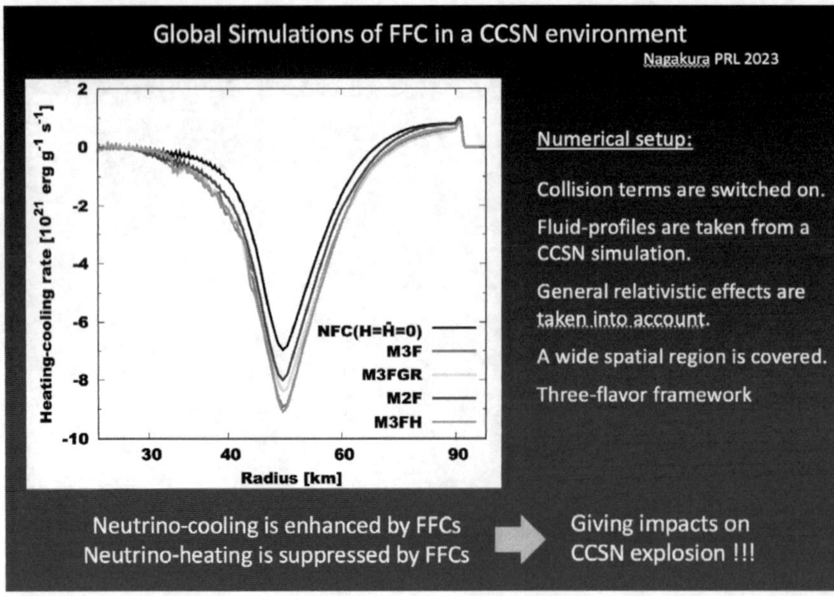

Figure 3. Net neutrino heating-cooling rate as functions of radius [20]. Black line represents the case without FFC, while others denote the result with FFC. Our result suggests that neutrino cooling is enhanced in optically thick region by FFC, whereas the heating is hampered in the gain region. See the text for more details.

kinetics of neutrinos; in fact, we need to extend the collision term from classical to quantum one (see, e.g., [14, 15]). Fourth, we employ an explicit time-integration scheme (WENO) to solve QKE in GRQKNT, whereas implicit treatments are mandatory for Boltzmann neutrino transport. This is attributed to the fact that the QKE is not a stiff equation, if flavor instabilities are triggered by neutrino self-interactions. In these cases, the neutrino advection needs to be solved with the same time scale as neutrino-oscillation one.

One thing we do notice here is that flavor conversions, in particular for fast mode (or FFC), occur in much shorter time- and length- scales than those in CCSNe and BNSMs. The disparity between the two scales reaches several orders of magnitude, indicating that global simulations are intractable in the current computational powers. In [18, 19], we proposed a novel approach to overcome this issue. We introduced an attenuation parameter, which reduces a coupling constant of neutrino oscillation Hamiltonian. Although the reduced coupling results in artificial outcomes, the realistic features can be extracted by convergence studies with respect to the attenuation parameter. In [18, 19], we showed that the small-scale structures of flavor conversions are sensitive to the attenuation parameter, whereas their time-averaged and global features are not affected (see Fig. 2). In the next section, we introduce some noticeable results in our recent study of FFCs in CCSN and BNSM environments, while we used the attenuation prescription in all simulations. We refer readers to [18, 19] for the detailed properties in their convergence studies.

3. Fast neutrino-flavor conversion in CCSN and BNSM

Figure 3 summarizes the most important result for the study of FFC in our recent paper [20]. We find that FFC can change neutrino radiation fields radically from those obtained from classical (Boltzmann) neutrino transport, which gives impacts on both neutrino cooling and heating in

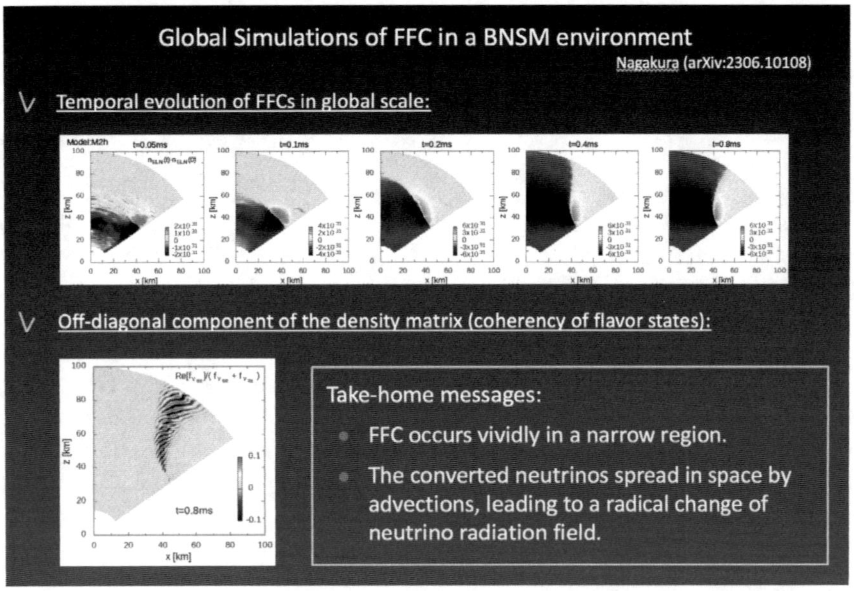

Figure 4. Global features of FFCs in BNSM environments [21]. In the top panels, the color contour displays the difference of electron-neutrino lepton number (ELN) number density from those in the initial (which corresponds to a steady state obtained from Boltzmann simulation) profile. In the bottom panel, we show the off-diagonal component of density matrix of neutrinos, which represents the coherency of flavor states; see the text for more details.

post-shock regions. In optically thick region, the colored lines (corresponding the cases with FFC; see [20] for the detail of color code) are always lower than the black one (the case with no FFC), indicating that the neutrino cooling is enhanced by FFC. At large radii, on the other hand, the net neutrino heating is suppressed by FFC.

One thing we should point out here is that these two effects would compete each other for CCSN explosions. It has been suggested that the large neutrino cooling in the gain region accelerates PNS cooling, resulting in fast PNS contraction. This also increases the average energy of neutrinos, which facilitates neutrino absorptions, i.e., heating in the gain region (it should be noted, however, that this effect is not taken into account in our simulations, since we need to include feedback effects from neutrino radiation field to matter, but we fix the matter background in time). However, the neutrino fluxes of electron- and antielectron type neutrinos are reduced by FFC, which reduces the neutrino heating in the gain region, as shown in Fig. 3. At the moment, we do not give a conlusive answer at the moment whether FFC gives a positive or negative effect on CCSN explosions. However, our result clearly shows that the fluid dynamics including shock revival must be impacted by FFC.

In Fig. 4, we show the main results for the study of FFC in BNSM environments [21]. It should be emphasized that this is the first axisymmetric global FFC simulation. We find that the dynamical features of FFCs are very different from those expected from local simulations. There are mainly two noticeable features. First, an ELN-XLN Zero Surface (EXZS), corresponding to a surface where ELN and heavy-leptonic one (XLN) number densities become the same, emerges, and then the neutrinos undergo flavor swaps when they pass through it. The flavor swap is strong evidence that FFC can give a large impact on neutrino radiation field, since ELN and XLN can be completely exchanged at this surface. Next, FFC occurs only in narrow regions, while linear stability analysis (based on the radiation field from Boltzmann neutrino transport) suggests that FFC occurs in almost everywhere above the accretion disk. This represents that non-local neutrino advection changes the overall property of FFC in BNSM environments. This finding presents an importance of including effects of global advection of neutrinos when we study actual impacts of FFC on BNSM.

4. Summary

There is a long history of neutrino transport simulations for the modeling of central engines of CCSN and BNSM phenomena. These numerical modelings rely on Boltzmann-base kinetic theory, in which the neutrinos are treated as classical particles and their flavor conversions have been neglected. This is due to the fact that matter potential is much stronger than the vacuum contribution, which can effectively reduce the difference between mass- and flavor eigenstates. On the other hand, effects of neutrino-self interactions can induce rapid flavor conversions even in dense matter environments. In fact, linear stability analysis of flavor instabilities suggest that these types of flavor conversions can occur in both CCSN and BNSM environments ubiquitously. Motivated by these facts, we developed a new code, GRQKNT, which has the ability to solve mean-field QKE for neutrino transport.

By using the code, we demonstrated QKE simulations for FFC in both CCSN and BNSM environments. Our results clearly show that FFC changes substantially the neutrino radiation field. We also find that global advections should be taken into account appropriately when we discuss impact of FFC on these phenomena. Last but not least, we would like to mention that the community of neutrino flavor conversion is rapidly evolving. Although there still remain many open issues in non-linear features of neutrino flavor instabilities, we will address all these issues one by one in collaborations with CCSN and BNSM modellers.

References

[1] A. Mezacappa et al. 2020, Physical, numerical, and computational challenges of modeling neutrino transport in core-collapse supernovae, Living Reviews in Computational Astrophysics, Volume 6, Issue 1, article id.4
[2] K. Sumiyoshi and S. Yamada, 2012, Neutrino Transfer in Three Dimensions for Core-collapse Supernovae. I. Static Configurations, The Astrophysical Journal Supplement, Volume 199, Issue 1, article id. 17, 32 pp.
[3] H. Nagakura et al., 2014, Three-dimensional Boltzmann Hydro Code for Core Collapse in Massive Stars. I. Special Relativistic Treatments, The Astrophysical Journal Supplement Series, Volume 214, Issue 2, article id. 16, 19 pp.
[4] H. Nagakura et al., 2018, Simulations of Core-collapse Supernovae in Spatial Axisymmetry with Full Boltzmann Neutrino Transport, The Astrophysical Journal, Volume 854, Issue 2, article id. 136, 13 pp.
[5] W. Iwakami et al., 2020, Simulations of the Early Postbounce Phase of Core-collapse Supernovae in Three-dimensional Space with Full Boltzmann Neutrino Transport, The Astrophysical Journal, Volume 903, Issue 2, id.82, 24 pp.
[6] R. Akaho et al., 2023, Protoneutron Star Convection Simulated with a New General Relativistic Boltzmann Neutrino Radiation Hydrodynamics Code, The Astrophysical Journal, Volume 944, Issue 1, id.60, 18 pp.
[7] S. Abbar et al., 2021, On the characteristics of fast neutrino flavor instabilities in three-dimensional core-collapse supernova models, PRD, 103, 063033
[8] H. Nagakura et al., 2019, Fast-pairwise Collective Neutrino Oscillations Associated with Asymmetric Neutrino Emissions in Core-collapse Supernovae, The Astrophysical Journal, Volume 886, Issue 2, article id. 139, 11 pp.
[9] H. Nagakura et al., 2021, Where, when, and why: Occurrence of fast-pairwise collective neutrino oscillation in three-dimensional core-collapse supernova models, PRD, 104, 083025
[10] S. Furusawa and H. Nagakura, 2023, Nuclei in core-collapse supernovae engine, Progress in Particle and Nuclear Physics, Volume 129, article id. 104018
[11] Z. Xiong et al., 2023, Evolution of collisional neutrino flavor instabilities in spherically symmetric supernova models, PRD, 107, 083016
[12] M. Wu and I. Tamborra, 2017, Fast neutrino conversions: Ubiquitous in compact binary merger remnants, PRD, 95, 103007
[13] Z. Xiong et al., 2023, Collisional flavor instability in dense neutrino gases, arXiv:2212.03750

[14] D. Blaschke and V. Cirigliano, 2016, Neutrino Quantum Kinetic Equations: The Collision Term, PRD, 94, 033009
[15] R. Sherwood, 2019, Neutrino quantum kinetics in compact objects, PRD, 99, 123014
[16] H. Nagakura, 2022, General-relativistic quantum-kinetics neutrino transport, PRD, 106, 063011
[17] M. Shibata et al. 2014, Conservative form of Boltzmann's equation in general relativity, PRD, 89, 084073
[18] H. Nagakura and M. Zaizen, 2022, Time-Dependent and Quasisteady Features of Fast Neutrino-Flavor Conversion, PRL, 129, 261101
[19] H. Nagakura and M. Zaizen, 2022, Connecting small-scale to large-scale structures of fast neutrino-flavor conversion, PRD, 107, 063033
[20] H. Nagakura, 2022, Roles of Fast Neutrino-Flavor Conversion on the Neutrino-Heating Mechanism of Core-Collapse Supernova, PRL, 130, 211401
[21] H. Nagakura, 2022, Global features of fast neutrino-flavor conversion in binary neutron star merger, arXiv:2306.10108

Journal of Physics: Conference Series **2742** (2024) 012018 doi:10.1088/1742-6596/2742/1/012018

An operator-splitting numerical scheme for relativistic magnetohydrodynamics

David Phillips and Serguei Komissarov
School of Mathematics, University of Leeds, UK
E-mail: mmdnhp@leeds.ac.uk, s.s.komissarov@leeds.ac.uk

Abstract. We describe a novel operator-splitting approach to numerical relativistic magnetohydrodynamics designed to expand its applicability to the domain of ultra-high magnetisation. In this approach, the electromagnetic field is split into the force-free component, governed by the equations of force-free degenerate electrodynamics (FFDE), and the perturbation component, governed by the perturbation equations derived from the full system of relativistic magnetohydrodynamics (RMHD). The combined system of the FFDE and perturbation equations is integrated simultaneously, for which various numerical techniques developed for hyperbolic conservation laws can be used. At the end of every time-step of numerical integration, the force-free and the perturbation components of the electromagnetic field are recombined and the result is regarded as the initial value of the force-free component for the next time-step, whereas the initial value of the perturbation component is set to zero. To explore the potential of this approach, we build a 3rd-order WENO code, which was used to carry out 1D and 2D test simulations. Their results show that this operator-splitting approach allows us to bypass the stiffness of RMHD in the ultra-high-magnetisation regime where the perturbation component becomes very small. At the same time, the code performs very well in problems with moderate and low magnetisation too.

1. Introduction

The strong gravity of astrophysical black holes and neutron stars creates some of the most extreme physical conditions in the Universe, which cannot be achieved in research laboratories. In particular, they naturally develop magnetospheres with extremely high plasma magnetisation. Highly-relativistic winds and jets emerging from this magnetospheres create spectacular structures on enormous scales, from parsecs (pulsar wind nebulae) to hundreds of kiloparsecs (jets of active galactic nuclei). These flows transport huge amount of energy in the electromagnetic form (Pointing flux) and this energy drives the observed phenomena either via magnetic reconnection and dissipation or via conversion into the kinetic energy of bulk motion with subsequent shock dissipation. For plasma flows on such huge scales, relativistic magnetohydrodynamics (RMHD) provides the most suitable framework.

In contrast to the non-relativistic problems, where the magnetisation is well described by the ratio of thermodynamic and magnetic pressures ($\beta = p/p_m$), the magnetisation of relativistic plasma is best described by the parameter $\sigma = \hat{B}^2/4\pi(e+p)$, where $\hat{B}$ is the magnetic field strength as measured in the rest frame of plasma, p is the thermal pressure and $e = \rho c^2 + e_t$, where ρ is the density of the rest mass of plasma particles (for a given species, $\rho = mn$, where n is the proper number density of particles and m is the rest mass of a single particle) and e_t is the proper thermal energy of plasma. In the magnetospheres of neutron stars, σ can reach the values of the order of $10^3 - 10^6$. Unfortunately, modern conservative schemes for RMHD begin to fail in multi-dimensional problems already when σ = 1-10, depending on the problem. Namely, the conservative variables cannot be converted to physically meaningful primitive variables. Higher values of σ can be handled for isentropic flows, where the energy equation can be eliminated [1]. A fix based on the entropy obtained via parallel integration of the adiabatic entropy transport equation can help to bypass crashes for adiabatic flows as well [2]. Obviously, this approach would not work for problems involving shocks, like the termination shock of pulsar winds [3], and, more importantly, magnetic reconnection, which involves kinetic and magnetic energy dissipation. Excessive artificial plasma heating due to numerical resistivity is another issue in simulations of highly magnetised flows. Recently, a kinetic approach based on the particle-in-cell (PIC) method has been used to model pulsar and black hole magnetospheres [4–6]. This approach has no difficulty in dealing with highly magnetised plasma but suffers from the scale-separation constraint. PIC simulations must resolve the microphysics scales, which severely limits the size of computational domain. While PIC is invaluable for studying the emission of relativistic plasma, the fluid approach remains the only realistic option for most macroscopic phenomena of relativistic plasma astrophysics.

In the limit of infinite magnetisation, $\sigma \to \infty$, the inertia of plasma, both in the form of the rest mass of particles and the mass-energy of heat, becomes vanishingly small and can be ignored. However, the number density of charged particles may still remain sufficiently high to ensure perfect conductivity. This is the regime of the so-called Force-Free Degenerate Electrodynamics (FFDE, [7]). In this limit, the system of RMHD becomes degenerate as only two out of the four components of the energy-momentum equation remain independent [8]. It has been suggested, that this is the main reason for the shortcomings of conservative schemes in high-magnetisation problems [9]. If so, it would make sense to apply a perturbation approach, where the FFDE solution is a leading order approximation and the plasma inertia enters only equations governing small perturbations to this solution.

A similar issue had been identified some time ago in Newtonian MHD simulations of the collision between highly magnetised ($\beta \ll 1$) Earth's magnetosphere and the solar wind. In that problem, the Earth's magnetic field increases by many orders of magnitude from the collision site to the troposphere, where it is largely stationary and dipolar, whereas the perturbation of this field remains of about the same magnitude and hence increasingly small relative to the undisturbed Earth's field on approach to the troposphere. As a result, the truncation errors become large in comparison to the perturbation amplitude and hence the numerical solution for the perturbation becomes corrupted. To overcome this problem, Tanaka [10] proposed to separate the strong stationary dipolar field from its perturbation and numerically integrate only the nonlinear equations governing the perturbation. This approach has been very effective and it is now standard in numerical modelling of planetary magnetospheres (e.g. [11]). Here we describe its generalisation to RMHD. Moreover, we no longer assume that the zero-order solution is time-independent, which significantly broadens the range of its applications.

2. Ideal relativistic magnetohydrodynamics

The covariant system of ideal RMHD consists of the Faraday equation

$$\nabla_\nu {}^*F^{\mu\nu} = 0\,, \tag{1}$$

the energy-momentum equation

$$\nabla_\nu T^{\mu\nu} = 0\,, \tag{2}$$

the continuity equation

$$\nabla_\nu \rho u^\nu = 0\,, \tag{3}$$

and the perfect conductivity condition

$$F_{\mu\nu} u^\nu = 0\,. \tag{4}$$

In these equations, $F_{\mu\nu}$ is the Maxwell tensor and ${}^*F_{\mu\nu}$ is the Faraday tensor, $T^{\nu\mu}$ is the total stress-energy-momentum tensor, and u^μ is the four-velocity vector of macroscopic motion. In highly ionised plasma the Faraday tensor is Hodge dual to the Maxwell tensor, and hence

$${}^*F^{\alpha\beta} = \frac{1}{2} e^{\alpha\beta\mu\nu} F_{\mu\nu}\,, \quad F^{\alpha\beta} = -\frac{1}{2} e^{\alpha\beta\mu\nu}\, {}^*F_{\mu\nu}, \tag{5}$$

where $e_{\alpha\beta\mu\nu} = \sqrt{-g}\,\epsilon_{\alpha\beta\mu\nu}$ is the Levi-Civita alternating tensor of space-time and $\epsilon_{\alpha\beta\mu\nu}$ is the four-dimensional Levi-Civita symbol. The total stress-energy momentum tensor,

$$T^{\mu\nu} = T^{\mu\nu}_{(m)} + T^{\mu\nu}_{(e)}\,, \tag{6}$$

is the sum of the stress-energy momentum tensor of electromagnetic field

$$T^{\mu\nu}_{(e)} = F^{\mu\gamma} F^{\nu}{}_{\gamma} - \frac{1}{4}(F^{\alpha\beta} F_{\alpha\beta}) g^{\mu\nu}\,, \tag{7}$$

and the stress-energy momentum tensor of matter

$$T^{\mu\nu}_{(m)} = w u^\mu u^\nu + p g^{\mu\nu}\,. \tag{8}$$

Here p is the thermodynamic pressure, $w(p,\rho)$ is the relativistic enthalpy per unit volume as measured in the rest frame of fluid, and $g^{\mu\nu}$ is the metric tensor of space-time.

3. Force-free and perturbation equations

Let us split the electromagnetic field into the force-free part $f^{\mu\nu}_{(0)}$ and the correction part $f^{\mu\nu}_{(1)}$

$$F^{\mu\nu} = f^{\mu\nu}_{(0)} + f^{\mu\nu}_{(1)}\,. \tag{9}$$

The corresponding expansion for the stress-energy-momentum tensor of the electromagnetic field is

$$T^{\mu\nu}_{(e)} = t^{\mu\nu}_{(0)} + t^{\mu\nu}_{(1)}\,, \tag{10}$$

where

$$t^{\mu\nu}_{(0)} = f^{\mu\gamma}_{(0)} f^{\nu}_{(0)\gamma} - \frac{1}{4}(f^{\alpha\beta}_{(0)} f^{(0)}_{\alpha\beta}) g^{\mu\nu} \tag{11}$$

and

$$t^{\mu\nu}_{(1)} = f^{\mu\gamma}_{(0)} f^{\nu}_{(1)\gamma} + f^{\mu\gamma}_{(1)} f^{\nu}_{(0)\gamma} + f^{\mu\gamma}_{(1)} f^{\nu}_{(1)\gamma} - \frac{1}{4}\left(2 f^{\alpha\beta}_{(0)} f^{(1)}_{\alpha\beta} + f^{\alpha\beta}_{(1)} f^{(1)}_{\alpha\beta}\right) g^{\mu\nu}\,. \tag{12}$$

The differential equations of the leading order system are the FFDE equations

$$\nabla_\nu\, {}^*f^{\mu\nu}_{(0)} = 0\,, \tag{13}$$

$$\nabla_\nu t^{\mu\nu}_{(0)} = 0\,. \tag{14}$$

Since the four velocity of plasma does not enter these equations, the perfect conductivity condition is written in the form

$${}^*f^{\mu\nu}_{(0)} f^{(0)}_{\mu\nu} = 0\,, \quad f^{\mu\nu}_{(0)} f^{(0)}_{\mu\nu} > 0\,, \tag{15}$$

Journal of Physics: Conference Series **2742** (2024) 012018 doi:10.1088/1742-6596/2742/1/012018

which ensures the existence of inertial frames where the electric field vanishes.

The perturbation equations are obtained from the full system of RMHD by removing the terms vanishing due to Eqs.13,14

$$\nabla_\nu {}^*f^{\mu\nu}_{(1)} = 0\,, \tag{16}$$

$$\nabla_\nu (t^{\mu\nu}_{(1)} + T^{\mu\nu}_{(m)}) = 0\,, \tag{17}$$

$$\nabla_\nu \rho u^\nu = 0\,. \tag{18}$$

It is important that this system is fully nonlinear with respect to its hydrodynamic component and hence allows the non-linear steepening of magnetosonic waves. In order to close this system we use the perfect conductivity condition

$$(f^{\mu\nu}_{(0)} + f^{\mu\nu}_{(1)}) u_\mu = 0\,. \tag{19}$$

By construction, the perturbation equations do not involve terms quadratic in the leading order electromagnetic field, like $f^{\mu\gamma}_{(0)} f^{\nu}_{(0)\gamma}$, which would be dominant in problems with high σ. This is the main goal behind our approach. The terms linear in $f^{\mu\gamma}_{(0)}$ in the perturbation stress-energy-momentum tensor (12) describe the interaction with the FFDE field.

4. Numerical scheme

Since the FFDE equations are independent from the perturbation equations, even if initially the electromagnetic component of the RMHD solution is close to the FFDE solution, they will eventually diverge. Once the solutions have diverged, the perturbation of the electromagnetic field, which is the difference between the RMHD and FFDE solutions, is no longer small and the splitting becomes useless. This is particularly severe in the case of shock solutions - whereas one may find a reference frame where a fast RMHD shock remains stationary, the fast waves of FFDE always propagate with the speed of light. To avoid this issue, the leading order and perturbation components of the electromagnetic field are recombined at the end of every full integration time-step, $F^{\mu\nu} = f^{\mu\nu}_{(0)} + f^{\mu\nu}_{(1)}$ and split at the beginning of the next time-step as $f^{\mu\nu}_{(0)} = F^{\mu\nu}$, $f^{\mu\nu}_{(1)} = 0$. In other words, the electromagnetic field is first advanced as force-free and then corrected for the interaction with plasma, making our approach a variant of the operator-splitting method.

We have explored the potential of this approach using a 3rd order WENO scheme for Special Relativistic MHD, similar to the one described in [12]. In this explicit scheme, the solution is advanced forward in time using a 3rd order Runge-Kutta method. The FFDE equations 13-14 and the perturbation equations 16-18 are integrated simultaneously, without recombination of the electromagnetic field at the Runge-Kutta sub-steps. To set up the Riemann problems at the cell interfaces, we developed a novel WENO-interpolation which ensures rapid onset of 3rd convergence even for smooth solutions with local extrema. To compute the interface fluxes, we used an HLL Riemann solver [13], with the speed of light as the fastest wave speeds for the FFDE sub-system, and the fast-magnetosonic speeds for the perturbation sub-system.

The conserved variables of the perturbation sub-system are converted into its primitive variables via the approach similar to the one described in [12]. The conserved variables of the FFDE are converted into its primitive variables as in [8], where the energy equation is not utilised at all. This makes the scheme not fully conservative and allows us to bypass the issue of RMHD degeneracy in the high-σ regime, so detrimental to the performance of traditional conservative schemes. However, if left unchecked, this lack of energy conservation in the FFDE sub-system may also lead to large errors, particularly at current sheets, where strong magnetic dissipation leads to significant plasma heating. The approach we use to overcome this problem involves 1) integrating the energy equation of FFDE sub-system together with the momentum equations, 2) computing the difference between the energy obtained this way and the energy computed from the $\boldsymbol{E}$ and B fields of the sub-system, and 3) if it is positive, adding it to the energy of the perturbation sub-system at the end of every timestep. In principle, the condition of step 3) could be strengthen to prevent numerical plasma heating in smooth regions away from current sheets. This remains to be explored, though. Full details of the scheme will be presented elsewhere.

5. Test simulations

Here we present the results of some test simulations. In these tests we use the ideal equation of state $w = \rho + \Gamma p/(\Gamma - 1)$ with the ratio of specific heats $\Gamma = 4/3$ and integrate the dimensionless perfect RMHD equations where 4π does not appear and the dimensionless speed of light $c = 1$ (e.g. [16]).

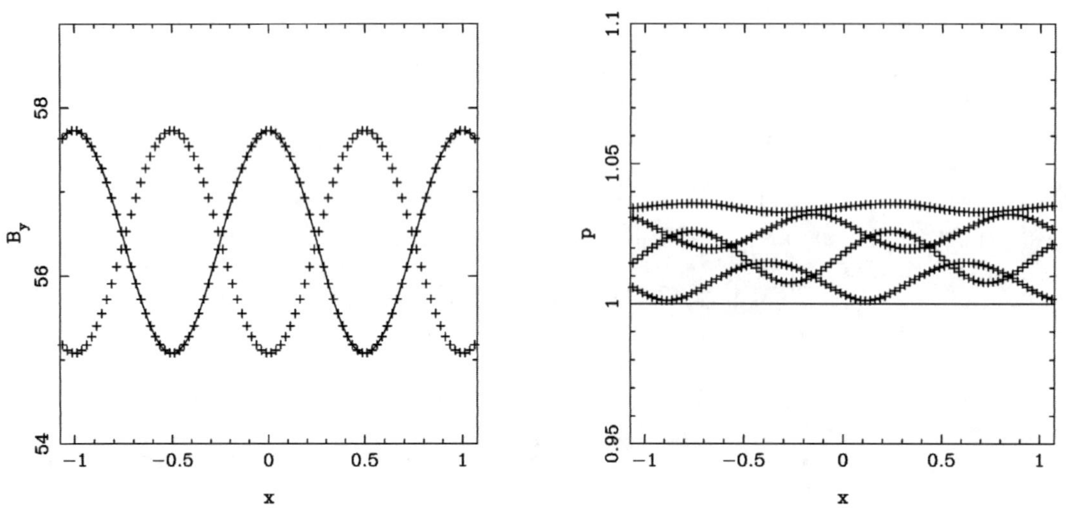

Figure 1. Alfvén wave test. In the exact solution, the wave moves to the left with the speed $v_a = 0.5$. On both panels, the solid line shows the initial solution ($t = 0$) and markers show the solutions for $t = 1, 2, 3, 4$. In the B_y-plots, the solutions for $t = 1, 3$ and $t = 0, 2, 4$ are indistinguishable. In the pressure plots, solutions for higher t have higher mean value of p. The resolution is n=80 and the Courant number $Cu = 0.4$.

5.1. 1D periodic Alfvén wave

The analytic solution to Alfvén waves of finite amplitude is given in [14]. In the exact solution used for the test problem, $\rho(x,t) = p(x,t) = 1$, $\sigma(x,t) = \sigma_0$, $B_x = 0.3B_0$ and the tangential magnetic field $B_t = \gamma_a B_0 e^{i\psi(x,t)}$, with $\psi(x,t) = \arcsin(0.3\sin(\pi(x + v_a t)))$, where $\gamma_a = (1 - v_a^2)^{-1/2}$ and the wavespeed $v_a = 0.5$. B_0 is the parameter determining the magnetic field strength in the Hoffmann-Teller frame, it is equivalent to the magnetisation parameter σ_0. Figure 1 illustrates the results for $\sigma_0 = 545$ and table 1 shows the results of the convergence study based on this problem, confirming the 3rd order accuracy of the scheme. For lower σ_0, the results are almost as good. The relative L_1 error for B_y increases only by the factor of about 4.5 for $\sigma_0 = 5.45 \times 10^{-4}$, suggesting that the approach is suitable for both the high- and low-σ regimes.

Journal of Physics: Conference Series **2742** (2024) 012018 doi:10.1088/1742-6596/2742/1/012018

Table 1. Convergence study based on the 1D Alfvén wave test. The numerical and analytic solutions are compared at $t = 4$. n is the number grid points in the computational domain.

n	20	40	80	160	320
$L_1(B_y)$	7.48E-2	8.89E-3	1.03E-3	1.31E-4	1.64E-5
$L_1(\rho)$	5.83E-1	1.11E-1	1.51E-2	1.90E-3	2.38E-4

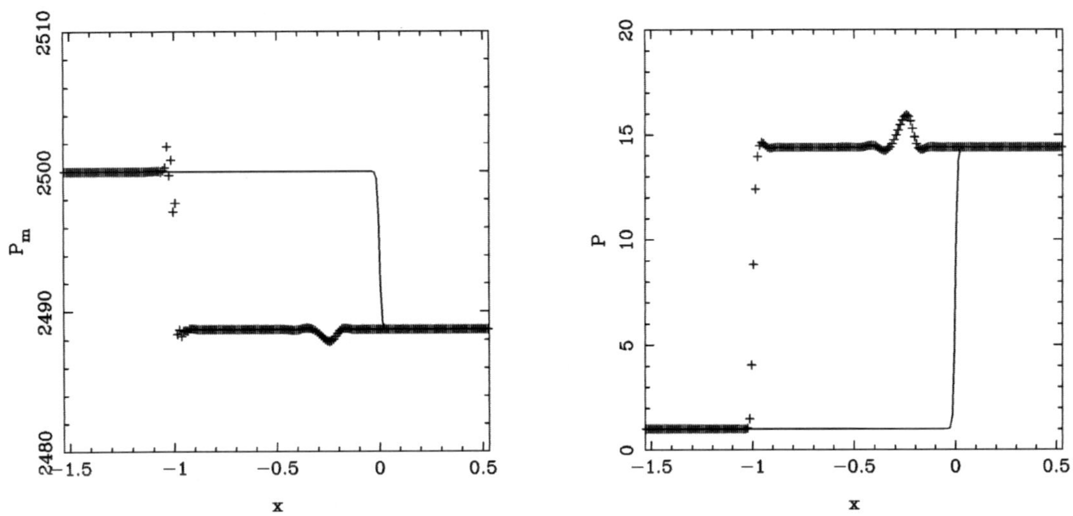

Figure 2. 1D slow shock test. The magnetic pressure (left panel) and the gas pressure (right panel) are shown for the initial solution ($t = 0$, solid lines) and the numerical solution at $t = 2$ (markers). The exact shock speed is $v_s = 0.5$. The disturbance located near $x = -0.3$ at $t = 2$ is caused by the evolution of the numerical shock structure at the beginning of the simulations.

5.2. 1D slow shock

The exact solutions for this and other shock tests were obtained using the method described in [15]. Left state: $p = \rho = 1$, $\boldsymbol{B} = (50, 51.03, 0)$, $\boldsymbol{v} = (0.1995, 0, 0)$, $\sigma = 1000$; Right state: $p = 14.41$, $\rho = 5.879$, $\boldsymbol{B} = (50, 50.82, 0)$, $\boldsymbol{v} = (-0.42.12, -0.6338, 0)$, $\sigma = 461$. The shock moves to the left with the speed $v_s = 0.5$. The angle between the shock front and magnetic field in the rest frame of the upstream state is 45°. The results at $t = 2$ are illustrated in figure 2.

5.3. 1D fast shocks

Here we present two tests, one for the high-σ regime and one for the low-σ regime. In both these cases, the angle between the shock front and the magnetic field in the rest frame of the upstream state is 45°. The fast magnetosonic Mach number of the shock in the same frame is $M_f = 2$, and relative to the grid the shock moves to the left with the speed $v_s = 0.5$. In the initial solution ($t = 0$) the shock is located at $x = 0$.

High-σ case. Left state: $p = \rho = 1$, $\boldsymbol{B} = (50, 1985., 0)$, $\boldsymbol{v} = (0.9997, 0, 0)$, $\sigma = 1000$. Right state: $p = 4.424$, $\rho = 2.618$, $\boldsymbol{B} = (50, 1989., 0)$, $\boldsymbol{v} = (0.9977, 0.0172, 0)$, $\sigma = 1119$. The numerical solution at $t = 2$ is illustrated in the right panel of figure 3.

Low-σ case. Left state: $p = \rho = 1$, $\boldsymbol{B} = (0.1581, 0.1765, 0)$, $\boldsymbol{v} = (0.4445, 0, 0)$, $\sigma = 0.01$. Right state: $p = 4.754$, $\rho = 3.032$, $\boldsymbol{B} = (0.1581, 0.4876, 0)$, $\boldsymbol{v} = (-0.1566, 0.0048, 0)$, $\sigma = 0.0116$. The numerical solution at $t = 2$ is illustrated in the left panel of figure 3.

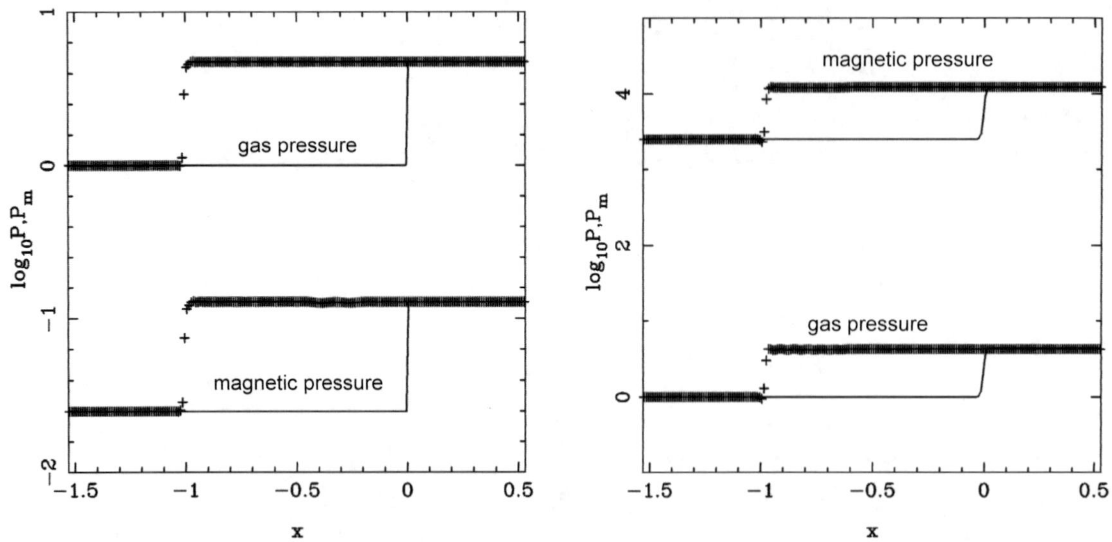

Figure 3. Low-σ (left panel) and high-σ (right panel) fast shock tests. The magnetic pressure (left panel) and the gas pressure (right panel) are shown for the initial solution ($t = 0$, solid lines) and for the numerical solution at $t = 2$ (markers). The exact shock speed is $v_s = 0.5$ in both the cases.

Our tests show that the code can handle all shocks in low-*sigma* plasma quite well. It also handles well shocks in high-σ plasma provided the jump of the tangential component of the magnetic field across the shock is relatively small. For larger jumps, the errors become substantial and the code may even crash. However, as illustrated by the strong explosion test, creating such strong shocks in high-*sigma* plasma is problematic.

5.4. 2D cylindrical explosion

This problem is set on a uniform Cartesian grid with 200 cells in each direction. The computational domain is $(-6, 6) \times (-6, 6)$. The initial solution describes a uniform ambient gas at rest with $p_a = 3 \times 10^{-5}$, $\rho_a = 10^{-4}$, $\boldsymbol{B} = (B_0, 0, 0)$, and $\boldsymbol{v} = (0, 0, 0)$. In the central region with the polar radius $r < 0.8$ the gas pressure and density are raised to $p_0 = 1$ and $\rho_0 = 0.01$ respectively. A tanh-profile is used to introduce a gradual transition between the two states in $0.8 < r < 1.0$. This is exactly the same setup as in [16], where $B_0 = 1$ ($\sigma_0 = 0.25$) was the highest value their conservative code could handle (Later, with a somewhat modified code, the author of that paper failed to reproduce this result, as the code crashed.). We have had successful runs for $B_0 = 0.01, 0.1, 1, 10, 100$, and 1000. For $B_0 = 0.1$ the results look very close to that in [16] and other papers repeating this test (e.g. [12]). Figure 4 illustrates the results for the extreme cases of $B_0 = 0.01$ and $B_0 = 1000$; the corresponding magnetisation in the centre of the explosion, $\sigma_0 = B_0^2/w(p_0, \rho_0)$, being $\sigma_0 = 2.5 \times 10^{-5}$ and $\sigma_0 = 2.5 \times 10^5$ respectively.

For $B_0 = 0.01$, the magnetic field has a very small effect on the flow up to $t = 4$, and so it is almost the same as in the case of unmagnetised fluid. The magnetic field is swept aside by

Journal of Physics: Conference Series **2742** (2024) 012018 doi:10.1088/1742-6596/2742/1/012018

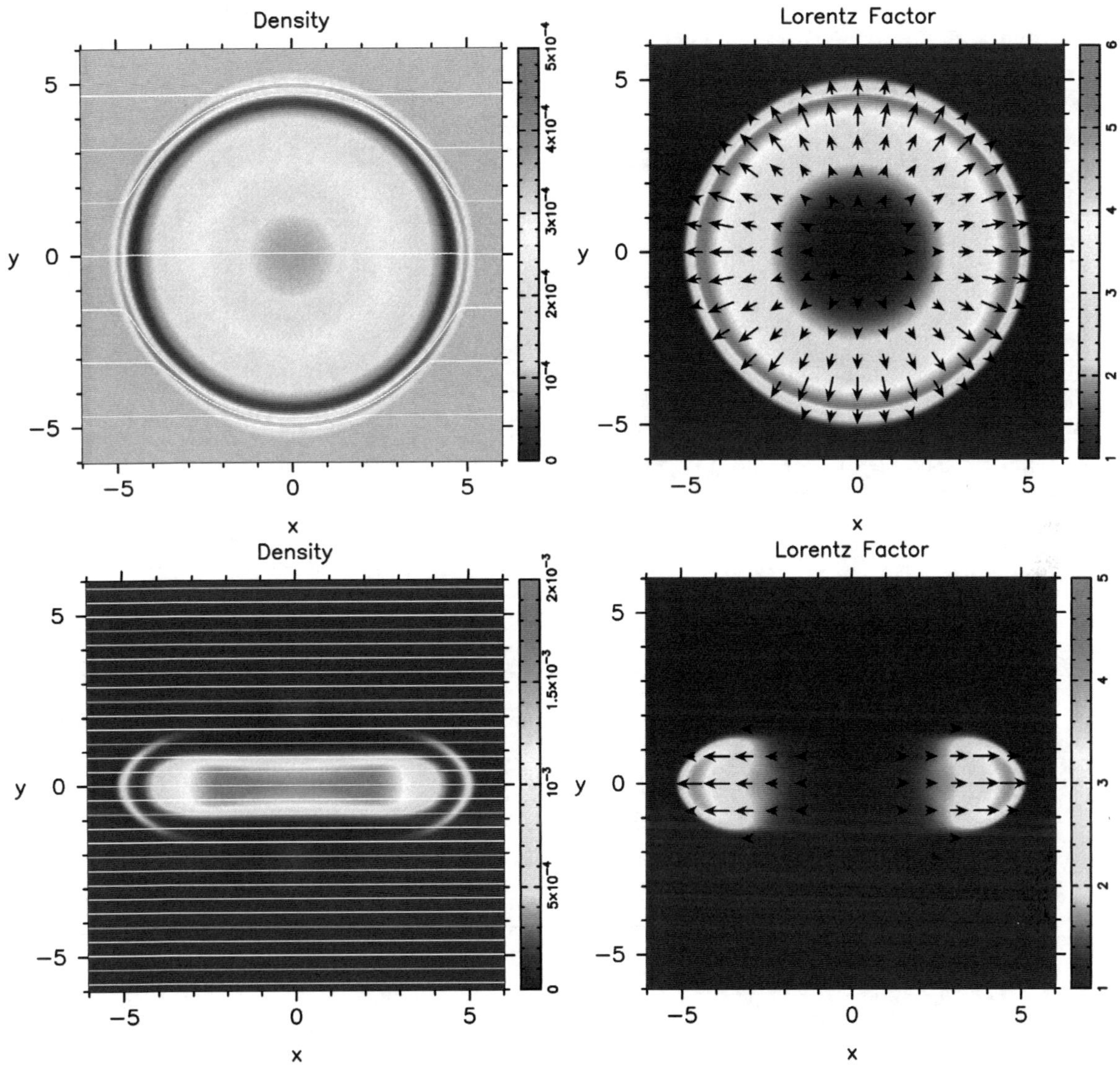

Figure 4. Cylindrical explosion test. The top panels show the solution for $\sigma_0 = 2.5 \times 10^{-5}$, whereas the bottom panels the one for $\sigma_0 = 2.5 \times 10^{5}$. In both cases, the integration time $t = 4$. The solid white lines are the lines of magnetic field, and the black arrows show the velocity field.

this flow. For $B_0 = 1000$, the magnetic field is so strong that the magnetic field lines are hardly deflected by the explosion (and hence the jump of the tangential component of magnetic field across the generated shocks is small) and the fluid velocity is perfectly aligned with the x-axis.

5.5. 2D ABC magnetic grid

The initial configuration of this problem describes a static uniform plasma distribution of density ρ_0 and pressure p_0 and the periodic force-free magnetic field

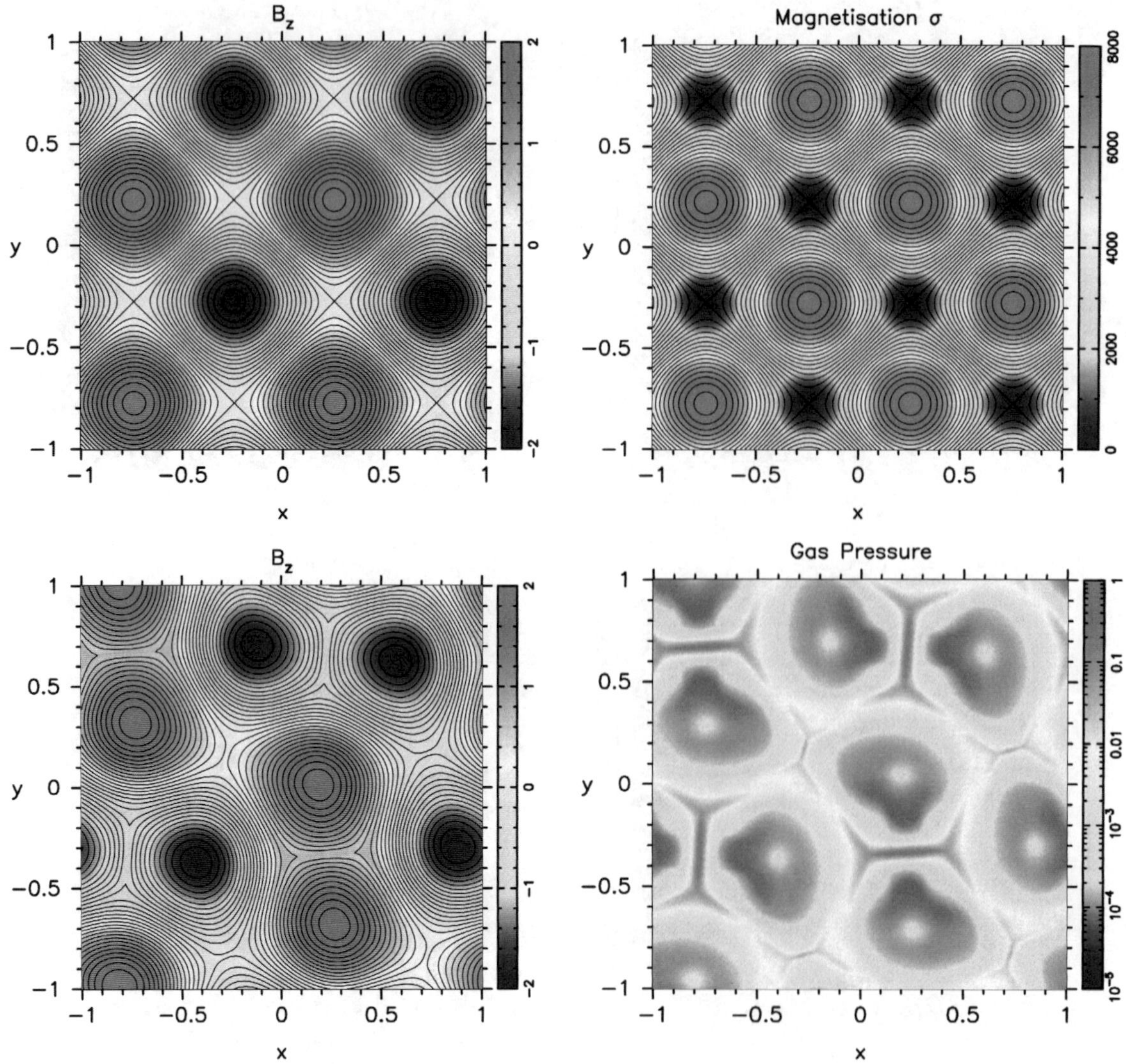

Figure 5. Collapse of the two-dimensional ABC grid. The top panels show the magnetic field lines and B_z (left) and σ (right) in the initial solution. The bottom panels show the solution at the onset of rapid merger of magnetic ropes with the same polarity ($t = 7$): magnetic field lines and B_z (left) and p (right).

$$\begin{cases} B_x = -B_0 \sin(2\pi y) \,, \\ B_y = B_0 \sin(2\pi x) \,, \\ B_z = B_0(\cos(2\pi x) + \cos(2\pi y)) \,, \end{cases}$$

which describes an equidistant grid of twisted magnetic field ropes aligned with the z-axis. In the xy-plane, these ropes appear as magnetic islands of opposite polarities (see the top-left pane of figure 5). Closest islands of the same polarities are separated via an x-point. This equilibrium configuration is unstable, leading to the eventual merger of islands with the same polarity via a collapse of the x-points into current sheets, followed by magnetic reconnection. Here we aimed to reproduce the FFDE and PIC simulations of this process described in [17].

The parameters of the simulations described here are $B_0 = 1.0$, $p_0 = \rho_0 = 10^{-4}$. The Cartesian computational domain is $(-1,1) \times (-1,1)$, with periodic boundary conditions and a 200×200 uniform grid. Random noise was used to trigger the instability.

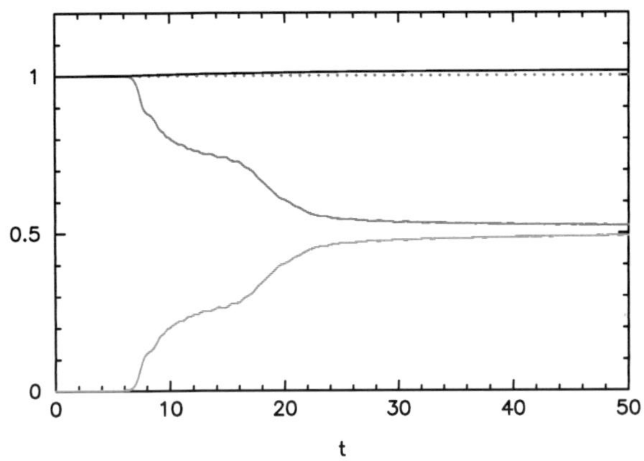

Figure 6. Energy balance in the ABC test. The black solid line shows the total energy in the box in the units of the initial total energy. The red line is the fraction of energy in the electromagnetic field and the green line is the fraction of energy in the plasma.

In the test simulations, the linear phase of the instability continues up to $t = 6$. This phase corresponds to the initial plateau of the magnetic energy curve in figure 6. By $t = 7$, the x-points separating the islands collapse into current sheets (see the bottom panels of figure 5). This opens the phase of rapid merger of neighbouring magnetic islands accompanied by strong magnetic dissipation. Once the number of the islands in the box reduces to two (of opposing polarities) the dissipation rate slows down significantly. By $t = 30$ almost 50 percent of the initial magnetic is converted into heat. The timescales of the evolution are similar to that seen in the PIC simulations (see figure 9 in [17]). Although we observe a somewhat smaller fraction of the dissipated magnetic energy, this could be related to the four-times larger size of the computational box in the PIC simulations, allowing it to trace the mergers to larger length scales. The total energy in the periodic box only slightly increases during the simulations.

6. Conclusions

In this contribution, we have described a rather unusual version of the operator-splitting method tailored to overcome the stiffness of RMHD equations in the regime of ultra-high magnetisation. The results of test simulations show that this approach allows us to obtain accurate solutions for a wide range of problems involving smooth flows, shock waves, and magnetic reconnection. It is suitable not only for highly magnetised flows but also for flows with low magnetisation, which is important for many astrophysical applications.

References

[1] Komissarov S S *et al.* 2007 *MNRAS* **380** 51
[2] Noble S C *et al.* 2009 *Ap.J* **692** 411
[3] Porth O *et al.* 2013 *MNRAS* **431** L48
[4] Philippov A and Spitkovsky A 2014 *Astrophys. J. Lett.* **785** L33
[5] Parfrey K *et al.* 2019 *Phys. Rev. Lett.* **122** 035101
[6] Crinquand B *et al.* 2020 *Phys. Rev. Lett.* **124** 145101
[7] Macdonald D and Thorne K 1982 *MNRAS* **198** 345
[8] Komissarov S S 2002 *MNRAS* **336** 759
[9] Komissarov S S 2006 *AIP Conference Proceedings* **856** 129
[10] Tanaka T 1994 *J.Comp.Phys.* **111** 381
[11] Eggington J W B *et al.* 2020 *J. Geophys. Research: Space Phys.* **125** e2019JA027510
[12] DelZanna L *et al.* 2007 *A&A* **473** 11
[13] Harten A *et al.* 1983 *SIAM Rev.* **5** 1

[14] Komissarov S S 1997 *Phys. Lett. A* **232** 435
[15] Majorana A and MAnile A 1987 *Phys. Fluids* **30** 3045
[16] Komissarov S S 1999 *MNRAS* **303** 343
[17] Lyutikov M *et al.* 2017 *J. Plasma Phys.* **83** 635830602

Journal of Physics: Conference Series **2742** (2024) 012019 doi:10.1088/1742-6596/2742/1/012019

Ookami: An A64FX Computing Resource

A C Calder[1,2], E Siegmann[2], C Feldman[1,2], S Chheda[2], D C Smolarski[3], F D Swesty[1], A Curtis[2], J Dey[2], D Carlson[2], B Michalowicz[4], and R J Harrison[2]

[1] Department of Physics and Astronomy, Stony Brook University, Stony Brook, NY 11794-3800, USA

[2] Institute for Advanced Computational Science, Stony Brook University, Stony Brook, NY 11794-5250, USA

[3] Department of Mathematics & Computer Science, Santa Clara University, Santa Clara, CA 95053, USA

[4] Department of Computer Science and Engineering, The Ohio State University, Columbus, OH 43210, USA

E-mail: `alan.calder@stonybrook.edu`

Abstract. We present a look at Ookami, a project providing community access to a testbed supercomputer with the ARM-based A64FX processors developed by a collaboration between RIKEN and Fujitsu and deployed in the Japanese supercomputer Fugaku. We provide an overview of the project and details of the hardware, and describe the user base and education/training program. We present highlights from previous performance studies of two astrophysical simulation codes and present a strong scaling study of a full 3D supernova simulation as an example of the the machine's capability.

1. Introduction

In this paper, we describe Ookami, a project providing access to a testbed computing system featuring the ARM-based A64FX processor developed by a collaboration between RIKEN and Fujitsu and deployed in Fugaku, what was, until June of 2022, the world's fastest supercomputer [1]. The project is run by Stony Brook University (SBU) in cooperation with the University at Buffalo [2] and provides open access along with training and resources to effectively use such hardware. Users have been able to port, analyze, and optimize the performance of many applications [3]. Below, we describe the hardware, our user base and education program, and provide highlights from previous performance studies of two astrophysical simulation codes, the astrophysical radiation hydrodynamics code V2D [4] and multi-application package FLASH [5, 6], here applied to thermonuclear supernovae. We also present a strong scaling study of a production 3D supernova simulation that demonstrates the capability of the hardware.

2. Overview of Project

The primary goal of the Ookami project is to provide access to a platform with the A64FX processors for testing and development. These processors offer an alternative to graphical programming units (GPUs) that power many contemporary supercomputers but may require considerable code development to make good use of. The expectation is that the reduced instruction set A64FX processors will provide high performance and reliability for applications with common programming models, particularly memory-intensive applications, while maintaining a good performance-to-power ratio similar to GPUs. Users with a variety of applications are able to explore and evaluate this new hardware and its potential for use in settings from a local cluster to the extreme scale of leadership-class supercomputing centers.

The Ookami project started in 2019 and will run for six years, providing researchers worldwide with access to this cutting-edge computing technology. For the first few years, access to Ookami was allocated via SBU. Researchers submitted allocation requests directly to the Ookami team, who reviewed them. Approved projects were granted access to the cluster for testbed projects, and once users could prove that their application performs well on this novel architecture, they could advance their projects to production status. In October 2022 Ookami became an ACCESS [7] resource provider, and now 90% of its resources are allocated via ACCESS. Users can submit requests to ACCESS and, if approved, get credits, which they can then exchange for resources (e.g. node hours, GPU hours, storage) on one or multiple of the ACCESS resource providers.

2.1. *The A64FX Hardware*

Ookami is an HPE (formerly Cray) Apollo 80 system with 174 + 2 debug Fujitsu A64FX-FX700 compute nodes, each of which has 48 cores divided into 4 core memory groups (NUMA regions), each with 8 GB of high-bandwidth memory (HBM) for a total of 32 GB per node. Each core has a 64 KB L1 cache, and an 8 MB L2 cache shared between the cores in each core memory group, and runs at 1.8 GHz. These processors use the ARMv8.2–A Scalable Vector Extension (SVE) SIMD instruction set with a 512-bit vector implementation, allowing for vector lengths from 128 to 2048 bits (in 128-bit increments) and enabling vector length agnostic programming [8].

The operating system for each processor resides on a node's local 512 GB SSD and processors communicate via an Infiniband HDR100 fat tree interconnect with 200 Gb/s switches. A high-performance Lustre file system provides about 800 TB of storage. At present, Ookami is running Rocky Linux 8.4, and compilers and toolchains from GNU, LLVM, ARM, HPE/Cray, NVIDIA, LLVM, and Fujitsu are installed.

2.2. *User Base*

Since Ookami was opened to researchers, 358 users have been onboarded. Those users work on 125 projects, 31 of which have been allocated via ACCESS. The experience of the users is very diverse as they range from undergrad students to professors and professionals with decades of High Performance Computing (HPC) experience. Figure 1 illustrates the distribution of users.

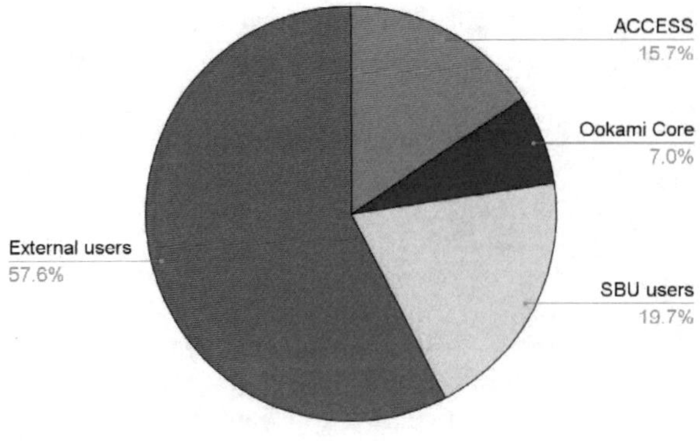

Figure 1. Pie chart with the distribution of users on Ookami, 358 as of November, 2023. SBU users are researchers from on-campus. External users are from other academic institutions. The core team is the personnel working on the project, including computational scientists, grad students, and system administrators. ACCESS users are the recent users added via the ACCESS program.

2.3. *Education and Training*

To cover the complex needs of the user community, the Ookami project has established a multi-modal support approach [9]. The extensive documentation, FAQ section, and getting-started guide on the project's website [2] allow users to learn everything about the system and its efficient use via self-study. However, because user support is especially important for novel technology, Ookami also offers a Slack channel, to which all new users are automatically added. There they can ask questions, get feedback, or just start an informal conversation with other users as well as with the project team. The average response time is normally within a few minutes, making this the perfect tool for quick help. Office hours, offered twice a week, enable users to ask more detailed questions. These are virtual meetings open to all users, led by Ookami team members. This is the perfect opportunity to do an interactive debugging or profiling session, or to just show applications and get feedback. On average these calls are attended by six to seven persons. Ookami also offers a traditional ticketing system, which is mainly used for installation requests, new user or project requests, or used by users who prefer this approach over the Slack channel.

Regular webinars spanning various topics, e.g. profilers, debuggers, and programming languages, are held to support the user community holistically and to provide updated input for the usage of HPC tools. The Ookami team also reaches out to communities that have potential interest in the cluster, e.g. Campus Champions [10], Science Gateways [11], and computational researchers with suitable applications, and introduces them to A64FX and the opportunity to use Ookami for their work. Other outreach activities include an introduction to HPC for high school students, who come to the institute without prior experience, learn the basics of HPC, get up and running on Ookami, and depart with a self-reported positive experience.

3. Performance Studies

Guided by the bountiful support of the Ookami team and user community, we were able to port and evaluate the performance of two different astrophysical simulation codes on Ookami: V2D and FLASH. Below, we report the results of our efforts including comparing the performance of different available compilers and considering the capacity for full production runs.

3.1. V2D

The V2D code solves the Euler equations of inviscid hydrodynamics and multi-species flux-limited diffusive radiation transport in two spatial dimensions via finite-difference methods. It was originally designed for the core-collapse supernovae problem of astrophysics but may be applied to other radiation-hydrodynamic problems. Details related to the underlying numerical methods can be found in [4]. V2D is written in modern Fortran, employs MPI for communication, and uses the the HDF5 library for parallel I/O.

The problem we chose for our study with V2D is a the propagation of a 2D Gaussian pulse of radiation, a diffusive radiation transport problem. The computational effort is in the solution of a large, sparse, memory-bandwidth-limited linear system that describes the time evolution of the radiation distribution. The linear system consists of $x1 \times x2 \times 2$ coupled linear equations, with $x1 = 200$, and $x2 = 100$ zones in the spatial dimensions, and there are two radiation species. Each time step thus requires the solution of a unique $x_1 \times x_2 \times 2$ linear system, and the solution is obtained via the BiCGSTAB algorithm. This is a relatively small test problem chosen to explore the performance of SVE optimization. Also, we make only limited use of the parallel capability of V2D when we vary the process topology to adjust the problem size on/ each processor. The linear system is sparse, but it has a regular structure. The method is matrix-free, but if the matrix corresponding to this system were stored with a dictionary-like ordering it would form a banded matrix with five bands.

We tested combinations of compilers and MPI implementations. As a sample of results, those presented below indicate test runs using the GNU (ver. 11.1.0), Fujitsu (ver. 4.5), and HPE/Cray (ver. 21.03) compilers. We compared the CPU times of simulations compiled with different compilers, both with and without SVE optimization. The Linux `perf stat` command of the Linux kernel performance monitor, with the `-e duration_time` flag, was used to measure the time of the simulations. This command measures the entire CPU time of the process. Each configuration (of the total number of processors used and the process topology in the $x1$- and $x2$- directions, determining how the linear system was partitioned) was run several times to confirm the timing results. Also, each test problem was run for 100 time steps.

All compilers available on Ookami are able to make use of SVE capabilities. In our tests, the SVE and optimization features were used, but we also turned off the SVE and other optimization features on some of the HPE/Cray tests for comparison purposes. When using a single processor, the executable compiled by GNU took the longest to complete the test run, around 363.91 seconds, while that compiled by the Fujitsu compiler, around 252.31 seconds, and the executable compiled using the HPE/Cray compiler (with optimization), around 181.26 seconds. Using more processors, however, with different domain decomposition to CPU core topologies, the executable compiled by Fujitsu performed better than the one compiled by HPE/Cray.

The following chart presents these results. The values in the N_p column refer to the total number of processors used for the run, with the numbers in the "Direction" `NX1` column indicating the number of domain decomposition tiles in the $x1$ direction and similarly for the numbers in the `NX2` column. Thus the product of the two values equals the total number of processors requested. The column labeled HPE/Cray (opt) indicates results obtained with an executable compiled both with both `-O3` optimization and SVE optimization enabled.

N_p	**Direction**		**Times by Compiler (seconds)**	
	`NX1`	`NX2`	Fujitsu	HPE/Cray (opt)
40	40	1	13.97	19.12
40	20	2	12.96	17.37
40	10	4	13.04	17.16
50	50	1	13.05	25.56
50	25	2	12.09	24.07
50	10	5	11.40	23.51

We note that the 50 core runs necessitated going to cores on a second node. The results varied by compiler, but in the case of the Cray compiler the "plateau" in the performance may follow from the overhead of communication across nodes. Further details may be found in [12].

3.2. FLASH

3.2.1. Overview and Initial Experiences FLASH is a simulation software package for addressing multi-scale, multi-physics applications. Initially developed at the University of Chicago to address thermonuclear flashes, stellar explosions powered by a thermonuclear runaway occurring on the surface or in the interiors of compact stars, FLASH continues to be developed for astrophysics [13] and high-energy-density physics [14], and a new code, FLASH-X, derived from FLASH and with a completely new infrastructure, is under development and will allow addressing more general problems [15].

At its heart, FLASH is a hydrodynamics plus additional physics (e.g. a stellar equation of state) method. FLASH uses the PARAMESH library to implement adaptive mesh refinement (AMR) to address problems with a wide range of physical scales on a block-structured mesh [16, 17]. FLASH is written primarily in modern Fortran and is parallelized primarily through MPI, although some solvers have been modified to take advantage of threaded approaches to parallelization [18] and development continues toward a more general design for better thread support [19–21].

PARAMESH manages a block-structured adaptive mesh, with the data typically in $16\times16\times16$ zone blocks (16×16 in 2D). Each block also includes four ghost zones in each direction around the block, and communication between blocks occurs through exchange of data in the ghost zones. Variables like density, temperature, internal energy, etc., are stored in

Journal of Physics: Conference Series **2742** (2024) 012019 doi:10.1088/1742-6596/2742/1/012019

a data container, `unk`, a Fortran array in the form `unk(nvar, il_bnd, iu_bnd, jl_bnd, ju_bnd, kl_bnd, ku_bnd, maxblocks)`, where `nvar` is the number variables, `il_bnd:iu_bnd, jl_bnd:ju_bnd, kl_bnd:ku_bnd` are the x, y, and z zone limits, and `maxblocks` is the maximum number of blocks allowed for a given processor element. Accordingly, block data is typically accessed block-by-block with the result being that there is a stride in memory for addressing variables in different zones or blocks.

As it was one of the marquee applications for the Ookami project, FLASH was ported as soon as Ookami was up and available. Sorting out the compilers and their options, versions of MPI, and requisite packages like the HDF5 library took some effort, but our initial experience with FLASH and other applications was overwhelmingly positive [3]. FLASH ran "right out of the box" with several compilers and MPI implementations, scaling reasonably well with no tuning. Profiling with Linaro (née Arm) MAP [22] indicated that thermonuclear supernova simulations, our problem of interest, spent considerable time in the hydrodynamics and equation of state (EOS) routines so we decided to focus our analysis on those routines while running 2D supernova simulations. We investigated use of SVE with the HPE/Cray, Fujitsu, and GNU compilers, but vectorization proved difficult due to significant branching in the main loops of the EOS routines. Details of our exploration with scaling results and our attempt to utilize the A64FX's SVE instructions and NUMA architecture may be found in [23].

To investigate other areas where we could improve performance, we instrumented the code with the Performance Application Programming Interface (PAPI) [24], and found that the number of data translation lookaside buffer (DTLB) misses were exceptionally high. The DTLB is a special cache that manages the mapping of virtual to physical memory. Given the stride in memory of the data layout of PARAMESH, a logical choice was to investigate memory management to study the high number of DTLB misses and we chose the use of Huge Pages.

3.2.2. Huge Pages and Compiler Comparison Huge Pages (hp) are a Linux kernel feature that allow blocks of memory larger than the default 64 KB page to be used when managing virtual memory, reducing the number of pages managed by the operating system (OS). This is useful for simulations with a large memory footprint. Hp come in two types, standard (explicitly managed) and transparent (automatically managed by the OS). Transparent hp are by default disabled on Ookami, so the results presented here use standard 2 MB hp.

For our hp study, we ran 2D supernova simulations and 3D pure hydrodynamics simulations of the Sedov explosion problem, one of the standard test problems provided with FLASH. We dubbed these "EOS" and "3-d Hydro" because the equation of state and 3D hydrodynamics routines were the parts of the code instrumented for performance testing. Details of the EOS and the Sedov explosion problem may be found in the original FLASH paper [5]. For the hp study, the EOS test ran a $\sim$ 1 GB 2D SN Ia simulation for 50 time steps and one refine ending with 624 blocks and the 3D Hydro test ran a $\sim$ 9 GB Sedov explosion simulation for 2 time steps and one refine ending with 3337 blocks. Both tests were run on 1 and 12 cores. FLASH Morton orders the blocks to be spatially located together, so to spread the blocks as much as possible the runs on 12 cores used the round robin distribution of processors. Details of the simulations, science, and results may be found in [25].

We began our study of hp with the Fujitsu compiler because it readily made use of hp (by default) and needed just a compiler flag to toggle between using hp or not. Subsequent investigation allowed us to utilize hp with other compilers by linking to the Fujitsu library `libmpg`. Using PAPI, we collected hardware counter data over the course of the run, and then used it to calculate derived quantities such as bandwidths and latencies following the calculations in the A64FX manual [26]. We refer to these counters and their derived metrics as "performance measures". The performance measures were then used to compare runs with hp, without hp, with the Fujitsu compiler, and with the GCC compiler, to better understand how different settings and compilers took advantage of the architecture. Runs on 1 core and 12 cores exhibited similar patterns in their performance measures, so we show results for only the 1 core runs below.

Figure 2 shows a comparison between simulations of the two test problems compiled with the Fujitsu 4.5 and GCC 12.2.0 compilers. Shown are ratios with and without hp of runtime, CPU cycles, DTLB misses, L1 and L2 DTLB misses, L1D and L2 cache misses, cycles waiting for memory, Bandwidth between L1D and L2 caches and L2 and HBM, and Latency for L1D cache misses and L2 cache misses.

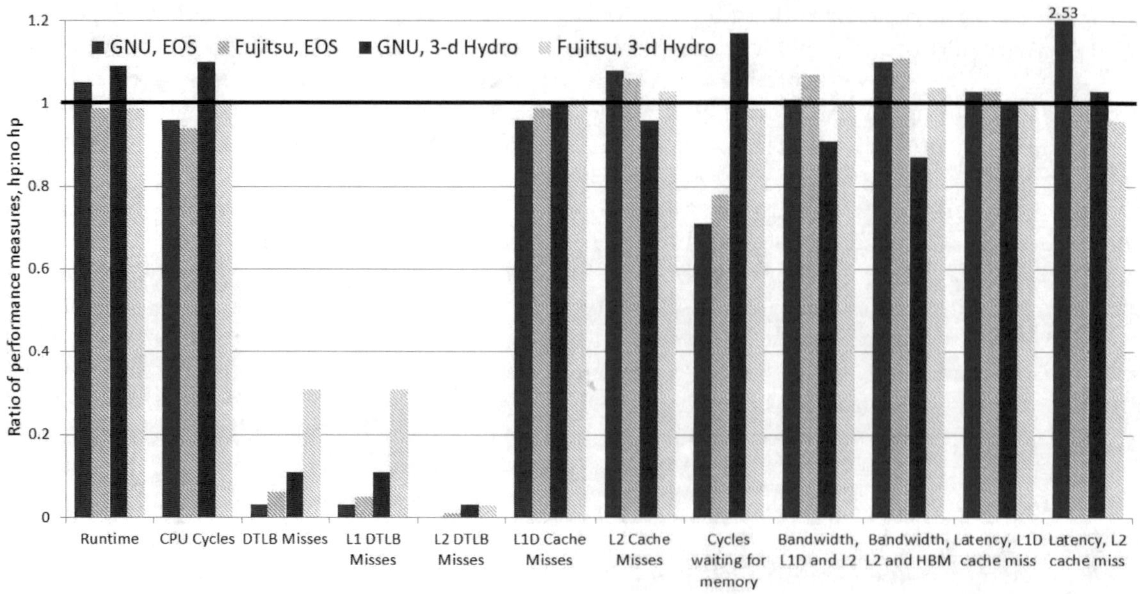

Figure 2. Bar chart showing the ratios of performance measures with and without hp for simulations of the two test problems with the Fujitsu and GNU compilers. Each bar is the ratio of each metric for simulations with and without hp. The DTLB misses show a drastic decrease when hp is enabled, regardless of compiler or test problem. For reasons unknown, the GNU executable for the EOS problem had 2.53× higher latency with hp enabled than disabled. The rest of the measures remain fairly close to 1.

Figure 3 presents the results of performance tests of FLASH for the two problems with and without hp. Shown are the same performance measures as the compiler test of Figure 2, except the ratios compare the compiler performance of the Fujitsu to the GNU compiler.

The results for the hp study may be summarized as follows. The use of hp decreased DTLB misses as expected, but this decrease did not affect performance as we expected. We attribute this finding to the performance of the translation table cache, but further research is needed to fully understand this result. The difference in compilers, however, had a large effect on runtime without making any changes to the code. FLASH performed faster with the Fujitsu compiler, requiring half the runtime and half the number of CPU cycles as with the GNU compiler. Fujitsu-compiled executables can access HBM 1.5 - 3× faster, so even though a similar number of cache misses as with the GNU compiler were seen, misses are not as expensive. Using the Fujitsu compiler, the code spends half as much time waiting for memory than when compiled with the GNU compiler. But because we also found that memory access is only 20-40% of the runtime (not illustrated in the plots), we conclude that the increased bandwidth can't completely account for the speedup. Complete results of our exploration of hp may be found in [27, 28].

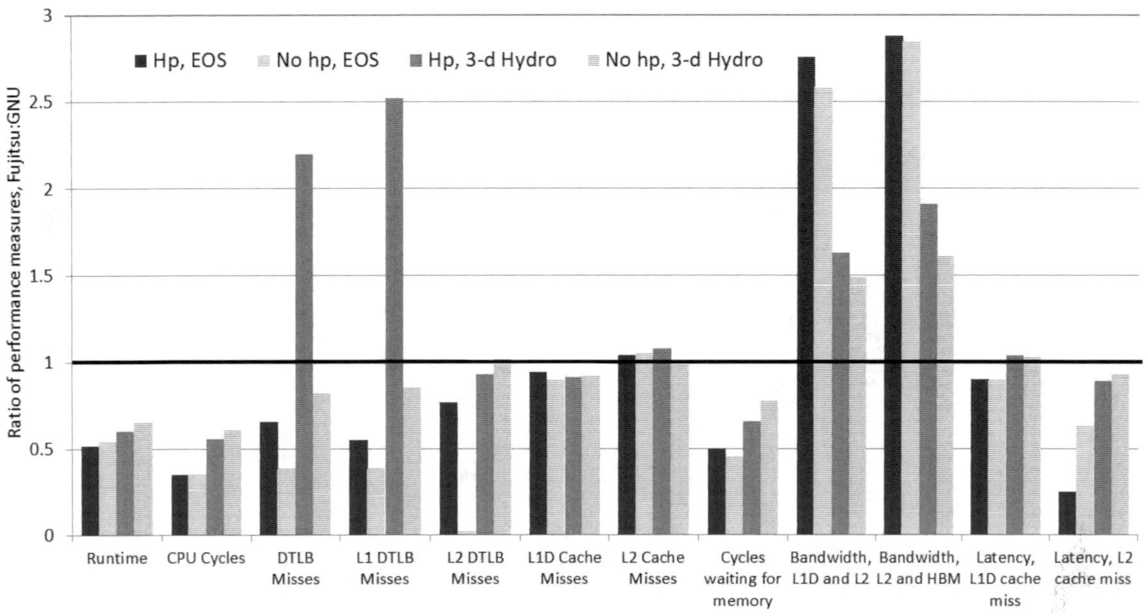

Figure 3. Bar chart showing the ratios of performance measures for Fujitsu and GNU compilers for simulations of the two test problems with (Hp) and without (No hp) hp. Each bar is the ratio of each metric for simulations compiled with the Fujitsu to the GNU compiler.

3.2.3. 3D Production Test As a way of testing the efficacy of Ookami for production runs, for this work we performed a scaling study with a full 3D simulation of a thermonuclear supernova, one of a suite of simulations from an ongoing science investigation. In this case, FLASH was compiled for $16 \times 16 \times 16$ zones per block and a maximum of 50 blocks per core. The simulations included the full physics of the flame model, equation of state, and multipole Poisson solve, and were run under production conditions in which the simulation was restarted from a checkpoint file and the mesh adapted as dictated by the fluid flow, typically every two time steps. FLASH was compiled with the GNU compiler v12.1.0 and were linked to the OpenMPI v4.1.4 MPI library and parallel hdf5 v1.12.1.

The simulations for the scaling study were started "in medias res" at about 0.5 s into the explosion when the flame is very active. The simulations consisted of approximately 33,000 blocks and ran for 20 time steps with 10 mesh refinements. The restart file was 30 GB, and the simulation required a minimum of 220 GB to run. The simulations were run on 21, 32, 64, and 128 nodes 7 times each, removing the minimum and maximum runtimes and averaging the remaining 5. The simulation ran on 36 of the 48 cores on each node, with 9 MPI ranks per NUMA region. This choice follows from experimentation with running simulations to fill much of the memory on a node. We discuss this issue in more detail below. The results of the scaling study are illustrated in Figure 4 and demonstrate reasonable strong scaling through 128 nodes. Also, the figure presents curves for the the total run time and the evolution time. The difference is the time required for initialization, reading the checkpoint file, and saving the final output, about 9 - 33 % of the run time. Although the input file is the same size in all cases, this percentage increases with the number of nodes as the amount of communication increases.

The choice of running on 36 of 48 possible cores follows from limitations of both FLASH and the A64FX memory. The number of blocks allocated must fit within the effective 27 GB node memory limit of Ookami. This is complicated by the adaptive mesh which needs more space to store its tree structure as the number of blocks grows, and the memory used by MPI which

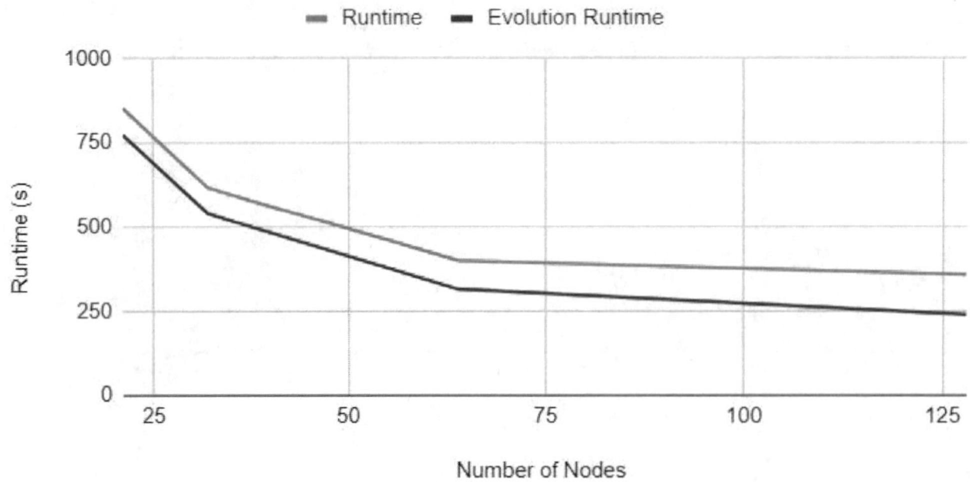

Figure 4. Strong scaling study for a full 3D simulation of a thermonuclear supernova with FLASH. The simulation had $\sim$ 33,000 blocks, and ran for 20 timesteps and 10 grid refinements. Shown are the full runtime and the evolution runtime which excludes I/O and initialization.

increases with the number of nodes in use. This sets an upper limit on the number of blocks that can be allocated to each processor. At the same time, for reasons we are still investigating, FLASH has issues with its guard blocks if too few blocks are allocated per processor. These opposing requirements meant that we needed to maximize the number of processors per node, and minimize the number of blocks allocated per process. We found a balance by allocating 50 blocks per process using 36 of the 48 available cores, which uses 26 GB of memory per node.

Using this configuration, we would be able to allocate a maximum of $\sim$316,000 blocks, or 4.7 TB, if we used the full machine. This limit is reached quickly in 3D, especially when we include the effects of a background turbulence field, a current work in progress, in our simulation. However, knowledge of this machine can lead to success on its larger counterpart, the supercomputer Fugaku, which has nearly 1000 times the number of nodes and memory than Ookami, along with a custom interconnect and additional assistant cores. This makes testing code on Ookami even more valuable, as understanding this testbed can lead to successful runs on more powerful machines with the same architecture.

4. Summary and Conclusions

Judging from the machine's use, the activity in Office Hours and the Slack channel, and the attendance at webinars, we conclude that the project has been successful in that it has indeed allowed many users with a variety of applications to meaningfully explore the Fujitsu A64FX processor. Our education and outreach efforts have been effective and have created a community of users that work well together.

Our performance studies for the two astrophysics codes produced mixed results. Both codes were readily ported to Ookami and ran with minimal adjustment. The use of SVE improved the performance of V2D but branching in the material EOS routines of FLASH prevented meaningful vectorization and thus there was no performance improvement. The use of hp with FLASH produced a dramatic decrease in DTLB misses but did not improve the performance of FLASH. This result suggests that DTLB misses do not adversely affect the performance of FLASH. We attribute this finding to the A64FX's translation table cache, which decreases the latency of virtual to physical address translation [26]. Finally, our results show that the best performance for V2D and FLASH applications on the A64FX architecture of Ookami occurs with the use of the Fujitsu compiler.

While the relatively small available memory per node of Ookami presented some challenges, our strong scaling study with FLASH under production conditions demonstrated reasonable scaling. We conclude that with experimentation and tuning, the machine is able to perform production simulations for multi-scale, multi-physics applications with a variety of solvers and provide valuable experience for porting larger and more detailed simulations to the supercomputer Fugaku.

Acknowledgments

Ookami is supported by the US NSF grant #1927880, and this research was supported in part by the US DOE under grant DE-FG02-87ER40317. FLASH was developed in part by the US DOE NSA-ASC and OSC-ASCR-supported Flash Center for Computational Science at the University of Chicago. The authors gratefully acknowledge the generous support of the Ookami community. The authors also thank Jens Domke at RIKEN for very helpful suggestions.

References

[1] Matsuoka S 2021 Fugaku and a64fx: the first exascale supercomputer and its innovative arm cpu *2021 Symposium on VLSI Circuits* pp 1–3

[2] IACS 2020 Ookami homepage URL `https://www.stonybrook.edu/commcms/ookami/`

[3] Burford A, Calder A, Carlson D, Chapman B, Coskun F, Curtis T, Feldman C, Harrison R, Kang Y, Michalowicz B, Raut E, Siegmann E, Wood D, DeLeon R, Jones M, Simakov N, White J and Oryspayev D 2021 Ookami: Deployment and initial experiences *Practice and Experience in Advanced Research Computing* PEARC '21 (New York, NY, USA: Association for Computing Machinery) ISBN 9781450382922 URL `https://doi.org/10.1145/3437359.3465578`

[4] Swesty F D and Myra E S 2009 *ApJS* **181** 1–52

[5] Fryxell B, Olson K, Ricker P, Timmes F X, Zingale M, Lamb D Q, MacNeice P, Rosner R, Truran J W and Tufo H 2000 *The Astrophysical Journal Supplement Series* **131** 273–334

[6] Dubey A, Antypas K, Calder A, Fryxell B, Lamb D, Ricker P, Reid L, Riley K, Rosner R, Siegel A, Timmes F, Vladimirova N and Weide K 2013 The software development process of flash, a multiphysics simulation code *2013 5th International Workshop on Software Engineering for Computational Science and Engineering (SE-CSE)* pp 1–8

[7] 2023 NSF ACCESS program URL `https://access-ci.org`

[8] Mann B 2017 Arm architecture - armv8.2-a evolution and delivery URL `https://community.arm.com/arm-community-blogs/b/architectures-and-processors-blog/posts/arm-architecture-armv8-2-a-evolution-and-delivery`

[9] Siegmann E, Calder A, Feldman C and Harrison R J 2021 Educating hpc users in the use of advanced computing technology *2021 IEEE/ACM Ninth Workshop on Education for High Performance Computing (EduHPC)* pp 16–23

[10] 2023 Campus Champions URL `https://campuschampions.cyberinfrastructure.org/`

[11] 2023 Science Gateways URL `https://sciencegateways.org/`

[12] Smolarski D C, Swesty F and Calder A C 2022 Performance of an astrophysical radiation hydrodynamics code under scalable vector extension optimization *2022 IEEE International Conference on Cluster Computing (CLUSTER)* (Los Alamitos, CA, USA: IEEE Computer Society) pp 545–548 URL `https://doi.ieeecomputersociety.org/10.1109/CLUSTER51413.2022.00071`

[13] Townsley D M, Calder A C and Miles B J 2019 Modeling subgrid combustion processes in simulations of thermonuclear supernovae *Journal of Physics Conference Series* (*Journal of Physics Conference Series* vol 1225) p 012004 (*Preprint* 1908.06176)

[14] 2023 The Flash Center for Computational Science URL https://flash.rochester.edu

[15] O'Neal J, Weide K and Dubey A 2018 Experience report: Refactoring the mesh interface in flash, a multiphysics software *WSSSPE6.1, colocated with eScience 2018, Amsterdam, Netherlands*

[16] MacNeice P, Olson K M Mobarry C, de Fainchtein R and Packer C 1999 *NASA Tech. Rep. CR-1999-209483*

[17] MacNeice P, Olson K M Mobarry C, de Fainchtein R and Packer C 2000 *Comput. Phys. Commun.* **126** 330–354

[18] Dubey A, Calder A, Fisher R, Graziani C, Jordan G, Lamb D, Reid L, Townsley D and Weide K 2013 *International Journal of High Performance Computing Applications* **27** 360–373 ISSN 1094-3420

[19] Daley C, Bachan J, Couch S, Dubey A, Fatenejad M, Gallagher B, Lee D and Weide K 2012 Adding shared memory parallelism to FLASH for many-core architectures *TACC-Intel Highly Parallel Computing Symposium* poster

[20] Dubey A 2019 Programming abstractions for orchestration of hpc scientific computing https://chapel-lang.org/CHIUW2019.html keynote, Chapel User's Group Meeting

[21] Dubey A 2019 Dynamic resource management, an application perspective https://project.inria.fr/resourcearbitration/program/ invited talk, RADR, co-located with IPDPS

[22] Linaro 2021 Linaro map URL https://www.linaroforge.com/linaroMap/

[23] Feldman C, Michalowicz B, Siegmann E, Curtis T, Calder A and Harrison R 2022 Experiences with porting the flash code to ookami, an hpe apollo 80 a64fx platform *International Conference on High Performance Computing in Asia-Pacific Region Workshops* HPCAsia 2022 Workshop (New York, NY, USA: Association for Computing Machinery) p 72–77 ISBN 9781450395649 URL https://doi.org/10.1145/3503470.3503478

[24] 2022 Performance Application Programming Interface URL http://icl.cs.utk.edu/papi/

[25] Feldman C, Gutierrez N, Eisenberg E, Willcox D E, Townsley D M and Calder A C 2023 *ApJ* **959** 112 (*Preprint* 2309.07283)

[26] Fujitsu 2023 A64fx microarchitecture manual URL https://github.com/fujitsu/A64FX/blob/master/doc/A64FX_Microarchitecture_Manual_en_1.3.pdf

[27] Calder A C, Feldman C, Siegmann E, Dey J, Curtis A, Chheda S and Harrison R J 2022 On using linux kernel huge pages with flash, an astrophysical simulation code *2022 IEEE International Conference on Cluster Computing (CLUSTER)* (Los Alamitos, CA, USA: IEEE Computer Society) pp 539–544 URL https://doi.ieeecomputersociety.org/10.1109/CLUSTER51413.2022.00070

[28] Feldman C, Chheda S, Calder A, Siegmann E, Dey J, Curtis T and Harrison R 2023 A further study of linux kernel hugepages on a64fx with flash, an astrophysical simulation code *Practice and Experience in Advanced Research Computing* PEARC '23 (New York, NY, USA: Association for Computing Machinery) ISBN 9781450382922 URL https://doi.org/10.1145/3569951.3597583

Journal of Physics: Conference Series **2742** (2024) 012020 doi:10.1088/1742-6596/2742/1/012020

Advancing parabolic operators in thermodynamic MHD models II: Evaluating a Practical Time Step Limit for Unconditionally Stable Methods

Ronald M. Caplan[1], Craig D. Johnston[2,3], Lars K. S. Daldoff[4,3], and Jon A. Linker[1]

[1] Predictive Science Inc., 9990 Mesa Rim Road Suite 170, San Diego, CA 92121

[2] Department of Physics and Astronomy, George Mason University, Fairfax, VA 22030

[3] Heliophysics Science Division, NASA Goddard Space Flight Center, Greenbelt, MD 20771

[4] Catholic University of America, 620 Michigan Ave., N.E. Washington, DC 20064

E-mail: caplanr@predsci.com, craig.d.johnston@nasa.gov, lars.daldorff@nasa.gov, linkerj@predsci.com

Abstract. Unconditionally stable time stepping schemes are useful and often practically necessary for advancing parabolic operators in multi-scale systems. However, serious accuracy problems may emerge when taking time steps that far exceed the explicit stability limits. In our previous work, we compared the accuracy and performance of advancing parabolic operators in a thermodynamic MHD model using an implicit method and an explicit super time-stepping (STS) method. We found that while the STS method outperformed the implicit one with overall good results, it was not able to damp oscillatory behavior in the solution efficiently, hindering its practical use. In this follow-up work, we evaluate an easy-to-implement method for selecting a practical time step limit (PTL) for unconditionally stable schemes. This time step is used to 'cycle' the operator-split thermal conduction and viscosity parabolic operators. We test the new time step with both an implicit and STS scheme for accuracy, performance, and scaling. We find that, for our test cases here, the PTL dramatically improves the STS solution, matching or improving the solution of the original implicit scheme, while retaining most of its performance and scaling advantages. The PTL shows promise to allow more accurate use of unconditionally stable schemes for parabolic operators and reliable use of STS methods.

1. Introduction

Thermodynamic magnetohydrodynamic (TMHD) models, like many other models of complex physical systems, contain processes spanning widely varying time scales. This makes numerical integration difficult, as the time steps required to describe the fastest processes can be quite small. When using explicit time-stepping methods, the stability requirements for these fast processes can further restrict the time steps (often a great deal), making the simulations computationally intractable. To combat this problem, unconditionally stable time-stepping algorithms can be used, effectively eliminating the numerical time step restrictions. However, such methods are prone to miss-use, as integrating them with very large time steps can cause fast processes of the system to not be captured, and/or introduce large errors to the system.

In our previous work ([1], referred henceforth as 'Paper I'), we compared two popular unconditionally stable time-stepping algorithms applied to the parabolic operators within the

Journal of Physics: Conference Series **2742** (2024) 012020 doi:10.1088/1742-6596/2742/1/012020

Magnetospheric Algorithm outside a Sphere (MAS) TMHD model, which is used to study the solar corona and heliosphere (see references therein). These were the L-stable [2] implicit Backward Euler (BE) method solved with a preconditioned conjugate gradient (PCG) solver, and an explicit super time stepping (STS) scheme, specifically, the A-stable [2] second-order Legendre Extended Stability Runge-Kutta method (RKL2) [3]. It was found that the RKL2 method computationally outperformed BE+PCG (especially in scalability across many CPUs), but suffered from solution artifacts in some isolated regions of the model domain. These artifacts appeared to be the result of using too large a time step combined with the poor high-mode damping of the A-stable RKL2 method when applied to the artificial kinematic viscosity term used to damp small spurious oscillations arising from other parts of the model.

This issue raised concerns about the robust use of STS methods in general, and in MAS, led to limiting its production use to a small selection of run types. While the BE+PCG's L-stable properties made it more robust (avoiding the solution artifacts), we have observed cases where it too can suffer from weak damping of oscillations when given too large of a time-step. One way to mitigate this issue is to 'cycle' the parabolic operator, effectively reducing the time-step, and allowing the high modes to damp through successive application of the operator. However, it is difficult to know how many cycles are required a-priory [4], and overestimation can significantly reduce the computational performance.

In this work, we follow-up Paper I by evaluating a novel, easy-to-implement method for dynamically calculating a practical time step limit (PTL) for unconditionally stable time stepping schemes applied to parabolic PDEs. The PTL is implemented in MAS, where we use it to automatically set the minimum number of cycles needed for the operator-split parabolic operators over one full MAS time step (set by the flow CFL condition). To improve performance, the PTL is re-evaluated dynamically each cycle, reducing the total number of cycles needed over the full time step. As in Paper I, we test the PTL using an implicit scheme and an STS scheme. We use two real-world cases to compare solution quality, performance, and scaling.

The paper is organized as follows. In Sec. 2, we describe the parabolic operators in the MAS model and present a brief derivation of the PTL. Then, in Sec.3, we describe the test setup, including the unconditionally stable methods, the test cases, and the computational environment. Solutions computed with and without using the PTL are evaluated in Sec. 4, including performance and scaling results. We conclude with a discussion in Sec.5.

2. A practical time step limit for integrating parabolic operators

Often, parts of a physical model that have smaller time scales than others are operator-split, such that each operator is integrated over the overall time step using the partial solution of the operator before it. The order and time steps of the splitting can be done in several ways, each with a cost of adding a splitting error to the system [5]. Each split operator is advanced over the time step independently by integrating

$$\frac{\partial \mathbf{u}}{\partial t} = \mathbf{F}(\mathbf{u}), \tag{1}$$

where $\mathbf{F}(\mathbf{u}) \equiv \mathbf{F}(\mathbf{x}, \mathbf{u}, ...)$ is the currently split operator.

The MAS model[1] described in Paper I (see references within) uses an overall time step set by a flow CFL condition, and operator splits several terms that would require very small time steps. These include the parabolic operators of artificial kinematic viscosity and Spitzer thermal conduction [6] given by

$$\mathbf{F}_{\text{visc}}(\mathbf{v}) = \frac{1}{\rho}\,\nabla\cdot\left(\nu(\mathbf{x})\,\rho\,\nabla\mathbf{v}\right), \text{ and } \mathbf{F}_{\text{tc}}(T) = \frac{(\gamma-1)\,m_p}{2\,k\,\rho}\,\nabla\cdot\Big(f_{\text{c}}(r)\,f_m(T_0)\,\kappa_0\,T_0^{5/2}\,\hat{\mathbf{b}}\hat{\mathbf{b}}\cdot\nabla T\Big), \tag{2}$$

[1] See predsci.com/mas for additional details and references

where $\mathbf{v}$, T, and ρ are the plasma flow velocity, temperature, and density respectively, $\nu(\mathbf{x})$ is the coefficient of viscosity, $\hat{\mathbf{b}} = |\mathbf{B}|/\mathbf{B}$ is the normalized direction of the magnetic field, $\gamma = 5/3$ is the adiabatic index, m_p is the proton mass, k is Boltzman's constant, κ_0 is the Spitzer coefficient for thermal conduction, $f_m(T)$ is used to increase the parallel thermal conductivity at low temperatures which, in conjunction with an inverse modification to radiative cooling, broadens the transition region [7], $f_C(r)$ is a profile that limits the radial extent that collisional thermal conduction is active, and T_0 is the previous step's temperature used to keep the operator linear through lagged diffusivity [8].

If the numerical stability time step limit of the operator-split parabolic term is smaller than the overall step, it can be cycled at its stable limit (at extreme computational cost), or an unconditionally stable scheme can be utilized. As described in Sec. 1, the latter can lead to problems when the overall time step is very large compared to the explicit limit. In such cases, the operator can be cycled (with many less cycles than an explicit scheme) to reduce solution artifacts. However, the number of cycles is typically set by experimentation (see the discussion in Ref. [4]) which can be time consuming to determine, and is therefore not practical for large MHD models.

A novel method to compute a practical time-step limit (PTL) for a parabolic operator is introduced in Ref [9]. Given a parabolic equation of the form of Eq. 1, the goal is to maintain the sign of the difference in adjacent cells when integrating from time t^n to $t^{n+1} = t^n + \Delta t$:

$$\Delta \mathbf{u}_{\vec{i}}^{n+1}\, \Delta \mathbf{u}_{\vec{i}}^{n} \geq 0, \tag{3}$$

where $\vec{i}$ is the grid point in question and $\Delta \mathbf{u}$ is the difference between the solution at $\vec{i}$ and its cell neighbor. In general, this condition can be considered in all directions and dimensions. For simplicity, we show one direction, where we define $\Delta \mathbf{u}_i^n = \mathbf{u}_i^n - \mathbf{u}_{i-1}^n$. Using the first-order Taylor expansion of the solution $\mathbf{u}^{n+1}$ in Δt and dropping higher order terms, Eq. 3 becomes

$$(\Delta \mathbf{u}_i^n)^2 + \Delta t\, \Delta \mathbf{u}_i^n\, \Delta \mathbf{F}_i^n \geq 0,$$

where $\Delta \mathbf{F}_i^n = \mathbf{F}_i(\mathbf{u}^n) - \mathbf{F}_{i-1}(\mathbf{u}^n)$. The only case where this condition is not unconditionally satisfied is when $\Delta \mathbf{u}_i^n \neq 0$, $\Delta \mathbf{F}_i^n \neq 0$, and $\Delta \mathbf{u}_i^n\, \Delta \mathbf{F}_i^n < 0$, in which case

$$\Delta t \leq -\frac{\Delta \mathbf{u}_i^n}{\Delta \mathbf{F}_i^n}. \tag{4}$$

The condition can be checked in all directions surrounding grid cell $\vec{i}$ that meet the criteria over every grid cell, and the minimum time step calculated. However, Ref. [9] found (in their cases) that the minimum Δt always occurred at the location of the maximum absolute value of the components of $\mathbf{F}$ (denoted as F_{max}). This allowed them to avoid needing to compute the conditions over the whole grid and sidestep numerical sensitivity issues near values very close to zero. Here, we make the assumption that their result extends to our cases and only compute the time step conditions at the grid location of F_{max}, which we denote as $\vec{k}$. The PTL can then be defined as

$$\Delta t_{\mathrm{PTL}} \leq \min \left[-\frac{u_{\vec{k}}^n - u_{\vec{k}+\vec{i}}^n}{F_{\vec{k}}(u^n) - F_{\vec{k}+\vec{i}}(u^n)} \right], \tag{5}$$

where the minimum is taken for all $\vec{i}$ one cell away in the 6 orthogonal grid directions.

Once the PTL is found, the number of cycles to integrate the operator at the PTL over the larger overall time-step can be set. However, since the operator is parabolic and smooths the solution with each application, the PTL can be dynamically re-evaluated and applied after each cycle. This can increase the PTL size after each step, leading to much fewer overall cycles, increasing performance. A detailed formulation and analysis of the PTL is discussed in Ref [9]. Here, we test applying the PTL to a full production code (MAS) on real-world problems.

3. Test setup

3.1. Unconditionally stable schemes

In Paper I, we described our implementation of the first-order accurate, L-stable BE+PCG scheme, which has been successfully used in the MAS code for many years. The implementation uses two choices of preconditioners (PC1 and PC2). PC1 (point-Jacobi) is inexpensive to formulate and apply, but limited in its effectiveness to reduce solver iterations, while PC2 (non-overlapping ILU0 factorization) is more expensive to formulate and apply, but also more effective at reducing iterations. PC1 is easily vectorized, making it simple to implement efficiently on vector-optimized hardware (e.g. GPUs), while the standard implementation of PC2 is not vectorizable, requiring the use of alternative algorithms and libraries for use on GPUs [10, 11]. The current version of MAS only supports PC1 on GPUs, while PC2 is used when running on CPUs (an important factor in comparing MAS's CPU and GPU performance). The BE+PCG has two important drawbacks - the requirement of global communications for the dot products which hinders scaling, and the necessity of using a linear operator.

The other scheme described and tested in Paper I was the explicit second-order extended stability Runge-Kutta Legendre super time stepping (RKL2) scheme [12, 3]. RKL2 is easy to implement, second-order accurate, can handle a non-linear operator, has no global synchronization points, and is vectorizable yielding efficient implementation on GPUs. However, it is only A-stable and does not damp high modes well for large time steps.

The second-order Gegenbauer method (RKG2) recently introduced in Ref. [13] has similar performance to the RKL2 scheme, but with a better damping amplification factor for high modes. In Fig. 1, we show the estimated speedups of RKL2 and RKG2, and compare the amplification factors of the BE, RKL2, and RKG2 with $\alpha = 3/2$ (see Ref. [14]) schemes to the exact solution of a simple 1D heat equation problem. We see that the RKG2 scheme's amplification factor

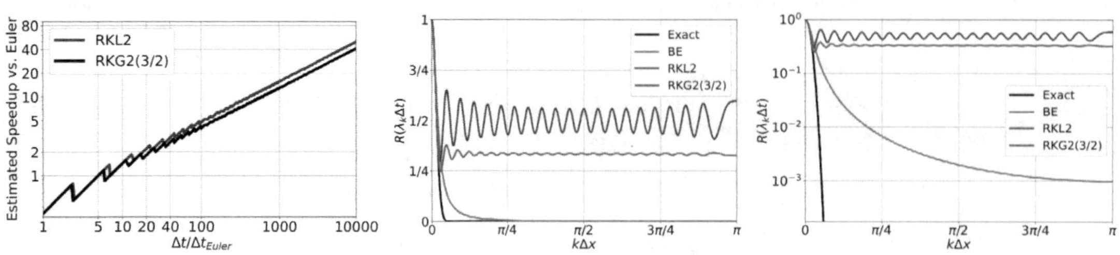

Figure 1: Left: Estimated speedup of the RKL2 and RKG2(3/2) schemes compared to explicit Euler (approximating one Euler step to have the same computational cost as one STS iteration). Middle: Amplification factors for the BE, RKL2, and RKG2 schemes when applied to a 1D uniform grid heat equation discretized with a second-order central finite difference at a time-step 500 times larger than the explicit Euler limit. Right: Same as middle, but on a log scale to show the difference in the BE scheme's damping rate compared to the exact solution.

is qualitatively similar to the RKL2, but saturates to a lower value and is less variable across modes. The RKG2 is also shown to be similar in performance to RKL2. We note that even the BE scheme's damping of modes diverges from the exact solution greatly, but overall it is much better than that of the STS schemes.

Since the inability to efficiently damp high modes is a key factor in the solution artifacts we observe in Paper I, we choose to use RKG2 over RKL2 in this work. We also note that it is critically important to ensure that the number of STS iterations of the RKG2 method be odd, otherwise the amplification factor approaches one at the highest modes. Details of the RKG2 scheme implementation are shown in Appendix A.

3.2. Test cases

To test the effects and performance of the PTL, we utilize two production MAS simulation runs. The first test (Test 1) is a lower resolution version of a thermodynamic MHD relaxation based on simulations used in Ref. [15]. This test has 3.4 million highly non-uniform grid points and uses a constructed lower radial magnetic field boundary combining a global dipole with a localized bi-pole representing an active region. The second test (Test 2) is a slightly modified (small changes in the viscosity values) version of the thermodynamic MHD relaxation used in Paper I with a size of 22.8 million grid points. This test uses real-world observed surface radial magnetic fields for the lower boundary.

For comparing solutions with and without using the PTL, we run each test's relaxation for a total of 8 hours of simulated physical time. For performance and scaling results, we only use Test 2, where we start the simulation with the final output of an 8-hour relaxation, and continue the run for another 6 simulated physical minutes. Since these performance runs are much shorter simulations than those used in production, we subtract off the loading time of the relaxation solution initial condition from the wall clock time.

3.3. Computational environment

Since Paper I, the MAS code has been updated to run on NVIDIA GPUs through the use of OpenACC [16] and Fortran's 'do concurrent' standard parallelism constructs [17]. We therefore test the methods on both CPU and GPU systems. For the solution comparison run of Test 1, we utilize an in-house workstation with an NVIDIA RTX 3090Ti GPU, while for Test 2 we use 32 nodes of the Expanse system at the San Diego Supercomputer Center (SDSC). For testing the parallel scaling performance on CPUs, we use between one and 32 nodes of Expanse, while on GPUs, we use between one and eight GPUs on a single node at the Delta system at the National Center for Supercomputing Applications (NCSA). The details of the hardware and software configurations used are shown in Appendix B.

We note that the MAS code is highly memory-bandwidth bound, making the maximum memory bandwidth the best indication of expected relative performance between systems, given the same algorithm. However, as discussed in Sec. 3.1, MAS uses a more effective preconditioner when run on CPUs than on GPUs, making the relative performance between CPU and GPU a practical time-to-solution comparison, and not an apples to apples hardware comparison.

4. Results

The key questions we want to address are:

(i) Can using the PTL with the RKG2 scheme obtain a solution similar (or better) to the original BE+PCG scheme?

(ii) Does using the PTL with the BE+PCG scheme improve *its* solution?

(iii) How does the PTL affect performance for both BE+PCG and RKG2 schemes?

(iv) Does using the PTL allow RKG2 to be competitive with the original BE+PCG scheme in both solution quality and performance?

4.1. Solution comparisons

To address the first two questions, we run the test cases with both BE+PCG and RKG2, each with and without using the PTL and compare the solutions. Like in Paper I, the overall global solution looks qualitatively the same using all methods (not shown here). Our focus here is on the locations and quantities within the solutions that exhibit oscillatory behavior, such as the zoomed views in Paper I's Fig. 7. For Test 1, this issue can be seen in the ϕ-component of the velocity field near the solar surface. A zoomed portion of the solution, along with 1D cuts through the view are shown in Fig. 2.

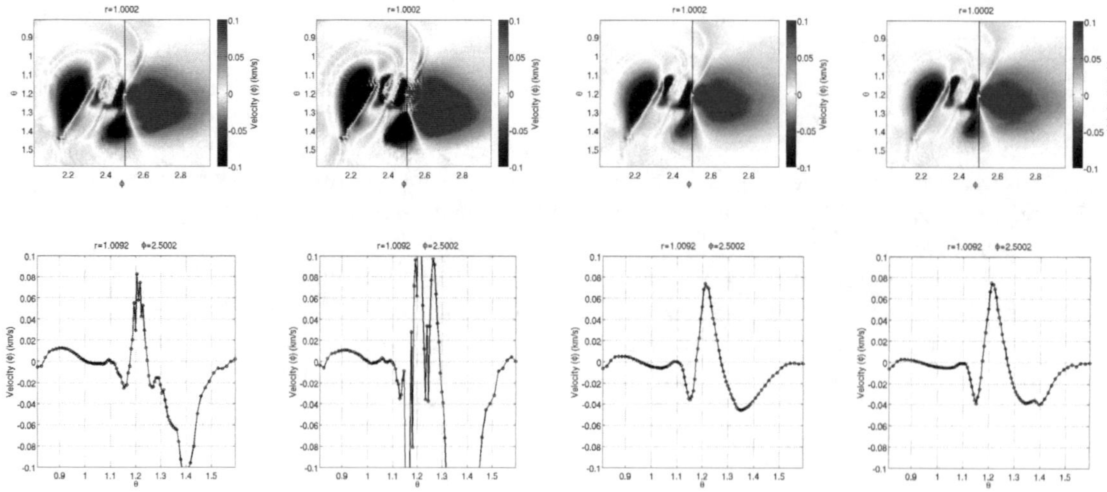

BE+PCG: Original RKG2: Original BE+PCG: PTL RKG2: PTL

Figure 2: Comparison run results for Test 1. A small section of the solution output of v_ϕ is shown in the $\theta - \phi$ plane (top) with the black line indicating the cut that is plotted in 1D (bottom). The PTL is shown to improve the quality of the solution when using either RKG2 and BE+PCG. The scheme used for each result is shown at the bottom of each column, where the text color corresponds to the performance result figures in the next section.

In this case, we notice that even the original BE+PCG scheme does not fully damp out the run's oscillations, implying that either the viscosity coefficient of the problem is not high enough, or that the time step is so large, not even the BE+PCG scheme can apply the specified viscosity correctly. The latter seems to be the case since the PTL run with BE+PCG eliminates the oscillations, answering question (ii) in the affirmative.

For the RKG2 scheme, we see that the original implementation does a poor job applying the viscosity to damp oscillations and/or is adding its own oscillations due to the large time step (see Ref. [9]). In contrast, when using the PTL with RKG2, the oscillations have been damped, and the resulting solution is qualitatively similar to the BE+PCG solution with PTL, with only a small 'notch' near $\theta = 1.4$ being a noticeable difference. The quality improvement of the solution when using the PTL is also seen clearly in the 2D cut planes, where the PTL has drastically cleaned up the original RKG2 solution. In this case, we have shown that the answer to question (i) is 'yes', as the PTL RKG2 is not only 'as good' as the original BE+PCG, but, in this case, is substantially better.

Moving to Test 2, in Fig. 3 we show zoomed-in portions of the ϕ-component of the current density for each solution, along with 1D cuts through the view. We see similar results as in Test 1. The original BE+PCG scheme does not have much oscillatory behavior in this case, but when using the PTL, the solution changes and becomes much smoother overall. The original RKG2 scheme once again suffers from oscillatory behavior, but is cleaned up substantially when applying the PTL. In this case, the PTL RKG2 solution is even closer qualitatively to the PTL BE+PCG than in Test 1.

We can therefore conclude (at least for these cases) that the answer to both questions (i) and (ii) is 'yes', as the PTL improves the solution of both schemes, and allows the RKG2 scheme to yield a very close solution to BE+PCG, with the PTL RKG2 solution being a significant improvement over the original BR+PCG scheme. However, to know whether it is practical to

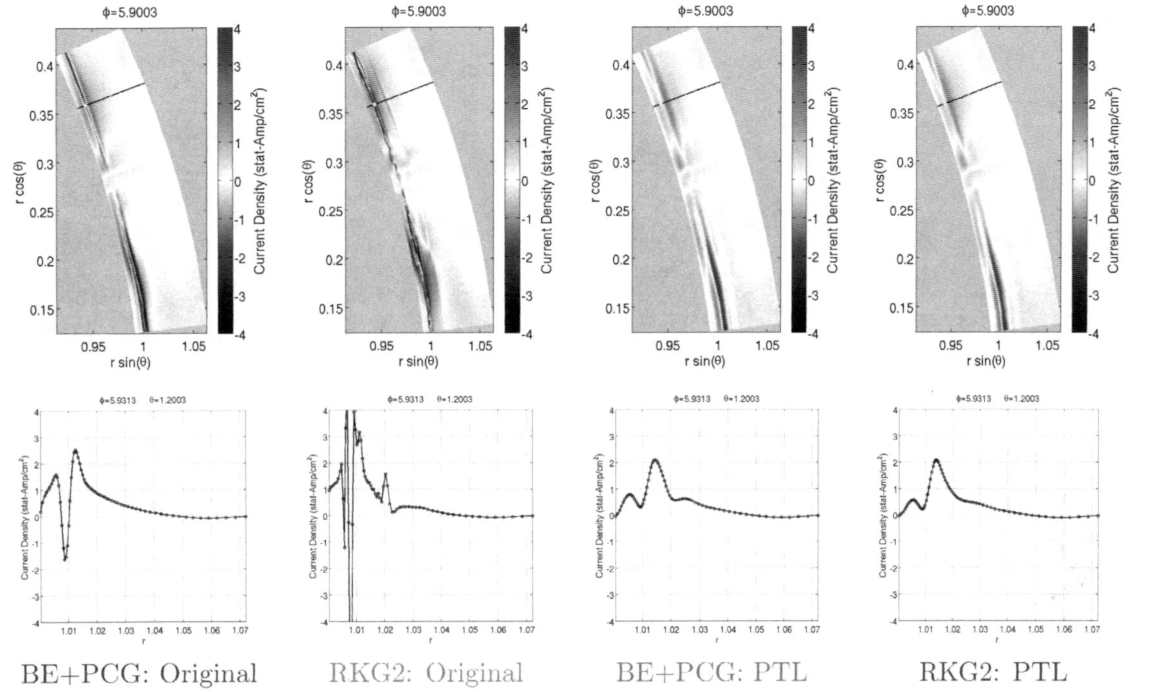

Figure 3: Comparison run results for Test 2. A small section of the solution output of j_ϕ is shown in the $r - \theta$ plane (top) with the black line indicating the cut that is plotted in 1D (bottom). The PTL is shown to improve the quality of the solution when using both RKG2 and BE+PCG. Notation is the same as Figure 2.

use the PTL for either scheme, these solution results need to be viewed in conjunction with each scheme's computational performance.

4.2. Performance

We start by looking at the performance results for the solution comparison runs shown in the previous section, which we show in Fig. 4. In both tests, we see that the original RKG2 scheme is faster than the original BE+PCG scheme as expected. Using the PTL on BE+PCG makes the code slower by up to a factor of two, with the slowdown for viscosity being a factor of around four. While this method did improve the solution over the original scheme, the improvement may not be worth the performance hit. Looking at the PTL RKG2 runs, we see that, while they are slower than the original RKG2 runs, they are comparable to the run time of the original BE+PCG scheme. Since the solution for the PTL RKG2 scheme is an improvement to the original BE+PCG solution, and comparable to the PTL BE+PCG scheme's solution, the small increase in run time is acceptable, and makes the RKG2 competitive with the original BE+PCG scheme, answering question (iv) in the affirmative for this case.

In Paper I, the performance advantages of the STS scheme over BE+PCG were even more pronounced when looking at parallel scaling across multiple CPUs. Here, we test the scaling with and without the PTL to see if the analysis above remains valid, and if the RKG2's advantage increases. The timing results for the two parabolic operators (thermal conduction and viscosity) portion of the runs are shown in Fig. 5. On the CPUs, we see that the BE+PCG exhibits expected scaling, and for the thermal conduction operator on large numbers of nodes, exhibits 'super-scaling'. This super scaling is likely due to the very large size and speed of the EPYC

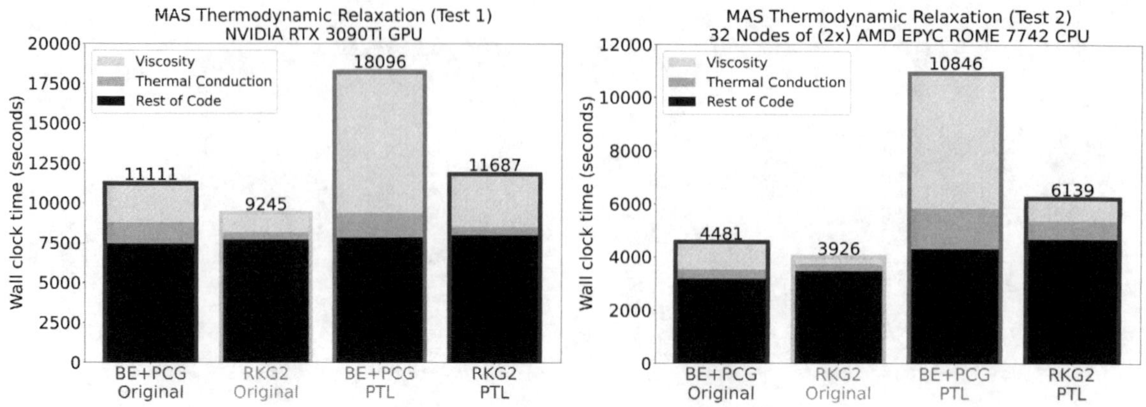

Figure 4: Timing results for the solution comparison runs in Sec. 4.1 for Test 1 (left) and Test 2 (right).

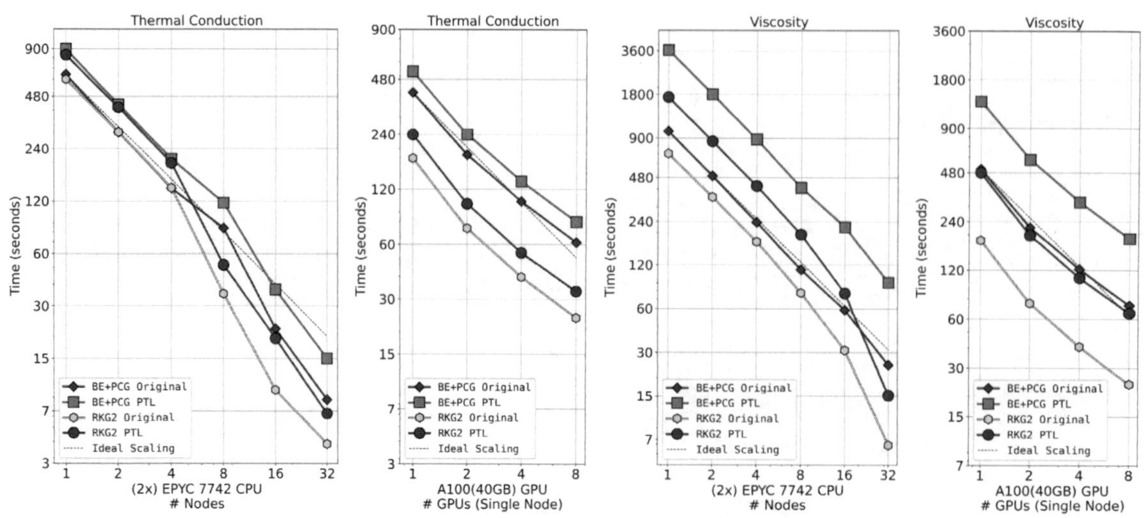

Figure 5: Test 2 scaling results for the thermal conduction (left two plots) and viscosity (right two plots) operators on CPUs (left) and GPUs (right).

CPU's cache, such that the local portion of the grid (which decreases with increasing number of CPU cores) is fitting into the fast cache. The RKG2 exhibits stronger super scaling for both the thermal conduction and viscosity operators, and is overall faster than the BR+PCG. When using the PTL, each method runs slower than their original implementations as expected, up to a factor of four. However, while the RKG2 with PTL runs slower than the original non-PTL BE+PCG method on small numbers of CPU nodes, when run on many CPU nodes, the super scaling causes it to be faster. Therefore, we see that the PTL RKG2 scheme can outperform the original BE+PCG scheme, while producing a better solution as shown in Sec. 4.1.

On the GPUs, all methods do not scale perfectly, likely caused by GPU-GPU communication overheads and the problem size not being large enough to fully saturate the GPUs. In this case, the RKG2 is always faster than the BE+PCG. This is expected since the GPU runs only use the less effective PC1 preconditoner for BE+PCG. For thermal conduction, the PTL RKG2 method outperforms the original BE+PCG method by a factor of 2, while for viscosity the two run times are almost the same. Given the superior solution quality of the PTL RKG2 over the original

BE+PCG method, these results strongly favor the use of PTL with RKG2 over BE+PCG.

In all the cases shown here, we see that using PTL with BE+PCG slows down the operator considerably, by up to a factor of 4. Therefore, although the PTL with BE+PCG yields a better solution than without PTL, the drop in performance may be unacceptable. This implies that the PTL has made STS methods competitive with BE+PCG, answering question (iv) in the affirmative.

The above analysis was for the two parabolic operators in isolation. However, since these operators are only part of the overall run time (around 30%, see Paper I) of the MAS code, we also look at the effects on the total wall clock time, which we show in Fig. 6. In the CPU case,

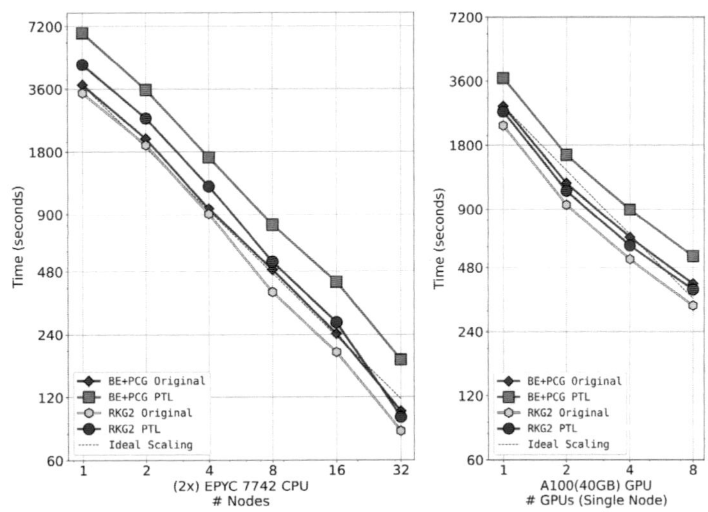

Figure 6: Test 2 scaling results for total Wall clock time (with startup time removed) on CPUs (left) and GPUs (right).

the overall code scales ideally, with a small amount of super-scaling when run on 32 CPU nodes. Given the amount of super scaling shown in Fig. 5, this shows that different sections of the code have varying levels of scaling efficiency.

We once again see that using the RKG2 scheme yields the best performance overall. The RKG2 scheme with PTL is slower than the original BE+PCG scheme for small numbers of CPU nodes, but has similar performance for more than 8 CPU nodes, and surpasses its performance for 32 CPU nodes. Using PTL with BE+PCG is once again much slower. On GPUs, the RKG2 with PTL closely matches the performance of the original BE+PCG scheme on all numbers of GPUs. Given the improvements of the solution using the PTL with RKG2 over the original RKG2 and BE+PCG, the new scheme is overall advantageous, where the performance loss on low numbers of CPU nodes is likely acceptable.

5. Discussion

In this paper, we have followed up our analysis in Paper I comparing unconditionally stable explicit STS schemes to implicit schemes for parabolic operators in a thermodynamic MHD model by applying an easy-to-use practical time step limit (PTL). This limit is designed to ensure a consistent solution structure from one time step to the next and is implemented as inner steps at a dynamically recalculated time step. Using the PTL was shown to improve the solution for both the RKG2 STS scheme and for the implicit BE+PCG scheme. Using the PTL reduces performance of the operators, in some cases by a large factor. However, it was found

Journal of Physics: Conference Series **2742** (2024) 012020 doi:10.1088/1742-6596/2742/1/012020

that the RKG2's original performance and scaling advantage over the BE+PCG scheme allow it to be competitive against the BE+PCG scheme even when using the PTL, all while exhibiting a much better solution than the original RKG2. The results presented imply the following answers to the questions in Sec. 4:

(i) Can using the PTL with the RKG2 scheme obtain a solution similar (or better) to the original BE+PCG scheme?
Yes, as shown in Figs. 2 and 3.

(ii) Does using the PTL with the BE+PCG scheme improve *its* solution?
Yes, as shown in Figs. 2 and 3.

(iii) How does the PTL affect performance for both BE+PCG and RKG2 schemes?
In both cases, the performance decreases, however due to the original performance advantage of the RKG2 scheme, the PTL applied to RKG2 exhibits similar or better performance than the original BE+PCG.

(iv) Does using the PTL allow RKG2 to be competitive with the original BE+PCG scheme in both solution quality and performance?
Yes, as shown in Figs. 2,3,4 and 6

We note that these answers are implied by the results obtained here, and are not guaranteed to be true in all cases. We have tested the PTL in other models, such as our flux evolution code HipFT[2], and our magnetic map smoothing tool called Diffuse [18]. In both these models, the PTL always avoided solution artifacts that were shown to occur when using very large time steps, with very little decline in performance. Work is proceeding on the theoretical formulation and analysis of the PTL, including careful testing, especially in nonlinear cases [9]. While the performance when applied to real-world simulations will be model and scheme dependent, the results shown here are encouraging that using the PTL can be used as a simple high performance way to avoid unwanted behavior of parabolic operators when using large time steps. The PTL also may be a great way to ensure proper behavior of STS methods, making them a more robust and practical option for the integration of parabolic operators in multi physics models such as thermodynamic MHD.

Appendix A.

The RKG2 scheme with $\alpha = 3/2$ [13, 14] is given by

$$u_0 = u^n$$
$$y_0 = F(u_0)$$
$$u_1 = u_0 + \tilde{\mu_1}\,\Delta t\, y_0$$

do k = 2 : s

$$u_k = \mu_k\, u_{k-1} + \nu_k\, u_{k-2} + (1-\mu_k-\nu_k)\, u_0 + \tilde{\mu_k}\,\Delta t\, F(u_{k-1}) + \gamma_k\,\Delta t\, y_0$$

enddo

$$u^{n+1} = u_s,$$

do k = 3 : s

$$b_k = \frac{4\,(k-1)\,(k+4)}{3\,k\,(k+1)\,(k+2)\,(k+3)}$$
$$\mu_k = \left(2+\frac{1}{k}\right)\frac{b_k}{b_{k-1}}$$
$$\nu_k = -\left(\frac{1}{k}+1\right)\frac{b_k}{b_{k-2}}$$
$$\tilde{\mu_k} = \mu_k\, w$$
$$\gamma_k = \left(\frac{k\,(k+1)}{2}\, b_{k-1} - 1\right)\tilde{\mu_k},$$

enddo

[2] Soon to be released at `https://www.github.com/predsci/hipft`

where u^n is the current solution, u^{n+1} is the solution after integrating the full time step Δt, F is the parabolic operator, u_k is the solution at STS iteration k, and

$$w = \frac{6}{(s+4)(s-1)}, \qquad b_0 = 1,\ b_1 = \tfrac{1}{3},\ b_2 = \tfrac{1}{15}, \qquad \mu_1 = 1,$$
$$\tilde{\mu_1} = w, \qquad \mu_2 = \frac{1}{2}, \qquad \nu_2 = -\frac{1}{10}, \qquad \tilde{\mu_2} = \mu_2 w, \qquad \gamma_2 = 0.$$

The number of required iterations s is given by

$$s = \left\lceil \frac{1}{2}\sqrt{25 + 24\,\frac{\Delta t}{\Delta t_{\mathrm{Euler}}}} - \frac{3}{2} \right\rceil,$$

where $\Delta t_{\mathrm{Euler}}$ is the explicit Euler stability time step. We note that it is critically important to ensure that the number of iterations be odd (by adding an iteration if needed), otherwise the amplification factor approaches one at the highest modes.

Appendix B.

The details of the hardware and software specifications used in this paper are shown in Table B1.

	SDSC Expanse CPU Node	**NCSA Delta GPU(8x) Node**	**Local GPU Workstation**
CPU	(2x) AMD EPYC Rome 7742	(2x) AMD EPYC Milan 7763	Intel Core-i5 Raptor Lake 13600KF
GPU	N/A	(8x) NVIDIA A100-40GB-SXM4	NVIDIA RTX 3090 Ti
CPU Cores	128	128	14
DRAM	128 GB	320 GB (GPUs)	24 GB (GPU)
Peak Memory Bandwidth	410 GB/s	12,440 GB/s (GPUs)	1,008 GB/s (GPUs)
Max Flops	9 TFlops	78 TFlops (GPUs)	0.625 TFlops (GPUs)
Network	HDR InfiniBand	HPE SlingShot	N/A
MPI Library	OpenMPI 4.1.3	OpenMPI 3.1.5	OpenMPI 3.1.5
Compiler	GCC 10.2.0	NV HPC SDK 23.5	NV HPC SDK 23.5
OS	Rocky 8.7	Red Hat Enterprise 8.8	Mint 21.2

Table B1: System hardware and software configuration of SDSC Expanse, NCSA Delta, and our local workstation used for the results shown in this paper.

Acknowledgments

Work at Predictive Science Inc. was supported by the NASA LWS Strategic Capabilities Program (grant 80NSSC22K0893), the NSF PREEVENTS program (grant ICER1854790), and the NSF/NASA SWQU program (grants AGS 2028154 and 80NSSC20K1582). CDJ and LKSD acknowledge support from the NASA GSFC Heliophysics Internal Scientist Funding Model program (PI Jim Klimchuk). CDJ also acknowledges support from the NASA LWS Focused Science Topic programs: NNH21ZDA001N-LWS "The Origin of the Photospheric Magnetic Field: Mapping Currents in the Chromosphere and Corona" (PI Pete Schuck) and NNH17ZDA001N-LWS "Investigating Magnetic Flux Emergence with Modeling and Observations to Understand the Onset of Major Solar Eruptions" (PI Mark Linton). Computational resources at the San Diego Supercomputer Center and the National Center for Supercomputing Applications were provided by the NSF-supported ACCESS program (allocation MCA03S014). We thank Puget Systems for providing the RTX 3090 Ti GPU for our local workstation.

References

[1] Caplan R M, Mikić Z, Linker J A and Lionello R 2017 *Journal of Physics: Conference Series* **837** 012016
[2] Bui T 1979 *Journal of the ACM (JACM)* **26** 483–493
[3] Meyer C D, Balsara D S and Aslam T D 2014 *Journal of Computational Physics* **257** 594–626
[4] Dawes A 2021 *Computers and Fluids* **214** 104762 ISSN 0045-7930
[5] Simpson M J and Landman K A 2008 *Mathematics and Computers in Simulation* **77** 9–21
[6] Priest E 2014 *Magnetohydrodynamics of the Sun* (Cambridge University Press)
[7] Linker J A, Lionello R, Mikić Z and Amari T 2001 *Journal of Geophysical Research: Space Physics (1978–2012)* **106** 25165–25175
[8] Galligani E 2012 *Advances in Difference Equations* **2012** 1–21
[9] Johnston C D, Daldoff L K S and Caplan R M 2024 *In Preparation*
[10] Dufrechou E and Ezzatti P 2018 *2018 26th Euromicro International Conference on Parallel, Distributed and Network-based Processing (PDP)* pp 196–203
[11] Ma W and Cai X C 2021 *The International Journal of High Performance Computing Applications* **35** 121–135
[12] Meyer C D, Balsara D S and Aslam T D 2012 *Monthly Notices of the Royal Astronomical Society* **422** 2102–2115
[13] O'Sullivan S 2019 *Journal of Computational Physics* **388** 209–223
[14] Skaras T, Saxton T, Meyer C and Aslam T D 2021 *Journal of Computational Physics* **425** 109879
[15] Reeves K K, Török T, Mikić Z, Linker J and Murphy N A 2019 *The Astrophysical Journal* **887** 103
[16] Caplan R M, Linker J A, Mikić Z, Downs C, Török T and Titov V S 2019 *Journal of Physics: Conference Series* **1225** 012012
[17] Caplan R M, Stulajter M M and Linker J A 2023 *2023 IEEE International Parallel and Distributed Processing Symposium Workshops (IPDPSW)* pp 582–590
[18] Stulajter M M, Caplan R M and Linker J A 2022 *Accelerator Programming Using Directives* ed Bhalachandra S, Daley C and Melesse Vergara V (Cham: Springer International Publishing) pp 3–21 ISBN 978-3-030-97759-7

A Framework for Exploring Nuclear Physics Sensitivity in Numerical Simulations

Zhi Chen[1], **Eric T. Johnson**[1], **Max Katz**[1], **Alexander Smith Clark**[1], **Brendan Boyd**[1], **Michael Zingale**[1]

[1]Department of Physics and Astronomy, Stony Brook University, Stony Brook, NY 11794-3800 USA

E-mail: `zhi.chen.3@stonybrook.edu`

Abstract. We describe the AMReX-Astrophysics framework for exploring the sensitivity of astrophysical simulations to the details of a nuclear reaction network, including the number of nuclei, choice of reaction rates, and approximations used. This is explored by modeling a simple detonation with the Castro simulation code. The entire simulation methodology is open-source and GPU-enabled.

1. Introduction

Nucleosynthesis is the process of forming new atomic nuclei via nuclear reactions. By studying both stellar and primordial nucleosynthesis, we gain better insights into stellar evolution, the chemical evolution of galaxies, and the element composition of the Universe. Moreover, enormous amounts of energy are released during nuclear fusion reactions in stellar nucleosynthesis, which is the power source leading to thermonuclear explosions such as Type Ia supernovae (SN Ia) [1, 2] and X-ray bursts (XRBs) [3]. Given the wide range of astrophysical environments in which nucleosynthesis occurs, each characterized by distinct thermodynamic conditions and dynamics, nuclear reaction networks are generally tailored to each case.

Due to limited computational resources, simulations of astrophysical explosions often use small to moderate-sized networks. Generally, the number of isotopes and reaction rates that can be included depends on the dimensionality of a simulation. In 1D, it is possible to incorporate more than a thousand isotopes along with thousands of nuclear reaction rates [4]. However, the situation becomes significantly more challenging in 2- and 3D, due to the increased memory and the need to advect each species separately. A contemporary 2D XRB simulation can include 25 isotopes easily [5], while a 3D XRB simulation can be restricted to ~ 10 isotopes [6, 7].

A reaction sensitivity study is essential to assess the reliability of a simulation where reactions provide a significant part of the energy budget. This can explore the significance of each individual reaction rate, the size of the network, the modifications to rates (like screening), and approximations used in the network. For example, through a sensitivity study, we can identify the most critical reaction rates that should be included in the network while eliminating unimportant ones to reduce cost. During this process, we also gain insights into how reaction rates impact the properties of the problem, such as the propagating speed of a thermonuclear flame in an XRB [5] and the light curve of thermonuclear explosions, leading to a better convergence between the numerical models and observations.

Here we present the AMReX-Astrophysics / pynucastro framework for exploring nuclear reaction physics in numerical simulations. AMReX [8] is the fundamental C++ framework designed to support the development of block-structured adaptive mesh refinement (AMR) algorithms for solving systems of partial differential equations. The reaction networks are generated via the pynucastro library [9, 10], which writes GPU-enabled C++ networks for use in the AMReX-Astrophysics Microphysics library [11]. In section 2, we describe the overall procedure and workflow for generating an arbitrary reaction network with its right-hand-side (RHS) and Jacobian using pynucastro. In section 3, we demonstrate a simple sensitivity study using the compressible hydrodynamics code Castro [12, 13] to model a 1D ^{4}He detonation. All of these codes are developed openly and are freely available on GitHub.

2. Arbitrary Network Generation

The first step in a reaction sensitivity study is the generation of the reaction network. We will use pynucastro for this purpose. pynucastro is a Python library that allows users to interactively explore nuclear reaction rates, construct any arbitrary reaction network, and export the RHS and Jacobian for the network for use in an integrator. The `ReacLibLibrary` class provides access to the collection of JINA ReacLib [14] rates and the `TabularLibrary` class provides tabulated weak rates from various sources [15, 16]. A detailed description of pynucastro can be found in [10], here we only emphasize essential points related to network generation.

To illustrate the process of network generation, we start by explaining the design philosophy and steps involved in building the `subch_simple` network, a network employed in our prior research studies [5, 17]. Subsequently, we showcase how this methodology can be seamlessly extended to develop more complex networks.

The first step in the network generation process is selecting the nuclei of interest. We will build a network that is appropriate for He burning in a detonation. Both `ReacLibLibrary` and `TabularLibrary` provide a method called `linking_nuclei()` that takes a list of nuclei as input and returns a `Library` object with all possible rates that connect the input nuclei. This is demonstrated in the following code:

```
import pynucastro as pyna

# Create a library object that contains all ReacLib rates
reaclib_lib = pyna.ReacLibLibrary()

# Create a list of nuclei
nuc_list = ["p", "he4", "c12", "o16", "ne20", "mg24", "si28",
            "s32", "ar36", "ca40", "ti44", "cr48", "fe52", "ni56",
            "al27", "p31", "cl35", "k39", "sc43", "v47", "mn51",
            "co55", "n13", "n14", "f18", "ne21", "na22", "na23"]

# Returns a subset of ReacLib Library
subch_lib = reaclib_lib.linking_nuclei(nuc_list)
```

Next, various fine-tuning procedures can be done on this `Library` to reduce the overall size of the network. It is often desirable to manually remove rates that have been determined as insignificant for energy generation or nucleosynthesis based on past studies. This can be accomplished using `Library` built-in method `remove_rate()` by supplying either the `Rate` object or the short name of the rate in the form of a string: `"A(x, y)B"`. Below we demonstrate this by removing the reverse rates of C and O reactions as well as some links involving heavy nuclei:

```
for r in subch_lib.get_rates():
    # Check if the products of the rate are the following
    if sorted(r.products) in [[pyna.Nucleus("c12"), pyna.Nucleus("c12")],
                              [pyna.Nucleus("c12"), pyna.Nucleus("o16")],
```

```
                                  [pyna.Nucleus("o16"), pyna.Nucleus("o16")]]:
        # Remove the rate
        subch_lib.remove_rate(r)

# A list of rates to remove by simply supplying the short-name of the rate
rates_to_remove = ["p31(p,c12)ne20", "si28(a,c12)ne20",
                   "ne20(c12,p)p31", "ne20(c12,a)si28",
                   "na23(a,g)al27", "al27(g,a)na23",
                   "al27(a,g)p31", "p31(g,a)al27"]

# Remove rates individually from the network
for r in rates_to_remove:
    subch_lib.remove_rate(r)
```

We can also directly modify the products of a reaction, to further simplify the network. This is useful, for example, in a neutron-capture sequence like A(B, n)Y(n, γ)C where the neutron capture on Y is significantly faster than the first reaction. In this case, a simple approximation involves modifying the product of the first rate to yield the final product of the two-reaction sequence, C, while retaining the rate of the first reaction and adjusting the overall Q-value accordingly. This is demonstrated in the following code:

```
# Select the reaction rates and end product to modify
modify_rates = [("c12(c12,n)mg23", "mg24"),
                ("o16(o16,n)s31", "s32"),
                ("o16(c12,n)si27", "si28")]

for r, mp in modify_rates:
    # Get the specifc Rate object using the short-name
    _r = reaclib_lib.get_rate_by_name(r)
    # Modify the product of that rate
    _r.modify_products(mp)
    # Add the modified rate into the customized Library.
    subch_lib.add_rate(_r)
```

This network describes α-captures up to the iron group (and is the starting point for the `subch_simple` network from [5]). However, for modeling SN Ia, the formation of iron-peak isotopes via electron capture during the explosion is important [18–21]. We can include these by adding tabulated weak rates. The code below obtains a new set of rates from both `ReacLibLibrary` and `TabularLibrary` in the iron-peak and combines them with our existing rates to create our custom rate library `all_lib`. Note that while the original set of nuclei did not include neutrons, we do include neutrons in the iron-group reactions.

```
# Define a new set of nuclei describing iron peak rates
iron_peak = ["n", "p", "he4", "mn51", "mn55",
             "fe52", "fe53", "fe54", "fe55", "fe56",
             "co55", "co56", "co57", "ni56", "ni57", "ni58",
             "cu59", "zn60"]

# Get a subset library with reaclib rates associated with these nuclei
iron_reaclib = reaclib_lib.linking_nuclei(iron_peak)

# Create the library that contains tabulated weak rates from Langanke
# and create a subset library with the rates associated with these nuclei
weak_lib = pyna.TabularLibrary()
iron_weak_lib = weak_lib.linking_nuclei(iron_peak)

# Combine all three libraries together.
all_lib = subch_lib + iron_reaclib + iron_weak_lib
```

There may be duplicated weak rates from the ReacLib library and the tabulated weak rates. To avoid errors when creating the network, these duplicates can be identified and removed via the `find_duplicated_links()` method:

```
# Find the duplicated links
dupes = all_lib.find_duplicate_links()

rates_to_remove = []
# Remove the ReacLibRate version in duplicates
for d in dupes:
    for r in d:
        if isinstance(r, pyna.rates.ReacLibRate):
            rates_to_remove.append(r)

# Remove rates
for r in rates_to_remove:
    all_lib.remove_rate(r)
```

After we've completed the construction of the customized rate libraries, pynucastro can output functions to compute both the righthand-side (RHS) of the ODE system of evolution equations and the Jacobian matrix. Several formats are possible, each provided by a different class: `PythonNetwork` for Python, `SimpleCxxNetwork` for a basic C++ network, and `AmrexAstroCxxNetwork` for C++ for use in the AMReX-Astrophysics suite. For our simulations, we want an `AmrexAstroCxxNetwork`, which can be used with Castro, and couples with hydrodynamics through a variety of integration techniques (e.g., our simplified-SDC method [22]). We create a network in this format simply as:

```
# Use previously defined customized library to create our network
net = pyna.AmrexAstroCxxNetwork(libraries=[all_lib])
```

Finally, pynucastro can also approximate rate sequences like $A(\alpha,p)X(p,\gamma)B$ rates into an effective as $A(\alpha,\gamma)B$ rate, which is used in the traditional "aprox" [23] networks. This approximation holds valid when the proton-capture rate happens at a much faster timescale compared to the (α, p) rate, causing an overall equilibrium in the proton flow. Similar to `modify_products()`, this approximation reduces the size of the network by removing the intermediate nuclei. This method is demonstrated in the following code:

```
# Perform the (a,p)(p,g) approximation by specifying the intermediate nuclei
net.make_ap_pg_approx(intermediate_nuclei=["cl35", "k39", "sc43", "v47"])])

# Remove intermediate nuclei since now they're approximated out
net.remove_nuclei(["cl35", "k39", "sc43", "v47"])])
```

Our final network has 36 isotopes and 149 rates including 12 `TabularRate` rates (compared to the original `subch_simple` with 22 isotopes and 57 rates). Note that `subch_simple` also uses the (α, p)(p, γ) approximation for ^{35}Cl, ^{39}K, ^{43}Sc, ^{47}V, ^{51}Mn, and ^{55}Co. Before writing out the actual RHS and the Jacobian, there are also methods that allow the user to explore the network they created and make adjustments if needed. For example, there is an easy interface to plot each reaction rate with a specified density, temperature, and mass abundances that mimic the simulation conditions to help determine the importance of each rate. An example output of the built-in plotting functions is shown in Figure 1.

Once we're satisfied with the current network, pynucastro can write out all the necessary files, including the RHS and the Jacobian of the network, compatible with Microphysics and Castro with a single function:

```
# Write all files needed by the package, Microphysics, to integrate the network
net.write_network()
```

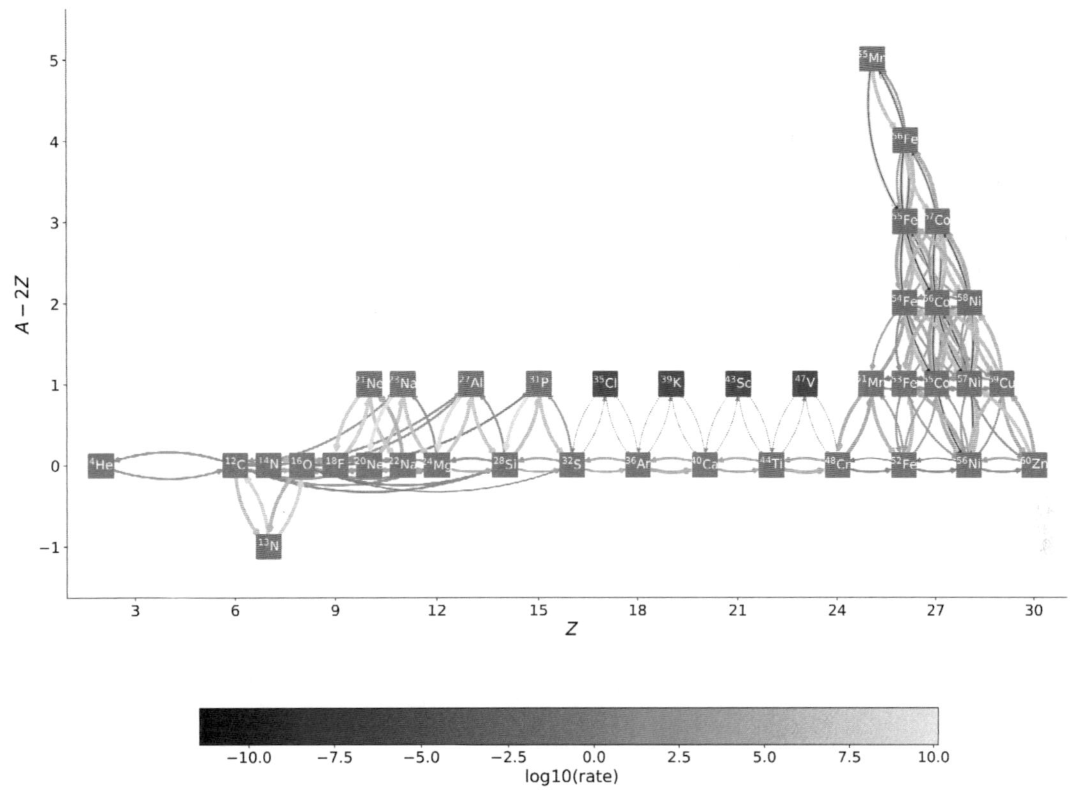

Figure 1: A visualization of the custom network that we created using the pynucastro package. The color bar shows the reaction rates with solar composition, $\rho = 9\times10^7$ g cm^{-3}, and $T = 6\times10^9$ K. The horizontal axis and the vertical axis show the atomic number, Z, and the extra number of neutrons compared to the number of protons, respectively. The gray nodes with dotted gray lines represent the $(\alpha,\mathrm{p})(\mathrm{p},\gamma)$ approximation, which are not directly in the network. All reactions that have the form, $A(X,\alpha)B$ or $A(\alpha,X)B$, are hidden for better clarity.

The AMReX-Astrophysics Microphysics library contains all of the code needed to support the integration of this network, including an equation of state (we'll use the Helmholtz EOS from [24]), integrators such as VODE [25], as well as routines for neutrino cooling and plasma screening. Microphysics can be used on its own to perform network evolution with self-heating or coupled to a simulation code, like Castro. It provides the necessary interfaces to support several different methods of coupling hydrodynamics and reactions.

Typically, a user creates a dedicated directory in Microphysics under `Microphysics/networks` (we'll use `custom_net` for the network described above). This directory serves as the repository containing both the network-generating script using pynucastro and all the associated auto-generated files.

3. Network Sensitivity Testing

We will now show how to do a network sensitivity study similar to the one recently performed for XRBs in [5]. Any Castro problem setup can be told to use our new network (in the Microphysics directory hierarchy) at compile time, via:

```
make NETWORK_DIR=custom_net
```

To demonstrate a simple sensitivity testing, we use the `Detonation` test problem in `Castro`. `Detonation` models a 1D propagating detonation wave, offering a way to replicate the thermodynamic conditions of the He detonation wave found in the shell-burning stage of the sub-Chandrasekhar double-detonation model [26], as discussed in previous work [17, 22]. This sensitivity testing on the `Detonation` problem helps us choose the appropriate network for a realistic sub-Chandrasekhar double-detonation model, especially when studying nucleosynthesis.

To demonstrate this idea, we will compare `subch_simple` and the classic 19-isotope network, `aprox19` [27], with the network we generated in Section 2, which will be referred to as `He-C-Fe`. Our detonation simulation spans a domain up to $x = 4.0 \times 10^7$ cm, with 256 coarse grid zones. By employing two refinement levels with ratios of 2, we reach 1024 zones at the finest level, providing a resolution of ~ 0.4 km. A zero-gradient outflow boundary condition is used on both boundaries. The initial thermodynamic conditions are taken from the base of the He layer from the simulations in [17]: a uniform $\rho = 1.1 \times 10^9$ g cm^{-3}, $T_{\text{pert}} = 1.1 \times 10^9$ K from $0 < x_{\text{pert}} < 1.2 \times 10^7$ cm, $T = 1.75 \times 10^8$ K for $x > 1.2 \times 10^7$ cm, and a composition of 99 % ^{4}He and 1 % ^{14}N.

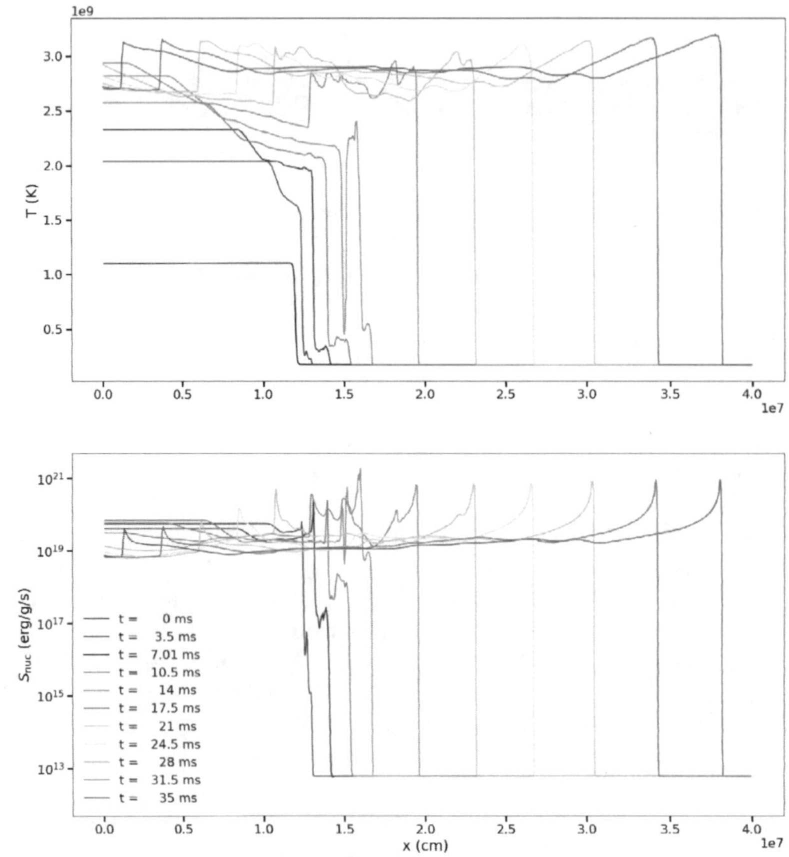

Figure 2: Temperature and specific nuclear energy generation evolution for `He-C-Fe`

Figure 2 shows the time evolution of the temperature and specific nuclear energy generation rate using the `He-C-Fe` network. The peak temperature and specific energy generation rate reach $\sim 3 \times 10^9$ K and $\sim 10^{21}$ erg g^{-1} s^{-1}, consistent with results of the shell-burning stage in the previous study [17].

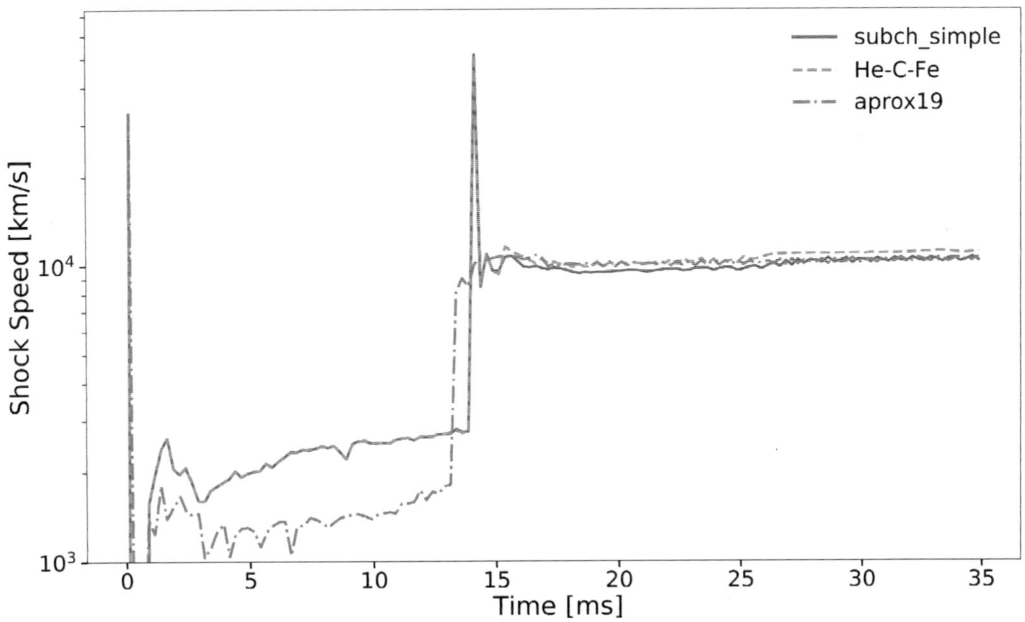

Figure 3: Time evolution of the shock speed for different networks.

We now look at the effect of the network. Figure 3 shows the time evolution of the shock speed for our different networks. It is observed that all models reach a shock speed of $\sim 10^4$ km s^{-1} after ~ 15 ms, while some noticeable differences for $t \lesssim 15$ ms. Specifically, `He-C-Fe` and `subch_simple` exhibit similar profiles, with roughly double the shock speed compared to `aprox19` for $t \lesssim 15$ ms. This difference in shock speed is due to the missing ^{12}C(p, γ)^{13}N(α, p)^{16}O rate in `aprox19`, resulting in weaker initial carbon-burning [28–30]. A similar effect was seen with flames in XRBs in the study by [5].

This carbon burning enhancement effect is evident when examining the total mass abundance of ^{12}C in Figure 4, showing the time evolution of ^{4}He, ^{12}C, and iron-peak isotopes available in the network. The enhanced carbon burning path in both `He-C-Fe` and `subch_simple` leads to a greater ^{12}C consumption rate, leading to a higher shock speed. After ~ 15 ms, the total mass of ^{12}C stabilizes, reaching a steady speed for all three networks. In terms of iron-peak nucleosynthesis, both `aprox19` and `subch_simple` indicate that ^{56}Ni is the primary product, whereas `He-C-Fe` predominantly yields ^{58}Ni. This insight is important for determining the correct mass abundance of ^{56}Ni following the explosion since the beta decay from $^{56}\mathrm{Ni} \rightarrow {}^{56}\mathrm{Co} \rightarrow {}^{56}\mathrm{Fe}$ primarily shapes the observed SN Ia light curve.

To demonstrate the computational cost associated with adding additional iron-peak isotopes, Figure 5 shows the CPU-hours used with each network. As expected, a larger network size corresponds to a higher CPU hour usage. Specifically, the CPU-hours used by `He-C-Fe` increased drastically after 15 ms, coinciding with the establishment of the steady propagating detonation wave. In the end, `He-C-Fe` required nearly 4 times the CPU-hours compared to `subch_simple`, which has $\sim 1/3$ of the total number of rates in `He-C-Fe`. This result suggests the need for further simplification and optimization of the network before employing it in actual simulations

Journal of Physics: Conference Series **2742** (2024) 012021 doi:10.1088/1742-6596/2742/1/012021

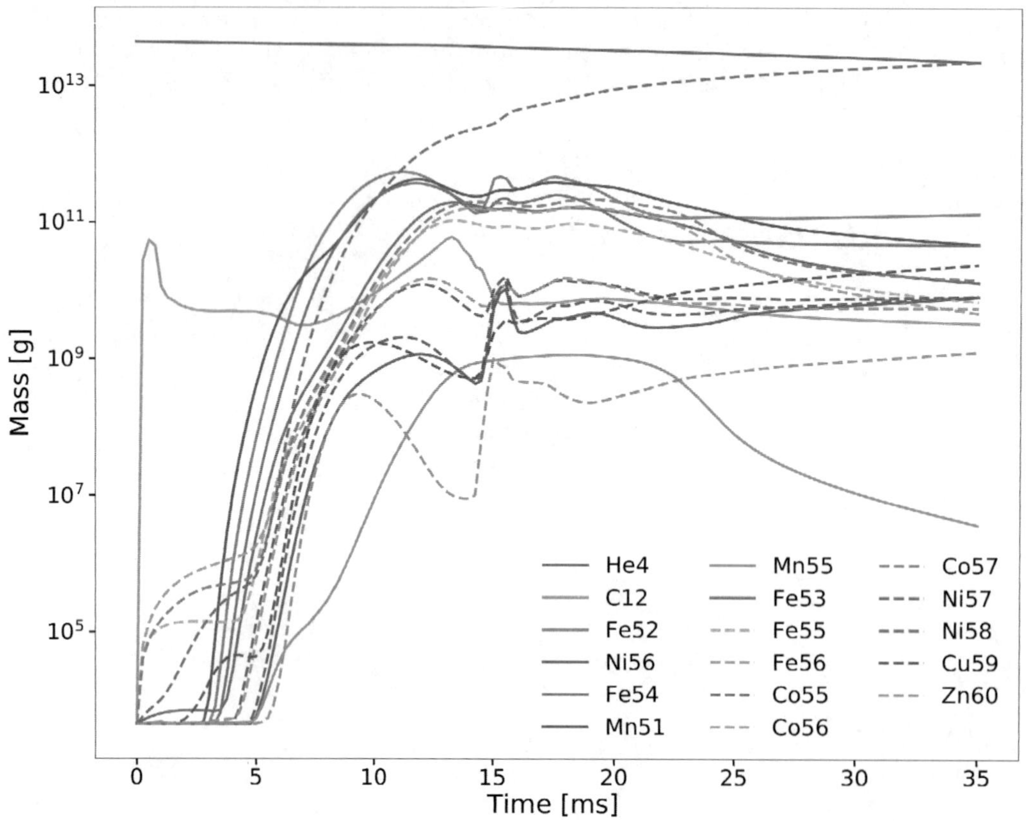

(a) He-C-Fe. Both solid and dashed lines are used to better distinguishability.

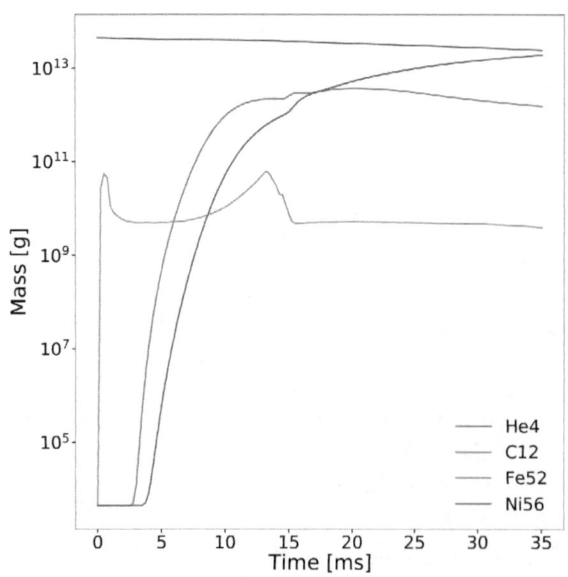

(b) subch_simple

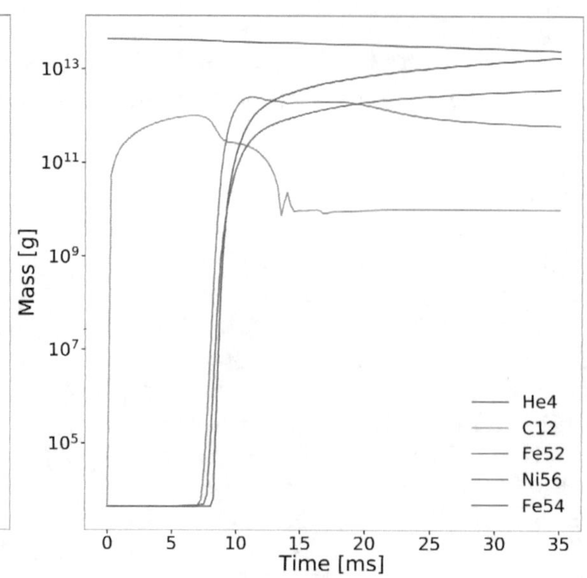

(c) aprox19

Figure 4: Time evolution of the total mass of ^{4}He, ^{12}C and iron-peak isotopes over time.

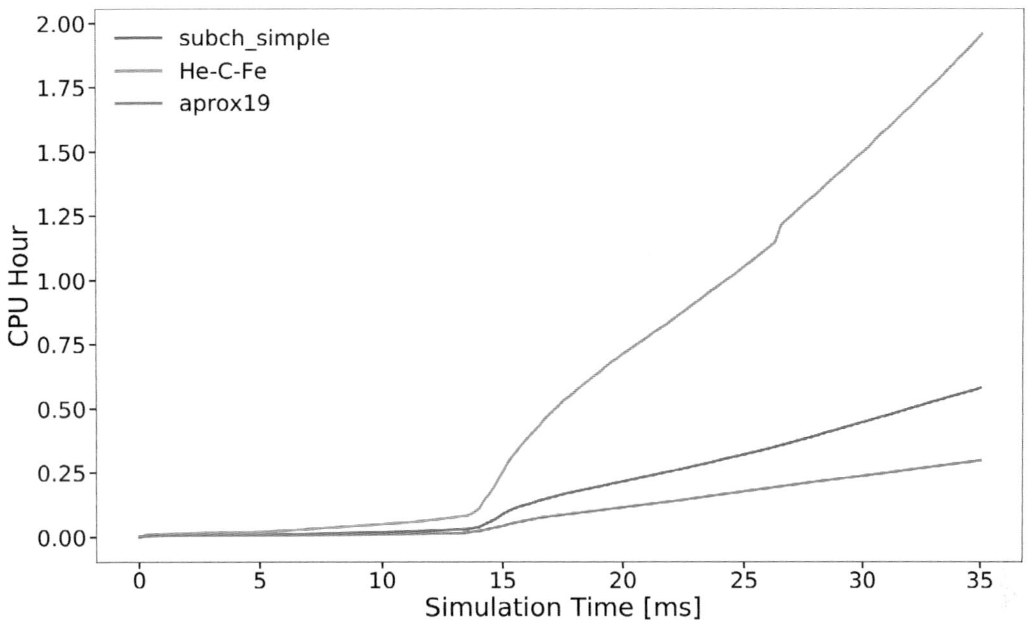

Figure 5: Total number of CPU hours used for each network.

such as a 2D sub-Chandrasekhar double detonation model.

4. Future Development

We've outlined the versatile approach for nuclear reaction sensitivity testing using pynucastro, Microphysics, and Castro. By leveraging pynucastro for network generation, Microphysics for network burning, and Castro for reaction-hydrodynamics coupling, we can effortlessly investigate any network of choice. Demonstrated through the 1D `Detonation` problem in Castro, we assess reaction network sensitivity and emulate shell-burning of the sub-Chandrasekhar double detonation model nucleosynthesis. In the future, this testing capability aims to identify an efficient network with sufficiently many iron-peak isotopes, similar to `He-C-Fe`, for studying nucleosynthesis in the actual sub-Chandrasekhar double detonation model.

Acknowledgments

Castro is open-source and freely available at `http://github.com/AMReX-Astro/Castro`. The work at Stony Brook was supported by DOE/Office of Nuclear Physics grant DE-FG02-87ER40317. This research was supported by the Exascale Computing Project (17-SC-20-SC), a collaborative effort of the U.S. Department of Energy Office of Science and the National Nuclear Security Administration. This research used resources of the National Energy Research Scientific Computing Center, a DOE Office of Science User Facility supported by the Office of Science of the U. S. Department of Energy under Contract No. DE-AC02-05CH11231. This research used resources of the Oak Ridge Leadership Computing Facility at the Oak Ridge National Laboratory, which is supported by the Office of Science of the U.S. Department of Energy under Contract No. DE-AC05-00OR22725, awarded through the DOE INCITE program. We thank NVIDIA Corporation for the donation of a Titan X and Titan V GPU through their academic grant program. This research has made use of NASA's Astrophysics Data System Bibliographic Services.

References

[1] Hillebrandt W and Niemeyer J C 2000 *Annual Review of Astronomy and Astrophysics* **38** 191–230 (*Preprint* https://doi.org/10.1146/annurev.astro.38.1.191) URL https://doi.org/10.1146/annurev.astro.38.1.191

[2] Maoz D, Mannucci F and Nelemans G 2014 *Annual Review of Astronomy and Astrophysics* **52** 107–170 (*Preprint* https://doi.org/10.1146/annurev-astro-082812-141031) URL https://doi.org/10.1146/annurev-astro-082812-141031

[3] Galloway D K and Keek L 2021 *Timing Neutron Stars: Pulsations, Oscillations and Explosions* (*Astrophysics and Space Science Library* vol 461) ed Belloni T M, Méndez M and Zhang C pp 209–262 (*Preprint* 1712.06227)

[4] Woosley S E, Heger A, Cumming A, Hoffman R D, Pruet J, Rauscher T, Fisker J L, Schatz H, Brown B A and Wiescher M 2004 *ASTROPHYS J SUPPL S* **151** 75–102 ISSN 0067-0049, 1538-4365 URL https://doi.org/10.1086/381533

[5] Chen Z, Zingale M and Eiden K 2023 *The Astrophysical Journal* **955** 128 URL https://dx.doi.org/10.3847/1538-4357/acec72

[6] Zingale M, Malone C M, Nonaka A, Almgren A S and Bell J B 2015 *ApJ* **807** 60 ISSN 1538-4357 (*Preprint* 1410.5796) URL https://doi.org/10.1088/0004-637x/807/1/60

[7] Zingale M, Eiden K and Katz M 2023 *ApJ* **952** 160 URL https://dx.doi.org/10.3847/1538-4357/ace04e

[8] Zhang W, Almgren A, Beckner V, Bell J, Blaschke J, Chan C, Day M, Friesen B, Gott K, Graves D, Katz M, Myers A, Nguyen T, Nonaka A, Rosso M, Williams S and Zingale M 2019 *JOSS* **4** 1370 ISSN 2475-9066 URL https://doi.org/10.21105/joss.01370

[9] E Willcox D and Zingale M 2018 *JOSS* **3** 588 ISSN 2475-9066 URL https://doi.org/10.21105/joss.00588

[10] Clark A I S, Johnson E T, Chen Z, Eiden K, Willcox D E, Boyd B, Cao L, DeGrendele C J and Zingale M 2023 *ApJ* **947** 65 URL https://dx.doi.org/10.3847/1538-4357/acbaff

[11] AMReX-Astro Microphysics Development Team, Bishop A, Fields C E, Chen Z, Harpole A, Jacobs A M, Johnson E, Katz M, Li X, Malone C, Timmes F, Willcox D and Zingale M 2023 Amrex-astro/microphysics: Release 23.08 URL https://doi.org/10.5281/zenodo.8206742

[12] Almgren A S, Beckner V E, Bell J B, Day M S, Howell L H, Joggerst C C, Lijewski M J, Nonaka A, Singer M and Zingale M 2010 *ApJ* **715** 1221–1238 ISSN 0004-637X, 1538-4357 (*Preprint* 1005.0114) URL https://doi.org/10.1088/0004-637x/715/2/1221

[13] Almgren A, Sazo M B, Bell J, Harpole A, Katz M, Sexton J, Willcox D, Zhang W and Zingale M 2020 *JOSS* **5** 2513 URL https://doi.org/10.21105/joss.02513

[14] Cyburt R H, Amthor A M, Ferguson R, Meisel Z, Smith K, Warren S, Heger A, Hoffman R D, Rauscher T, Sakharuk A, Schatz H, Thielemann F K and Wiescher M 2010 *ApJS* **189** 240–252 ISSN 0067-0049, 1538-4365 URL https://doi.org/10.1088/0067-0049/189/1/240

[15] Suzuki T, Toki H and Nomoto K 2016 *ApJ* **817** 163 URL https://dx.doi.org/10.3847/0004-637X/817/2/163

[16] LANGANKE K and MARTÍNEZ-PINEDO G 2001 *Atomic Data and Nuclear Data Tables* **79** 1–46 ISSN 0092-640X URL https://www.sciencedirect.com/science/article/pii/S0092640X01908654

[17] Zingale M, Chen Z, Rasmussen M, Polin A, Katz M, Clark A S and Johnson E T 2023 Sensitivity of simulations of double detonation type ia supernova to integration methodology (*Preprint* 2309.01802)

[18] Nomoto K, Iwamoto K, Nakasato N, Thielemann F K, Brachwitz F, Tsujimoto T, Kubo Y and Kishimoto N 1997 *Nuclear Physics A* **621** 467–476 (*Preprint* astro-ph/9706025)

[19] Parikh A, José J, Sala G and Iliadis C 2013 *Prog. Part. Nucl. Phys.* **69** 225–253 ISSN 0146-6410 (*Preprint* 1211.5900) URL https://doi.org/10.1016/j.ppnp.2012.11.002

[20] Bravo, E 2019 *A&A* **624** A139 URL https://doi.org/10.1051/0004-6361/201935095

[21] Lach, F, Röpke, F K, Seitenzahl, I R, Coté, B, Gronow, S and Ruiter, A J 2020 *A&A* **644** A118 URL https://doi.org/10.1051/0004-6361/202038721

[22] Zingale M, Katz M P, Nonaka A and Rasmussen M 2022 *ApJ* **936** 6 URL https://dx.doi.org/10.3847/1538-4357/ac8478

[23] Timmes F X 2019 aprox13 network URL https://cococubed.com/code_pages/burn_helium.shtml

[24] Timmes F X and Swesty F D 2000 *ApJS* **126** 501–516 ISSN 0067-0049, 1538-4365 URL https://doi.org/10.1086/313304

[25] Brown P N, Byrne G D and Hindmarsh A C 1989 *SIAM J. Sci. and Stat. Comput.* **10** 1038–1051 ISSN 0196-5204, 2168-3417 URL https://doi.org/10.1137/0910062

[26] Fink M, Hillebrandt W and Röpke F K 2007 *Astronomy and Astrophysics* **476** 1133–1143 (*Preprint* 0710.5486)

[27] Weaver T A, Zimmerman G B and Woosley S E 1978 *ApJ* **225** 1021 ISSN 0004-637X, 1538-4357 URL https://doi.org/10.1086/156569

[28] Weinberg N N, Bildsten L and Schatz H 2006 *Astrophysical Journal* **639** 1018–1032 (*Preprint* astro-ph/0511247)

[29] Fisker J L, Schatz H and Thielemann F 2008 *ApJS* **174** 261–276 ISSN 0067-0049, 1538-4365 URL https://doi.org/10.1086/521104

[30] Shen K J and Bildsten L 2009 *ApJ* **699** 1365–1373 URL https://doi.org/10.1088%2F0004-637x%2F699%2F2%2F1365

Challenges Modeling the Low-Luminosity Type Iax Supernovae

C Feldman[1,2], E Eisenberg[3], D M Townsley[4] and A C Calder[1,2]

[1] Department of Physics and Astronomy, Stony Brook University, Stony Brook, NY 11794-3800, USA

[2] Institute for Advanced Computational Science, Stony Brook University, Stony Brook, NY 11794-5250, USA

[3] Department of Physics, Yale University, New Haven, CT 06520, USA

[4] Department of Physics and Astronomy, The University of Alabama, Tuscaloosa, AL 35487-0324, USA

E-mail: catherine.feldman@stonybrook.edu

Abstract. Numerical models allow the investigation of phenomena that cannot exist in a laboratory. Computational simulations are therefore essential for advancing our knowledge of astrophysics, however, the very nature of simulation requires making assumptions that can substantially affect their outcome. Here, we present the challenges faced when simulating dim thermonuclear explosions, Type Iax supernovae. This class of dim events produce a slow moving, sparse ejecta that presents challenges for simulation. We investigate the limitations of the equation of state and its applicability to the expanding, cooling ejecta. We also discuss how the"fluff", i.e. the low-density gas on the grid in lieu of vacuum, inhibits the ejecta as it expands. We explore how the final state of the simulation changes as we vary the character of the burning, which influences the outcome of the explosion. These challenges are applicable to a wide range of astrophysical simulations, and are important to discuss and overcome as a community.

1. Introduction

Numerical modeling is the bridge between theory and observation, allowing experimentation in a controlled virtual laboratory. Astrophysics heavily relies on numerical modeling, as it depends on processes than span length scales from the quantum level to galaxies, and timescales of milliseconds to billions of years and most environments are difficult to reproduce in the laboratory. With simulations, we can collide galaxies, explode stars, and form planets, which allows us to test our theories of the Universe. But as British statistician George Box reminds us, "All models are wrong, but some are useful." Choices for initial conditions, simplifying assumptions, and approximations of large- or small-scale processes need to be made in any numerical model for it to be computationally feasible. These choices can determine the entire outcome of a simulation, and it is therefore important to know a simulation's limitations and quirks, and determine quantitatively how the outcome depends on them. Despite their importance, the reasoning and selection of these choices are more often than not passed over in the literature, but are nonetheless considered common knowledge in the field.

Our application, thermonuclear supernovae, is no different. These events, known as Type Ia supernovae (SNe Ia), are created when a white dwarf (WD) star gains enough mass to reignite nuclear fusion and subsequently explode. While advances have been made in both observational and computational capabilities, both the characteristics of the initial system and the method of explosion are still a matter of debate [1]. Modeling SNe Ia involves processes from sub-centimeter scales (the initial "flame", where nuclear burning runs away) to thousands of kilometers (the size of the star). No simulation can simultaneously resolve both of these scales, so approximations must be made, usually by the introduction of a burning model that is tuned to reproduce the energy release of small-scale flame experiments. The initial star is often parameterized, as is the location and size of the ignition.

This study focuses on the dimmest SNe Ia, a subclass known as Type Iax supernovae (SNe Iax). These events are thought to occur when a WD undergoes a deflagration (subsonic burning) without a subsequent detonation (supersonic burning) [2]. In simulation, this deflagration does not fully unbind the star and produces low-energy (10^{48} erg), low-density (10^1 g/cm^3) material. As we show below, the behavior of this low density material sensitively depends on the initial conditions and care must be taken to accurately model it. This paper seeks to explicitly lay out and discuss the challenges we faced while performing an exploratory study of SNe Iax. Section 2 briefly describes our simulation; Section 3 describes the challenges we faced; and Section 4 summarizes our findings and considers the next steps.

2. Methods

Our simulation for Type Iax events is briefly described here. Full details about our exploratory 2D deflagration study and its results can be found in [3].

To simulate the deflagration, we use an adapted version of FLASH, a highly parallelizable, multi-scale, multi-physics simulation instrument used to model the dynamics and evolution of astrophysical and high-energy phenomena [4]. FLASH uses the PARAMESH adaptive mesh library to decompose a volume into grid blocks where features of interest can have a higher resolution (i.e. have smaller sized blocks) than the rest of the grid. This ability allows FLASH to accurately model radiation magneto-hydrodynamic flows while putting heavy computation only where needed. FLASH's modular structure allows it to be extended to model different problems. For our study, we use FLASH version 4.6.2 with a custom burning model developed specifically for thermonuclear supernovae [5].

Our investigative 2D simulations of a deflagration are performed in FLASH on a cylindrically symmetric grid. First, a realistic progenitor [6] created with MESA [7] was mapped to the grid in FLASH [8]. FLASH's hydrodynamics solver requires gas on the entire grid so, instead of the vacuum of space, low density "fluff" material was placed everywhere outside of the star. The initial conditions of the fluff were chosen with care, as the fluff can introduce artifacts and affect the results of the simulation. A deflagration was then artificially ignited by placing a circular 150 km fully-burned region at the temperature peak at the center of the star. To create conditions for plumes to form, random sinusoidal perturbations were placed on the circumference of this burned region. Thirty sets of random perturbations, each of which is called a "realization", were used to create a sample.

The equation of state (EOS) used for our supernova application is the Helmholtz EOS from [9] with Coulomb corrections, that is distributed with FLASH [4] and described in Appendix A.2 of [3]. This EOS is meant to describe a degenerate, fully ionized, electron-positron plasma such as that inside a WD. The hydrodynamics solver uses the split piecewise parabolic method; gravity uses a multipole Poisson solver; the deflagration front is artificially thickened to a 4-grid-zone wide layer and tracked by a scalar variable; and nuclear burning is approximated by 3 distinct stages tracked by scalar progress variables – the first of which (C-burning) is evolved with a modified advection-diffusion-reaction (ADR) scheme and the second (C-burning ash to Si-group elements) and third (Si-group to Fe-group elements) of which are evolved at tuned timescales – plus a dynamic final burned state in nuclear statistical equilibrium.

Journal of Physics: Conference Series **2742** (2024) 012022 doi:10.1088/1742-6596/2742/1/012022

Each of the thirty realizations were run until 20 s. As found in other studies, the deflagration alone is not strong enough to unbind the entire WD [10, 11]. The resulting system is a hot, dense, gravitationally bound remnant surrounded by sparse plumes of ash. Some parts of the plumes have enough kinetic energy to escape the pull of the remnant; this material will be ejected from the system, while the remaining material will fall back onto the remnant. Our simulation produces ^{56}Ni yields and ejecta velocities that are in agreement with observations, but ejecta and kinetic energy yields that are 10 times smaller than observations suggest. These findings propel us to continue work in a 3D simulation that can model the interaction between the flame and background turbulence [12].

3. Simulation Challenges

This paper focuses on the challenges faced by simulating this slow evolving phenomenon that produces low-density, low-temperature material in a partially gravitationally bound environment. These challenges include reaching the limitations of the EOS (Section 3.1), slowdown of ejecta caused by the fluff (Section 3.2), results dependent on our choice of included or excluded physics modules (Section 3.3), and variability of the results due to variation of the initial conditions, in this case the initial flame radius (Section 3.4). While discussed here in the context of our Type Iax application, these challenges are applicable to many simulations involving a grid-based approach to stellar phenomena.

3.1. Equation of State

Our simulation would sometimes return a negative pressure and internal energy at the edges of the rising and expanding plumes. Finding that the Coulomb corrections would overwhelm the pressure in these cases, we decided to investigate the EOS. We called the FLASH EOS with intervals in density-temperature (ρ-T) space as input, and examined the resulting pressure and internal energy. The latter is plotted in Figure 1 as a function of the density and temperature.

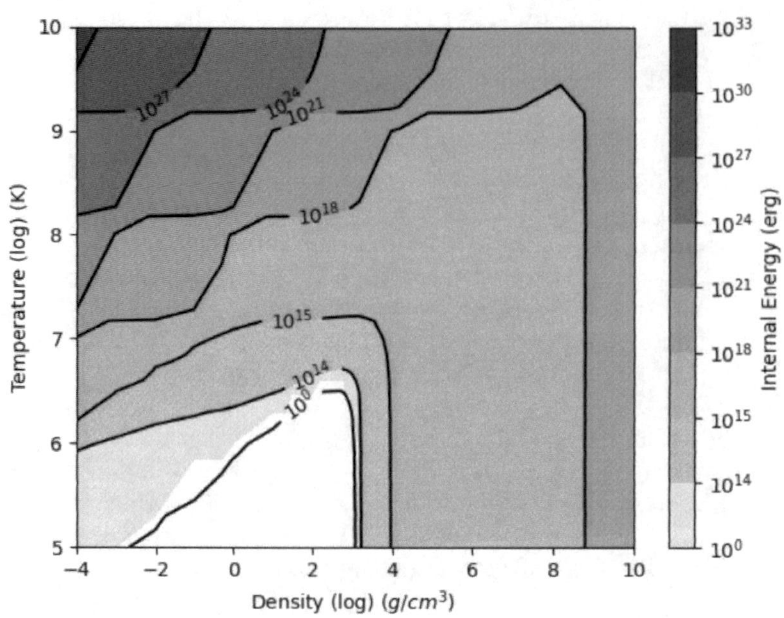

Figure 1. Contour plot of internal energy in ρ-T space. Data was collected by sampling from the FLASH EOS using density and temperature as inputs. The white triangular region shows where the internal energy is negative, and therefore the EOS is invalid.

At temperatures lower than 10^7 K, as the density decreases from 10^4 to 10^3 g/cm^3 the internal energy sharply drops by over ten orders of magnitude. The internal energy then becomes negative in a triangular-shaped region in ρ-T space that begins at densities on the order of 10^3 g/cm^3 and temperatures on the order of 10^6 K. At even lower densities, the internal energy remains negative with a decrease in temperatures. This triangular region of negative pressure and internal energy can also be seen in the newest version of the Helmoltz EOS, denoted the Skye EOS [13]. In this region, the assumption that the gas is fully ionized breaks down, and this EOS becomes invalid.

Our "workaround" to this issue was to set the temperature floor to the relatively high value of 2×10^6 K to prevent the material from entering the invalid region of the EOS. As can be seen in Figure 2, the edges of the ejecta and the surrounding material all reach and are held at the temperature floor. As the simulation continues, more and more material is held at the temperature floor, which means that energy was added to the simulation in these regions. This artificial increase of temperature, and therefore pressure and internal energy, did not significantly affect our results. In the future, an internal energy floor could be used instead, as long as the results also remain unaffected by the artificial increase in energy.

The next step is to determine why our simulation ventures into this problematic area of ρ-T space. Does the expanding ejecta naturally fall into this region, implying a physically motivated cause that will require the coupling of the current EOS to another? Or, do the numerics of the FLASH solvers, perhaps when coupled to our burning model, artificially lead the solution astray?

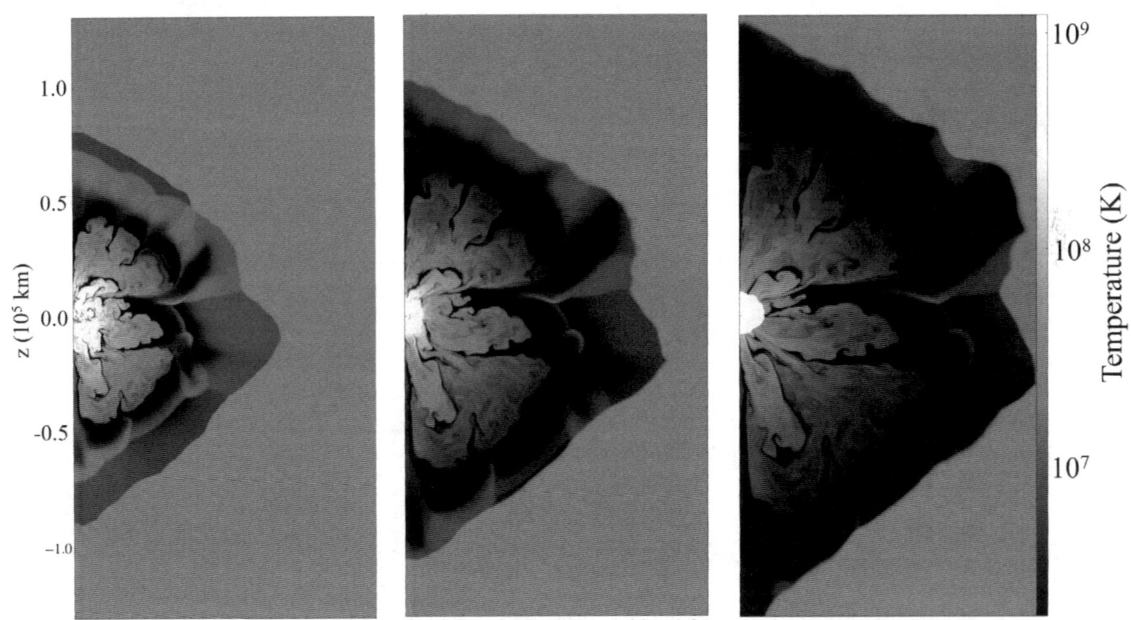

Figure 2. Temperature of our simulation at (from left to right) 10.1 s, 15 s, and 19.8 s. Regions of low temperature can be seen in front of the expanding plumes of burned material. The temperature floor of 2×10^6 K is shown in black on the plots, and is reached as the ejecta expand and cool.

3.2. Fluff-ejecta Interaction

The fluff is another feature of the simulation that inhibits running out to long (>20 s) times. At the start of the simulation, everything is at rest. As the deflagration commences, the star expands into the fluff, colliding and interacting with this low density gas. As the ejecta plows through the fluff, a reverse shock is sent into the ejecta, slowing it down. This effect has been seen in DDT and double detonation simulations [14] and is also shown here, where the energy release and expansion velocities are much lower.

The properties (density, temperature, pressure, etc.) of the fluff must therefore be chosen to prevent an imbalance at the fluff-star boundary. To prevent a composition gradient, the composition of the fluff is set to match the composition of the outermost zone of the star. Here, the pressure in the fluff is set to 5×10^{14} dyne/cm^2, 25% higher than the pressure at the edge of the star, to prevent any shocks at the beginning of the simulation where the outer edge of the star expands in reaction to the sudden introduction of the flame. The density of the fluff must be also chosen with care; the higher the fluff density, the more material that will get in the way of any expanding star material. But if the fluff density is too low, the sound-crossing time of the fluff material will limit the timestep of the entire simulation. Ideally, the density of the fluff will be able to be orders of magnitude lower than that of the ejecta, but for a Type Iax this is difficult as parts of the sparse ejecta reach 10^{-2} g/cm^3. Here, we choose a fluff density (ρ) of 10^{-4} g/cm^3, and use a refinement scheme based on the location of the fluff, rather than density.

One potential issue is that, since the fluff parameters are constant, it is not in hydrostatic equilibrium (HSE). Therefore, the fluff could fall onto the surface of the star, dampening the explosion and changing its character. Putting the fluff in HSE is only valuable if its free-fall time ($t_{\rm ff}$) is much shorter than the time of explosion. Assuming HSE, the scale height (H) of the fluff can be calculated as $H = \frac{P}{\rho g}$, where P is the fluff pressure and g is the gravitational acceleration at the WD's surface. Using H, $t_{\rm ff}$ can be calculated as $\sqrt{\frac{\frac{4}{3}\pi H^3}{GM}}$, where G is the gravitational constant and M is the mass of the WD, 1.36 $M_\odot$. For our simulation, H = 1.68 $\times 10^4$ km, nearly seven times the radius of the star, and $t_{\rm ff}$ = 10.4 s. For this initial study, it is therefore not worth putting the fluff in HSE, because the deflagration begins expanding the star (0.2s) and burns to completion (3s) far sooner than $t_{\rm ff}$. However, it is important to note that for a longer simulation time (100s of seconds) it would be necessary to do so. Next, we examine the potential interaction between the ejected material and the fluff at later times.

Figure 3 shows the radial velocity profile of one deflagration simulation from 10-20 s. At small radii, the bound remnant pulls material back towards the center, giving it an increasingly smaller radial velocity. At larger radii, the material becomes unbound and will be ejected from the system. Notably, in the ejecta, the characteristic linear relationship between the radial velocity and the radius can be seen, demonstrating that the ejecta are in free expansion. Past the point of the ejecta, the ejecta and fluff interact, leading to a noisy radial velocity profile.

At the ejecta-fluff boundary, the ejecta pushes forward and sends a shock into the fluff, which responds by piling up at the boundary and sending a reverse shock back into the ejecta, slowing it down. The ejecta has enough energy that it is able to overtake the fluff and continue pushing it aside, moving outwards. Therefore, unlike true free expansion where over time the velocity of the ejecta would remain unchanged, the ejecta loses speed. This is most apparent at the boundary between the ejecta and the fluff, denoted by the black points in Figure 3, where the velocity of the ejecta at the boundary decreases with time. The slope of the line, which is the inverse of the expansion time, also decreases, signifying that the expansion time is increasing. Therefore, interaction with the fluff is a limiting factor in how long our simulations are accurate.

One way to ameliorate the effects of the fluff is the implementation of a "roll-off", where the density and temperature of the fluff is linearly decreased with radius. This strategy is implemented and discussed in [15]. The higher fluff density at the star's surface will both

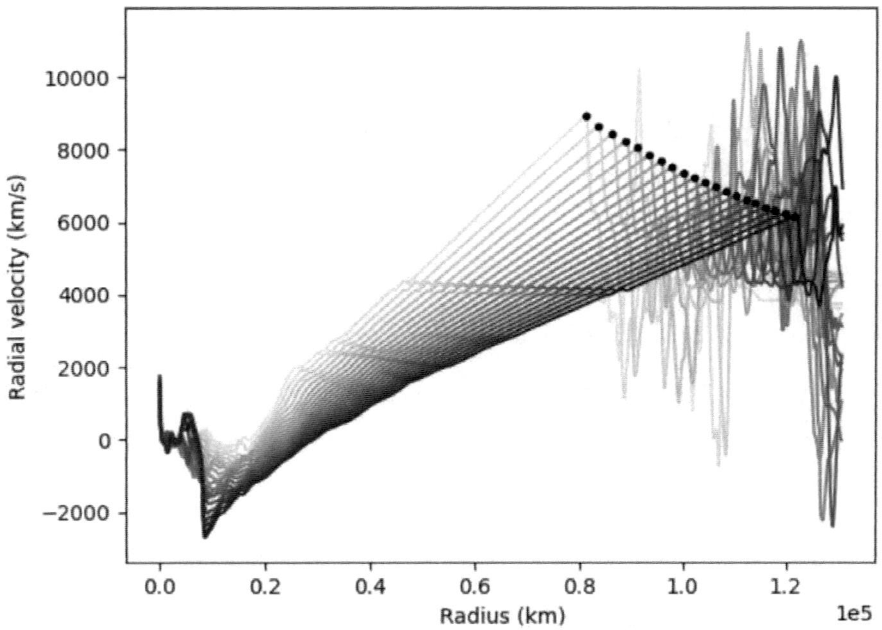

Figure 3. Radial velocity vs. radius of our simulation over time. Each line represents an individual timestep, from 10.1 s (yellow) to 20 s (purple) in intervals of 0.5 s. Time follows the darkening colors. The black points mark the highest velocity that the ejecta reaches at each timestep, which decreases with time due to interaction with the fluff.

prevent the timestep from being constrained and make it easier to keep the star-fluff boundary in pressure equilibrium without the need to drastically increase the fluff temperature. At the same time, the roll-off will provide overall less material for the expanding ejecta to run into, reducing the strength of the reverse shock and enabling the ejecta to be followed further in time. [10] and other studies follow the ejecta out to 100 s or more, and a fluff roll-off could make this feasible for our and other future studies.

3.3. Reignition

As we ran out to the relatively long time of 20 s, some of our realizations reignited by means of a pulsational detonation caused by material falling back onto the dense, mostly unburned remnant. This process is visually demonstrated in Figure 4. To prevent reignition from happening in our deflagration study, we manually turned off burning at 5 s, by which time the deflagration had quenched. This manual setting is also used by other deflagration studies [11], and a separate, accompanying study of pulsational detonations is sometimes performed [16].

This reignition is interesting and prompts many questions. Why do some, but not all, the realizations reignite before 20 s? Or do all realizations reignite, but at different times? Does reignition depend on the amount of mass ejected from the system? To probe these questions, we artificially reduced the flame speed to 0 after the deflagration completed so that the flame would essentially be deactivated; the ADR scheme is tuned to approximate the initial deflagration phase of an SN Ia, and not suited for modeling reignition. Instead, the thermally-activated C-burning reaction rate is used in place of the ADR equation to evolve the first (C-burning) stage of our burning model. Figure 5 shows results from a few selected realizations that span the space of ejected mass.

Realizations that ejected less mass in our deflagration study reignite sooner, most likely

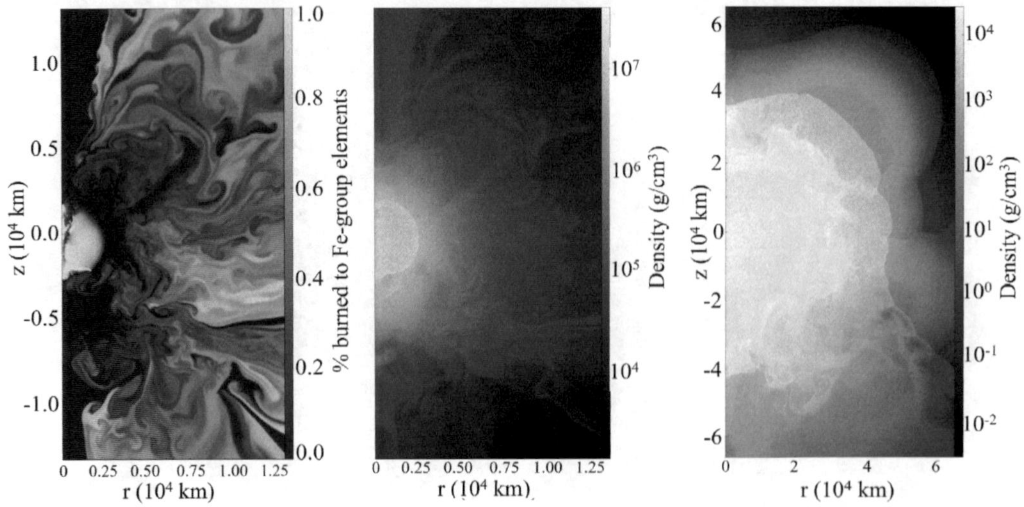

Figure 4. Reignition of the remnant caused by material fallback in realization 28. (Left) The percent of material burned to Fe-group elements at 7.3 s. The detonation can be seen in the center of the remnant, which is mostly unburned fuel surrounded by plumes of burned material from the deflagration phase. (Center) The density at 7.3 s. The detonation can be seen in the center of the remnant, which is much denser than the surrounding plumes. (Right) The density at 12 s, where the expanding burned material and the detonation shock can be seen.

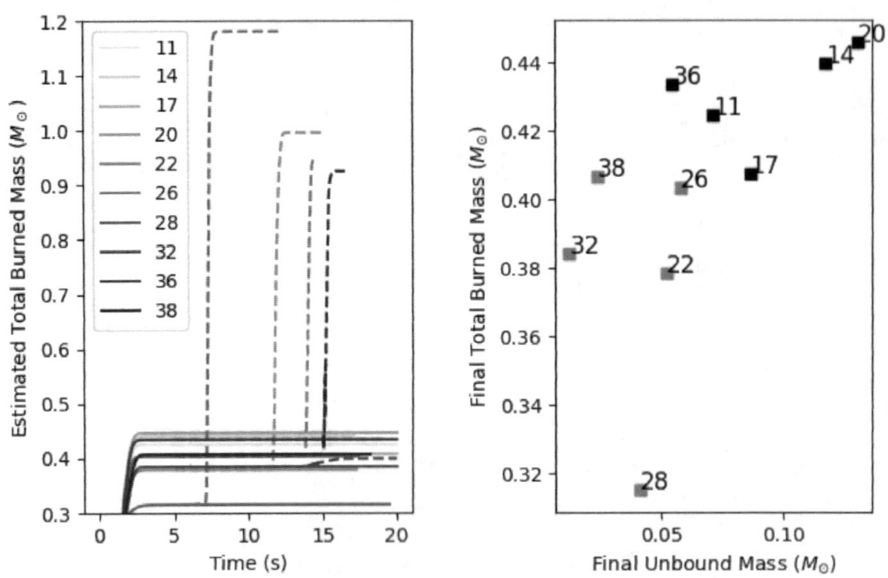

Figure 5. (Left) Burned mass vs. time for selected realizations run to test reignition. The solid lines show the results from our deflagration study (burning manually turned off at 5 s), and the dotted lines show the results for thermally-activated burning turned on after the deflagration completes. (Right) Total burned mass vs. total ejected mass of selected realizations from our deflagration study. Marked in red are the realizations that reignited, and black are the ones that did not (yet).

because the remnant is more massive, denser, and hotter, which makes it easier for the fallback to create a spark. In general, they also burn more mass in the reignition, because there is more mass at higher densities available to consume. It cannot be definitively determined whether or not all realizations will reignite; the process may just take much longer for some than for others.

It is also important to note that our simulation is unsuited to determine conditions for initiating a pulsational detonation, and therefore this process has a great deal of uncertainty. These results are interesting, but absolutely not methodical; they simply showcase an idea to be rigorously tested. First, it would need to be determined what resolution properly describes the uneven surface of the bound, compact remnant, the structure that detonates. Secondly, our three-stage burning model was created for a flame front moving through relatively homogeneous regions of fuel. In this case, the detonation occurs in a heterogeneous environment – the C/O fuel of the initial WD is mixed with burning ashes – and it would need to be determined if our burning model is still applicable. An additional, but certainly not the last, question is whether the C-burning reaction alone is enough to accurately describe the ignition, or if the burning is enhanced or suppressed by another reaction that would have to be included.

3.4. Initial Flame

As the sub-cm scales of a realistic flame cannot be resolved in our star-scale simulation, a deflagration is often ignited by inserting a volume of fully-burned material into the star. The size, location, and number of the initial flame bubbles are often varied to determine its effect on the simulation results. A smaller flame has longer to evolve and be affected by fluid instabilities before it reaches the size of a larger flame. However, it is unclear whether in reality ignition occurs in one location or multiple merging locations. Despite the uncertainties, it is important to vary this initial condition to determine its influence on the outcome of the simulation.

The initial flame in our simulation is placed at the temperature peak in the center of the star. It was given a radius of 150 km, a choice prompted by previous deflagration to detonation transition (DDT) studies [17, 18]. An 150 km initial flame allowed the star to expand enough during the deflagration phase so that the subsequent detonation produced yields consistent with observations. For our deflagration study, we investigated how the initial flame radius affected the results. We selected ten realizations that spanned the space of ejected ^{56}Ni mass and adjusted their initial flames to 75 km and 300 km. Examples of these three flame radii for the same realization are shown in Figure 6. Note that the radii and not the perturbations were scaled.

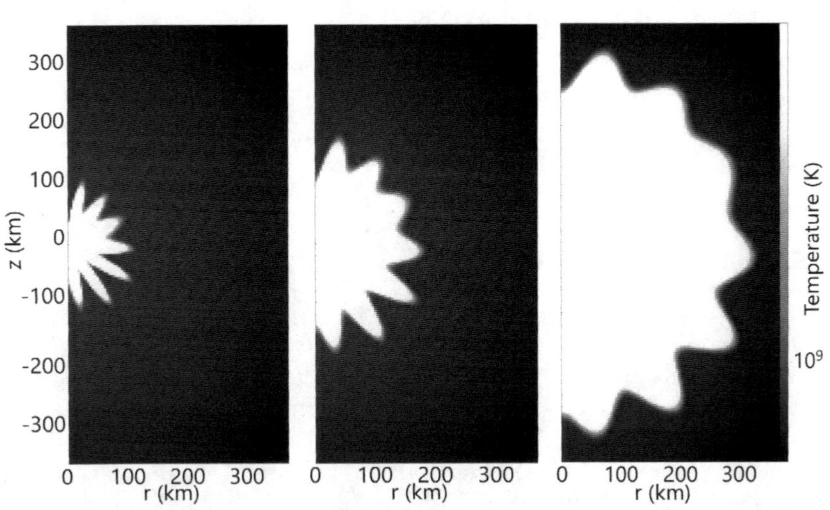

Figure 6. Three initial flames from the same realization with (from left to right) 75 km, 150 km, and 300 km initial radii.

Figure 7 shows the total ejected mass and the total ejected ^{56}Ni mass for the three initial

Journal of Physics: Conference Series **2742** (2024) 012022 doi:10.1088/1742-6596/2742/1/012022

flame radii. There is a clear linear relationship between the ejected mass and the ejected ^{56}Ni mass; roughly 10% of the ejecta is ^{56}Ni. In general, a larger initial flame radius produces less ejected mass, and less burned mass overall. This is because the larger flame injects more energy into the simulation at its start, causing the star to expand faster. Therefore, the burning occurs at lower densities and extinguishes faster, leading to a smaller burned yield and less overall energy release. Perhaps most interestingly, eight of the ten 300 km radius realizations produced no ejected mass – all of the mass remained gravitationally bound. Based on the findings of Section 3.3, the 300 km radius realizations are likely to all reignite, while the 75 km ones are more likely to not (at least within 15-20 s).

Importantly, decreasing the flame radius still produces ejected ^{56}Ni masses within the range of observations of SNe Iax, meaning that the deflagration model remains a viable hypothesis. If reignition does not occur, a deflagration and resulting SN Iax could also be a potential explanation for hypervelocity WDs such as LP 40-365 [19].

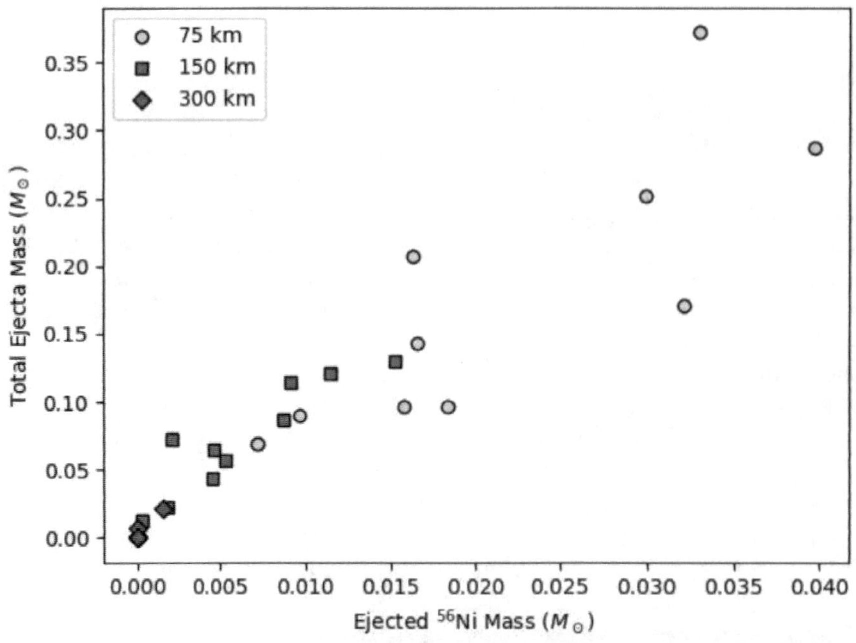

Figure 7. Total ejected mass vs. ejected ^{56}Ni mass for ten realizations using three different initial flame radii: 75 km (yellow ○), 150 km (purple □), and 300 km (pink ◇). In general, there is a direct relationship between the amount of ejected mass and the amount of ejected ^{56}Ni. Smaller initial flames produce more ejecta. Eight of the ten realizations with the 300 km radius initial flame produced no ejecta.

4. Summary and Conclusions

Numerical models provide a means to experiment on astrophysical systems. These tools comes with challenges including choosing initial conditions, assumptions, and approximations. We discuss the challenges faced when simulating the dim SNe Iax as a deflagration in a WD. We find that the slow-moving, sparse, and cool ejecta are sensitive to the choices made in simulation: the EOS ventures into an invalid region when describing it; the fluff impedes its progress; and the choice of the burning model and the initial flame size determines its mass. Despite all of these uncertainties, our simulation demonstrates that a deflagration in a WD is a viable mechanism to produce an SN Iax, encouraging future exploration in 3D.

It is important to understand the variability of simulations under different initial conditions and assumptions, especially for non-linear processes. Performing an extensive parameter study for a detailed simulation might be computationally prohibitive, however testing more abstracted simulations such as the 2D suite performed here can provide insight into a model's sensitivity.

Acknowledgments
The authors would like to thank Stony Brook Research Computing and Cyberinfrastructure, and the Institute for Advanced Computational Science at Stony Brook University (SBU) for access to the SeaWulf computing system, which was made possible by National Science Foundation (NSF) grant #1531492. This research was also supported in part by the US Department of Energy (DOE) under grant DE-FG02-87ER40317, in part by the NSF under grant OAC 1927880, and in part by the Simons Summer Research Program at SBU.

References

[1] Liu Z W, Röpke F K and Han Z 2023 *Research in Astronomy and Astrophysics* **23** 082001 URL `https://dx.doi.org/10.1088/1674-4527/acd89e`
[2] Foley R J, Challis P J, Chornock R, Ganeshalingam M, Li W, Marion G H, Morrell N I, Pignata G, Stritzinger M D, Silverman J M, Wang X, Anderson J P, Filippenko A V, Freedman W L, Hamuy M, Jha S W, Kirshner R P, McCully C, Persson S E, Phillips M M, Reichart D E and Soderberg A M 2013 *The Astrophysical Journal* **767** 57 URL `https://doi.org/10.1088/0004-637x/767/1/57`
[3] Feldman C, Gutierrez N, Eisenberg E, Willcox D E, Townsley D M and Calder A C 2023 *The Astrophysical Journal* **959** 112 URL `https://dx.doi.org/10.3847/1538-4357/acf658`
[4] Fryxell B, Olson K, Ricker P, Timmes F X, Zingale M, Lamb D Q, MacNeice P, Rosner R, Truran J W and Tufo H 2000 *The Astrophysical Journal Supplement Series* **131** 273–334 URL `https://doi.org/10.1086/317361`
[5] Townsley D M, Miles B J, Timmes F X, Calder A C and Brown E F 2016 *The Astrophysical Journal Supplement Series* **225** 3 URL `https://doi.org/10.3847/0067-0049/225/1/3`
[6] Brooks J, Schwab J, Bildsten L, Quataert E and Paxton B 2017 *The Astrophysical Journal* **834** L9 URL `https://doi.org/10.3847/2041-8213/834/2/19`
[7] Paxton B, Bildsten L, Dotter A, Herwig F, Lesaffre P and Timmes F 2011 *ApJS* **192** 3 (*Preprint* `1009.1622`)
[8] Willcox D E, Townsley D M, Calder A C, Denissenkov P A and Herwig F 2016 *The Astrophysical Journal* **832** 13 URL `https://doi.org/10.3847/0004-637x/832/1/13`
[9] Timmes F X and Swesty F D 2000 *ApJS* **126** 501–516
[10] Kromer M, Ohlmann S T, Pakmor R, Ruiter A J, Hillebrandt W, Marquardt K S, Röpke F K, Seitenzahl I R, Sim S A and Taubenberger S 2015 *Monthly Notices of the Royal Astronomical Society* **450** 3045–3053 ISSN 0035-8711 URL `https://doi.org/10.1093/mnras/stv886`
[11] Lach F, Callan F, Bubeck D, Röpke F, Sim S, Schrauth M, Ohlmann S and Kromer M 2022 *A&A* **658** 27 URL `https://doi.org/10.1051/0004-6361/202141453`
[12] Jackson A P, Townsley D M and Calder A C 2014 *The Astrophysical Journal* **784** 174 URL `https://doi.org/10.1088/0004-637x/784/2/174`
[13] Jermyn A S, Schwab J, Bauer E, Timmes F X and Potekhin A Y 2021 *The Astrophysical Journal* **913** 72 URL `https://dx.doi.org/10.3847/1538-4357/abf48e`
[14] Townsley D M, Miles B J, Shen K J and Kasen D 2019 *The Astrophysical Journal Letters* **878** L38 URL `https://dx.doi.org/10.3847/2041-8213/ab27cd`
[15] Boos S J, Townsley D M, Shen K J, Caldwell S and Miles B J 2021 *The Astrophysical Journal* **919** 126 URL `https://dx.doi.org/10.3847/1538-4357/ac07a2`
[16] Lach F, Callan F P, Sim S A and Röpke F K 2022 *A&A* **659** A27 (*Preprint* `2111.14394`)
[17] Townsley D M, Calder A C, Asida S M, Seitenzahl I R, Peng F, Vladimirova N, Lamb D Q and Truran J W 2007 *The Astrophysical Journal* **668** 1118–1131 URL `https://doi.org/10.1086/521013`
[18] Townsley D M, Jackson A P, Calder A C, Chamulak D A, Brown E F and Timmes F X 2009 *The Astrophysical Journal* **701** 1582–1604 URL `https://doi.org/10.1088/0004-637x/701/2/1582`
[19] El-Badry K, Shen K J, Chandra V, Bauer E B, Fuller J, Strader J, Chomiuk L, Naidu R P, Caiazzo I, Rodriguez A C, Nagarajan P, Yamaguchi N, Vanderbosch Z P, Roulston B R, Gänsicke B, Han J J, Burdge K B, Filippenko A V, Brink T G and Zheng W 2023 *The Open Journal of Astrophysics* **6** 28 (*Preprint* `2306.03914`)

Achieving Higher Order Accuracy in Space in Hydrodynamic Simulations of Self-Gravitating Gas

Tomoyuki Hanawa[1] **and Patrick Dean Mullen**[2]

[1]Center for Frontier Science, Chiba University, 1-33 Yayoi-cho, Inage-ku, Chiba 263-8522, Japan

[2]CCS-2, Los Alamos National Laboratory, Los Alamos, NM 87545, USA

E-mail: hanawa@faculty.chiba-u.jp

Abstract. Modern astrophysical simulation codes employ a variety of numerical algorithms capable of achieving higher-order accuracy in both space and time. Albeit they succeed in achieving an effective higher spatial resolution and in suppressing the numerical damping of waves, to our knowledge, all current astrophysical simulations invoking self-gravity are limited to second-order accuracy in space. If we can devise an algorithm to evaluate self-gravity with a higher-order spatial accuracy, we can better the evaluation of the gravitational acceleration and gravitational energy release which dictate the evolution of many astrophysical systems. Herein, we present a numerical algorithm for self-gravitating hydrodynamics capable of achieving fourth-order accuracy for a given density distribution on a Cartesian uniform grid. First, we derive the cell-averaged gravitational potential at fourth-order accuracy from the cell-averaged density by solving the Poisson equation. Next, we obtain the cell average of the product of the density and gravitational acceleration, which differs from the cell-averaged density multiplied by the cell-averaged gravitational acceleration. We then show the verification of the algorithm by applying it to critical test problems: (1) maintaining equilibria of self-gravitating slabs, even upon advection, (2) evolving a polytropic sphere with a massive power-law envelope, and (3) conservation of specific entropy during the propagation of a sound wave.

1. Introduction

Gravity plays a primary role in many astronomical objects and phenomena. Thus, it is important to evaluate the gravity accurately when solving the hydrodynamic equations for modeling them. The accuracy depends on the spatial and temporal resolution and numerical scheme employed. When a numerical scheme is of n-th order accuracy in space, the numerical error is proportional to h^n where h denotes the cell width in the simulation. To our knowledge, all current numerical codes available are restricted to second order accuracy in space when invoking self-gravity, even when the remainder of the physics are treated at fourth or higher order accuracy. In this, we introduce a method to evaluate the gravitational acceleration ($\rho \boldsymbol{g}$) and gravitational energy release ($\rho \boldsymbol{v} \cdot \boldsymbol{g}$) for a given density ($\rho$) distribution and mass flux ($\rho \boldsymbol{v}$). We show some numerical examples to demonstrate that we have achieved fouth-order accuracy in space, including (1) the evolution of a polytrope sphere and (2) the maintaining of self-gravitating slabs in equilibrium, (3) the advection of the slabs, and (4) the propagation of a sound wave modified by self-gravity.

We adopt MP5[1] to solve the hydrodynamic equations. MP5 reconstructs cell surface values of fifth-order accuracy in space from cell averages while preserving extrema. It employs the third order Runge-Kutta method for time stepping. The temporal accuracy inherits that of the hydrodynamic equation solver employed. We use the Roe scheme as an approximate Riemann solver, although any other upwind scheme should work.

Our scheme ensures the conservation of linear momentum, i.e., the updates to momentum can be recast as the divergence of a gravitational stress tensor. As an option, we can provide a scheme to conserve total energy (including gravitational potential energy) as in Mullen et al[8]. Moreover, we can evolve the the specific entropy of a gas element at third order accuracy in time, hence eliminating spurious heating or cooling exhibited by lower order schemes.

This paper is organized as follows. Section 2 describes our method and its philosophy. Our methods consist of three parts: the Poisson equation solver, the evaluation of the gravitational acceleration, and the evaluation of the gravitational energy release. Section 3 shows some numerical examples to showcase the accuracy achieved with the scheme. Section 4 summarizes the results and describes implications.

2. Methods

2.1. Problem Definition

We seek a fourth-order algorithm to solve the equations of self-gravitating hydrodynamics,

$$\frac{\partial \rho}{\partial t} + \boldsymbol{\nabla} \cdot (\rho \boldsymbol{v}) = 0, \tag{1}$$

$$\frac{\partial}{\partial t}(\rho \boldsymbol{v}) + \boldsymbol{\nabla} \cdot (P\boldsymbol{I} + \rho \boldsymbol{v}\boldsymbol{v}) = \rho \boldsymbol{g}, \tag{2}$$

$$\frac{\partial}{\partial t}(\rho E) + \boldsymbol{\nabla} \cdot [(\rho E + P)\, \boldsymbol{v}] = \rho \boldsymbol{v} \cdot \boldsymbol{g}, \tag{3}$$

on a uniform Cartesian grid, where

$$E = \frac{\boldsymbol{v}^2}{2} + \frac{P}{(\gamma - 1)\,\rho}, \quad \boldsymbol{g} = -\boldsymbol{\nabla}\phi. \tag{4}$$

Here, ρ, P, and ϕ denote the density, pressure and gravitational potential, respectively, while $\boldsymbol{v}$ and $\boldsymbol{g}$ the velocity and gravity, respectively. We assume that the flow obeys an ideal gas equation of state wherein the specific heat ratio adiabatic index γ is constant. The gravitational potential is related to the density by the Poisson equation,

$$\boldsymbol{\nabla}^2 \phi = 4\pi G \rho, \tag{5}$$

where G denotes the gravitational constant.

For simplicity, we ignore heating and cooling processes. Therefore, the specific entropy, $s = \ln P - \gamma \ln \rho$, should change in time only through numerical errors. We use this property as a proxy for the accuracy of our numerical methods.

We consider a uniform rectangular grid in Cartesian coordinates where the center of cell center is designated

$$(x_i, y_j, z_k) = \left[h\left(i - \frac{1}{2}\right), h\left(j - \frac{1}{2}\right), h\left(k - \frac{1}{2}\right)\right]. \tag{6}$$

Here, the indices i, j, and k denote the cell number in the x-, y-, and z-directions, respectively. The grid spacing, h, is equal in all the directions. In the following, we use the half indices such as $x_{i+1/2} = hi$, $y_{j+1/2} = hj$, and $z_{k+1/2} = hk$ to denote the position of the cell boundaries.

2.2. Poisson Equation
Using a fourth order accurate central difference, we discretize the Poisson Equation (Equation(5)) as

$$\begin{aligned}
&\frac{-\phi_{i+2,j,k} + 16\phi_{i+1,j,k} - 30\phi_{i,j,k} + 16\phi_{i-1,j,k} - \phi_{i-2,j,k}}{12h^2} \\
&\frac{-\phi_{i,j+2,k} + 16\phi_{i,j+1,k} - 30\phi_{i,j,k} + 16\phi_{i,j-1,k} - \phi_{i,j-2,k}}{12h^2} \\
&\frac{-\phi_{i,j,k+2} + 16\phi_{i,j,k+1} - 30\phi_{i,j,k} + 16\phi_{i,j,k-1} - \phi_{i-2,j,k-2}}{12h^2} = 4\pi G \rho_{i,j,k},
\end{aligned} \tag{7}$$

where $\rho_{i,j,k}$ and $\phi_{i,j,k}$ denote cell averaged values. The truncation error of $\phi_{i,j,k}$ is proportional to h^4.

We can solve Equation (7) by either multi-grid iterations[5] or the Fast Fourier Transform (FFT). Since both of these methods are established, we omit their details here.

When we apply the periodic boundary condition in all the directions, we replace the right hand side of Equation (7) with $4\pi G\left(\rho_{i,j,k} - \bar{\rho}\right)$, where $\bar{\rho}$ denotes the average density in the computation domain. We adopt this Jeans' swindle in the test problems when invoking periodic boundary conditions.

For later convenience, we rewrite Equation (7) as

$$\begin{aligned}
-\frac{g^{(4)}_{x,i+1/2,j,k} - g^{(4)}_{x,i-1/2,j,k}}{h} - \frac{g^{(4)}_{y,i,j+1/2,k} - g^{(4)}_{x,i,j-1/2,k}}{h} - \frac{g^{(4)}_{z,i,j,k+1/2} - g^{(4)}_{x,i,j,k-1/2}}{h} \\
= 4\pi G \rho_{i,j,k}, \qquad (8) \\
g^{(4)}_{x,i+1/2,j,k} = \frac{\phi_{i+2,j,k} - 15\phi_{i+1,j,k} + 15\phi_{i,j,k} - \phi_{i-1,j,k}}{h}, \qquad (9) \\
g^{(4)}_{y,i,j+1/2,k} = \frac{\phi_{i,j+2,k} - 15\phi_{i,j+1,k} + 15\phi_{i,j,k} - \phi_{i,j-1,k}}{h}, \qquad (10) \\
g^{(4)}_{z,i,j,k+1/2} = \frac{\phi_{i,j,k+2} - 15\phi_{i,j,k+1} + 15\phi_{i,j,k} - \phi_{i,j,k-1}}{h}, \qquad (11)
\end{aligned}$$

where $g^{(4)}_{x,i+1/2,j,k}$, $g^{(4)}_{y,i,j+1/2,k}$, and $g^{(4)}_{z,i,j,k+1/2}$ denote the cell surface average gravity of the fourth-order accuracy in space.

2.3. Gravitational Acceleration
Next we derive the gravitational acceleration $\rho \boldsymbol{g}$ from the gravitational tensor,

$$\boldsymbol{T}_g = \frac{1}{4\pi G}\left(\boldsymbol{g}\boldsymbol{g} - \frac{\boldsymbol{g}^2}{2}\right). \tag{12}$$

Since $\rho \boldsymbol{g} = -\boldsymbol{\nabla} \cdot \boldsymbol{T}_g$, we obtain

$$\begin{aligned}
\left(\rho g_x\right)_{i,j,k} h^3 = &\left(-T_{g,xx,i+1/2,j} + T_{g,xx,i-1/2,j,k} - T_{g,yx,i,j+1/2,k} + T_{g,yx,i,j-1/2,k}\right. \\
&\left.-T_{g,zx,i,j,k+1/2} + T_{g,zx,i,j,k-1/2}\right) h^2,
\end{aligned} \tag{13}$$

where the indices specify the components and cell surface on the right hand side. When each term on the right hand side denotes the cell surface average at fourth-order accuracy in space, the left hand side denotes the cell average gravitational acceleration of the fourth-order accuracy. Mullen et al.[8] derived the second order accurate gravitational tensor,

$$
\begin{aligned}
T^{(2)}_{g,xx,i+1/2,j,k} &= \frac{g^2_{x,i+1/2,j,k}}{8\pi G} - \frac{g_{y,i+1,j+1/2,k} g_{y,i,j+1/2,k} + g_{y,i+1,j-1/2,k} g_{y,i,j-1/2,k}}{16\pi G} \\
&\quad - \frac{g_{z,i+1,j,k+1/2} g_{z,i,j,k+1/2} + g_{z,i+1,j,k-1/2} g_{z,i,j,k-1/2}}{16\pi G}, && (14) \\
T^{(2)}_{yx,i,j+1/2,k} &= \frac{g_{y,i,j+1/2,k}\left(g_{x,i+1/2,j+1,k} + g_{x,i+1/2,j,k} + g_{x,i-1/2,j+1,k} + g_{x,i-1/2,j,k}\right)}{16\pi G}, && (15) \\
T^{(2)}_{zx,i,j,k+1/2} &= \frac{g_{z,i,j,k+1/2}\left(g_{x,i+1/2,j,k+1} + g_{x,i+1/2,j,k} + g_{x,i-1/2,j,k+1} + g_{x,i-1/2,j,k}\right)}{16\pi G}, && (16) \\
g_{x,i+1/2,j,k} &= -\frac{\phi_{i+1,j,k} - \phi_{i,j,k}}{h}, && (17) \\
g_{y,i,j+1/2,k} &= -\frac{\phi_{i,j+1,k} - \phi_{i,j,k}}{h}, && (18) \\
g_{z,i,j,k+1/2} &= -\frac{\phi_{i,j,k+1} - \phi_{i,j,k}}{h}. && (19)
\end{aligned}
$$

We can derive the other components of the tensor by changing x, y, z and indices cyclically. This second order gravitational tensor provides

$$
\begin{aligned}
\boldsymbol{\nabla}\cdot\boldsymbol{T}^{(2)}_g &= -\frac{\rho^{(2)}_{i,k,k}}{2}\begin{pmatrix} g_{x,i+1/2,j,k} + g_{x,i-1/2,j,k} \\ g_{y,i,j+1/2,k} + g_{y,i,j-1/2,k} \\ g_{z,i,j,k+1/2} + g_{z,i,j,k-1/2} \end{pmatrix}, && (20) \\
\rho^{(2)}_{i,j,k} &= -\frac{g_{x,i+1/2,j,k} - g_{x,i-1/2,j,k}}{4\pi G h} - \frac{g_{y,i,j+1/2,k} - g_{y,i,j-1/2,k}}{4\pi G h} \\
&\quad - \frac{g_{z,i,j,k+1/2} - g_{z,i,j,k-1/2}}{4\pi G h}. && (21)
\end{aligned}
$$

Since the Poisson equation is symmetric with respect to inversion, the truncation error of $\boldsymbol{T}^{(2)}_g$ should be of the even-order. Then the fourth order accurate gravitational tensor has the form,

$$
\boldsymbol{T}^{(4)}_g = \boldsymbol{T}^{(2)}_g + \Delta\boldsymbol{T}_g, \tag{22}
$$

where $\Delta\boldsymbol{T}_g$ denotes the correction term proportional to h^2.

We derive the correction term, $\Delta\boldsymbol{T}_g$, by expanding the gravitational potential around each cell surface center up to the third order such as

$$
\begin{aligned}
\phi(x,y,z) &= \phi\left(x_{i+1/2}, y_j, z_k\right) + \frac{\partial\phi}{\partial x}\left(x - x_{i+1/2}\right) + \frac{\partial\phi}{\partial y}\left(y - y_j\right) + \frac{\partial\phi}{\partial z}\left(z - z_k\right) \\
&\quad + \frac{1}{2}\frac{\partial^2\phi}{\partial x^2}\left(x - x_{i+1/2}\right)^2 + \frac{1}{2}\frac{\partial^2\phi}{\partial y^2}\left(y - y_j\right)^2 + \frac{1}{2}\frac{\partial\phi^2}{\partial z^2}\left(z - z_k\right)^2 \\
&\quad + \frac{\partial^2\phi}{\partial x\partial y}\left(x - x_{i+1/2}\right)\left(y - y_j\right) + \frac{\partial^2\phi}{\partial y}\left(y - y_j\right)\left(z - z_k\right) + \frac{\partial^2\phi}{\partial z\partial x}\left(z - z_k\right)\left(x - x_{i+1/2}\right) \\
&\quad + \frac{1}{6}\frac{\partial^3\phi}{\partial x^3}\left(x - x_{i+1/2}\right)^3 + \frac{1}{6}\frac{\partial^3\phi}{\partial y^3}\left(y - y_j\right)^3 + \frac{1}{6}\frac{\partial^3\phi}{\partial z^3}\left(z - z_k\right)^3 \\
&\quad + \frac{1}{2}\frac{\partial^3\phi}{\partial x^2\partial y}\left(x - x_{i+1/2}\right)^2\left(y - y_j\right) + \frac{1}{2}\frac{\partial^3\phi}{\partial y^2\partial z}\left(y - y_j\right)^2\left(z - z_k\right) \\
&\quad + \frac{1}{2}\frac{\partial^3\phi}{\partial z^2\partial x}\left(z - z_k\right)^2\left(x - x_{i+1/2}\right) + \frac{\partial^3\phi}{\partial x\partial y\partial z}\left(x - x_{i+1/2}\right)\left(y - y_j\right)\left(z - z_k\right). && (23)
\end{aligned}
$$

The cell surface averaged gravitational tensor is defined as

$$T_{g,xx,i+1/2,j,k} = \frac{1}{8\pi G h^2}\int_{y_j-h/2}^{y_j+h/2}\int_{z_k-h/2}^{z_k+h/2}\left[\left(\frac{\partial\phi}{\partial x}\right)^2-\left(\frac{\partial\phi}{\partial y}\right)^2-\left(\frac{\partial\phi}{\partial z}\right)^2\right]dydz, \quad (24)$$

$$T_{g,xy,i+1/2,j,k} = \frac{1}{4\pi G h^2}\int_{y_j-h/2}^{y_j+h/2}\int_{z_k-h/2}^{z_k+h/2}\frac{\partial\phi}{\partial x}\frac{\partial\phi}{\partial y}dydz, \quad (25)$$

$$T_{g,xz,i+1/2,j,k} = \frac{1}{4\pi G h^2}\int_{y_j-h/2}^{y_j+h/2}\int_{z_k-h/2}^{z_k+h/2}\frac{\partial\phi}{\partial x}\frac{\partial\phi}{\partial z}dydz, \quad (26)$$

where we evaluate the integrand on $x = x_{i+1/2}$ as a function of y and z. Substituting Equation (23) into Equations (24) through (26) we obtain

$$\begin{aligned}4\pi G T_{g,xx,i+1/2,j,k} &= \frac{1}{2}\left[\left(\frac{\partial\phi}{\partial x}\right)^2-\left(\frac{\partial\phi}{\partial y}\right)^2-\left(\frac{\partial\phi}{\partial z}\right)^2\right]+\frac{1}{24}\left(\frac{\partial^2\phi}{\partial x\partial y}\right)^2h^2+\frac{1}{24}\left(\frac{\partial^2\phi}{\partial z\partial x}\right)^2h^2\\&+\frac{1}{24}\frac{\partial\phi}{\partial x}\frac{\partial^3\phi}{\partial x\partial y^2}h^2+\frac{1}{24}\frac{\partial\phi}{\partial x}\frac{\partial^3\phi}{\partial x\partial z^2}h^2-\frac{1}{24}\left(\frac{\partial^2\phi}{\partial y}\right)^2h^2-\frac{1}{24}\left(\frac{\partial^2\phi}{\partial z}\right)^2h^2\\&-\frac{1}{12}\frac{\partial\phi}{\partial y}\frac{\partial^3\phi}{\partial y^3}h^2-\frac{1}{12}\frac{\partial\phi}{\partial z}\frac{\partial^3\phi}{\partial z^3}h^2-\frac{1}{12}\left(\frac{\partial^2\phi}{\partial y\partial z}\right)^2h^2+\mathcal{O}\left(h^4\right), \quad (27)\end{aligned}$$

$$\begin{aligned}4\pi G T_{g,xy,i+1/2,j,k} &= \frac{\partial\phi}{\partial x}\frac{\partial\phi}{\partial y}+\frac{1}{12}\frac{\partial^2\phi}{\partial x\partial y}\frac{\partial^2\phi}{\partial y^2}h^2+\frac{1}{24}\frac{\partial\phi}{\partial x}\frac{\partial^3\phi}{\partial y^3}h^2+\frac{1}{24}\frac{\partial\phi}{\partial y}\frac{\partial^3\phi}{\partial x\partial y^2}h^2\\&+\frac{1}{12}\frac{\partial^2\phi}{\partial x\partial z}\frac{\partial^2\phi}{\partial y\partial z}h^2+\mathcal{O}\left(h^4\right), \quad (28)\end{aligned}$$

$$\begin{aligned}4\pi G T_{g,xz,i+1/2,j,k} &= \frac{\partial\phi}{\partial x}\frac{\partial\phi}{\partial z}+\frac{1}{12}\frac{\partial^2\phi}{\partial x\partial z}\frac{\partial^2\phi}{\partial z^2}h^2+\frac{1}{24}\frac{\partial\phi}{\partial x}\frac{\partial^3\phi}{\partial z^3}h^2+\frac{1}{24}\frac{\partial\phi}{\partial z}\frac{\partial^3\phi}{\partial x\partial z^2}h^2\\&+\frac{1}{12}\frac{\partial^2\phi}{\partial x\partial y}\frac{\partial^2\phi}{\partial y\partial z}h^2+\mathcal{O}\left(h^4\right). \quad (29)\end{aligned}$$

We also obtain

$$g_{x,i+1/2,j,k} = -\frac{\partial\phi}{\partial x}-\frac{1}{12}\frac{\partial^3\phi}{\partial x^3}h^2-\frac{1}{24}\frac{\partial^3\phi}{\partial x\partial y^2}h^2-\frac{1}{24}\frac{\partial^3\phi}{\partial x\partial z^2}h^2+\mathcal{O}\left(h^4\right), \quad (30)$$

$$\begin{aligned}g_{y,i+1,j\pm1/2,k} &= -\frac{\partial\phi}{\partial y}-\frac{1}{2}\frac{\partial^2\phi}{\partial x\partial y}h\mp\frac{1}{2}\frac{\partial^2\phi}{\partial y^2}h-\frac{1}{6}\frac{\partial^3\phi}{\partial x^2\partial y}h^2\mp\frac{1}{4}\frac{\partial^3\phi}{\partial x\partial y^2}h^2\\&-\frac{5}{24}\frac{\partial^3\phi}{\partial y^3}h^2-\frac{1}{24}\frac{\partial^2\phi}{\partial y\partial z^2}h^2+\mathcal{O}\left(h^3\right), \quad (31)\end{aligned}$$

$$\begin{aligned}g_{y,i,j\pm1/2,k} &= -\frac{\partial\phi}{\partial y}+\frac{1}{2}\frac{\partial^2\phi}{\partial x\partial y}h\pm\frac{1}{2}\frac{\partial^2\phi}{\partial y^2}h-\frac{1}{6}\frac{\partial^3\phi}{\partial x^2\partial y}h^2\pm\frac{1}{4}\frac{\partial^3\phi}{\partial x\partial y^2}h^2\\&-\frac{5}{24}\frac{\partial^3\phi}{\partial y^3}h^2-\frac{1}{24}\frac{\partial^2\phi}{\partial y\partial z^2}h^2+\mathcal{O}\left(h^3\right), \quad (32)\end{aligned}$$

$$\begin{aligned}g_{z,i+1,j,k\pm1/2} &= -\frac{\partial\phi}{\partial z}-\frac{1}{2}\frac{\partial^2\phi}{\partial x\partial z}h\mp\frac{1}{2}\frac{\partial^2\phi}{\partial z^2}h-\frac{1}{6}\frac{\partial^3\phi}{\partial x^2\partial z}h^2\mp\frac{1}{4}\frac{\partial^3\phi}{\partial x\partial z^2}h^2\\&-\frac{5}{24}\frac{\partial^3\phi}{\partial z^3}h^2-\frac{1}{24}\frac{\partial^2\phi}{\partial y^2\partial z}h^2+\mathcal{O}\left(h^3\right), \quad (33)\end{aligned}$$

$$\begin{aligned}g_{z,i,j,k\pm1/2} &= -\frac{\partial\phi}{\partial z}+\frac{1}{2}\frac{\partial^2\phi}{\partial x\partial z}h\pm\frac{1}{2}\frac{\partial^2\phi}{\partial z^2}h-\frac{1}{6}\frac{\partial^3\phi}{\partial x^2\partial z}h^2\pm\frac{1}{4}\frac{\partial^3\phi}{\partial x\partial z^2}h^2\\&-\frac{5}{24}\frac{\partial^3\phi}{\partial z^3}h^2-\frac{1}{24}\frac{\partial^2\phi}{\partial y^2\partial z}h^2+\mathcal{O}\left(h^3\right), \quad (34)\end{aligned}$$

by substituting Equation (23) into Equations (17) through (19). Although $g_{y,i+1,j\pm 1/2,k}$, $g_{y,i,j\pm 1/2,k}$, $g_{z,i+1,j,k\pm 1/2}$, and $g_{z,i,j,k\pm 1/2}$ have not only even-order but also odd-order truncation errors, we can eliminate the odd-order truncation errors by symmetrizing $\Delta \boldsymbol{T}_g$. Thus we omit the details of the third-order truncation errors in Equations (31) to (34).

After some manipulation we obtain

$$\begin{aligned} 4\pi G \Delta T_{g,xx,i+1/2,j,k} &= -\frac{1}{12}\frac{\partial \phi}{\partial x}\frac{\partial^3 \phi}{\partial x^3}h^2 - \frac{1}{12}\left(\frac{\partial^2 \phi}{\partial x \partial y}\right)^2 h^2 - \frac{1}{12}\left(\frac{\partial^2 \phi}{\partial x \partial z}\right)^2 h^2 \\ &+ \frac{1}{12}\left(\frac{\partial^2 \phi}{\partial y^2}\right)^2 h^2 + \frac{1}{12}\left(\frac{\partial^2 \phi}{\partial z^2}\right)^2 h^2 - \frac{1}{12}\left(\frac{\partial^2 \phi}{\partial y \partial z}\right)^2 h^2 \\ &+ \frac{1}{6}\frac{\partial \phi}{\partial y}\left(\frac{\partial^3 \phi}{\partial x^2 \partial y} + \frac{\partial^3 \phi}{\partial y^3}\right) h^2 + \frac{1}{6}\frac{\partial \phi}{\partial z}\left(\frac{\partial^3 \phi}{\partial x^2 \partial z} + \frac{\partial^3 \phi}{\partial z^3}\right) h^2, \quad (35) \\ 4\pi G \Delta T_{xy,i,j+1/2,k} &= \frac{1}{12}\frac{\partial^2 \phi}{\partial x^2}\frac{\partial^2 \phi}{\partial x \partial y}h^2 + \frac{1}{12}\frac{\partial^2 \phi}{\partial x \partial z}\frac{\partial^2 \phi}{\partial y \partial z}h^2 - \frac{1}{12}\frac{\partial \phi}{\partial x}\frac{\partial \phi}{\partial y^3}h^2 \\ &- \frac{1}{6}\frac{\partial \phi}{\partial y}\left(\frac{\partial^3 \phi}{\partial x^3}h^2 + \frac{\partial^3 \phi}{\partial x \partial y^2}h^2\right), \quad (36) \\ 4\pi G \Delta T_{xz,i,j,k+1/2} &= \frac{1}{12}\frac{\partial^2 \phi}{\partial x^2}\frac{\partial^2 \phi}{\partial x \partial z}h^2 + \frac{1}{12}\frac{\partial^2 \phi}{\partial x \partial y}\frac{\partial^2 \phi}{\partial y \partial z}h^2 - \frac{1}{12}\frac{\partial \phi}{\partial x}\frac{\partial \phi}{\partial z^3}h^2 \\ &- \frac{1}{6}\frac{\partial \phi}{\partial z}\left(\frac{\partial^3 \phi}{\partial x^3} + \frac{\partial^3 \phi}{\partial x \partial z^2}\right) h^2. \quad (37) \end{aligned}$$

Here we evaluate all the derivatives and higher derivatives on the right hand side at the cell surface center. We can express them as the quadrature of $g_{x,i+1/2,j,k}$, $g_{y,i,j+1/2,k}$, and $g_{z,i,j,k+1/2}$. We can derive the other components of the gravitational tensor, $\boldsymbol{T}_g$, by changing x, y, z, and their indices cyclically.

2.4. *Gravitational Energy Release*

The gravitational energy release $\rho \boldsymbol{v} \cdot \boldsymbol{g}$ denotes the energy transfer rate from gravitational to other forms. The change in the gravitational energy is due to changes in the density distribution. The change in the density is described by the discretized continuity equation,

$$\begin{aligned} \frac{\partial \rho_{i,k,k}}{\partial t} &+ \frac{(\rho v_x)_{i+1/2,j,k} - (\rho v_x)_{i-1/2,j,k}}{h} + \frac{(\rho v_y)_{i,j+1/2,k} - (\rho v_y)_{i,j-1/2,k}}{h} \\ &+ \frac{(\rho v_z)_{i,j,k+1/2} - (\rho v_z)_{i,j,k-1/2}}{h} = 0, \quad (38) \end{aligned}$$

where (ρv_x), (ρv_y), and (ρv_z) denote the numerical mass flux at the cell surface. Thus, it is reasonable to use the mass flux to evaluate the gravitational release. We evaluate the gravitational energy release $\rho \boldsymbol{v} \cdot \boldsymbol{g}$ to be the product of the mass flux and gravity, both of which can be evaluated on the cell surface (See Equations (9), (10), (11) and (38)). The gravitational energy release consists of three components,

$$(\rho \boldsymbol{v} \cdot \boldsymbol{g})_{i,j,k} = (\rho v_x g_x)_{i,j,k} + (\rho v_y g_z)_{i,j,k} + (\rho v_z g_z)_{i,j,k}. \quad (39)$$

The first component defined as

$$(\rho v_x g_x)_{i,j,k} = \frac{1}{24}\left[-(\rho v_x)_{i+3/2,j,k}\, g^{(4)}_{x,i+3/2,j,k} + 13\,(\rho v_x)_{i+1/2,j}\, g^{(4)}_{x,i+1/2,k}\right.$$

$$
\begin{aligned}
&+13\,(\rho v_x)_{i-1/2,j,k}\, g^{(4)}_{x,i-1/2,j,k} - (\rho v_x)_{i-3/2}\, g^{(4)}_{x,i-3/2}\Big] \\
&+\frac{1}{96}\left[(\rho v_x)_{i+1/2,j+1,k} - (\rho v_x)_{i+1/2,j-1,k}\right]\left[g^{(4)}_{x,i+1/2,j+1,k} - g^{(4)}_{x,i+1/2,j-1,k}\right] \\
&+\frac{1}{96}\left[(\rho v_x)_{i-1/2,j+1,k} - (\rho v_x)_{i-1/2,j-1,k}\right]\left[g^{(4)}_{x,i-1/2,j+1,k} - g^{(4)}_{x,i-1/2,j-1,k}\right] \\
&+\frac{1}{96}\left[(\rho v_x)_{i+1/2,j,k+1} - (\rho v_x)_{i+1/2,j,k-1}\right]\left[g^{(4)}_{x,i+1/2,j,k+1} - g^{(4)}_{x,i+1/2,j,k-1}\right] \\
&+\frac{1}{96}\left[(\rho v_x)_{i-1/2,j,k+1} - (\rho v_x)_{i-1/2,j,k-1}\right]\left[g^{(4)}_{x,i-1/2,j,k+1} - g^{(4)}_{x,i-1/2,j,k-1}\right].) \qquad (40)
\end{aligned}
$$

is the fourth-order accurate cell average. We can derive the other components similarly. Note that Equation (40) takes account of the transverse gradients of mass flux and gravity in the last four lines. PPM employs a similar correction when evaluating a cell surface average (see, e.g., Equation (17) of Colella et al.[2].

2.5. *Time Integration*

As shown in the previous subsections, we can compute both the gravitational acceleration and gravitational energy release once the density and velocity are specified. Thus we can apply our method to each integration stage within a step when advancing the hydrodynamic equations in time. The temporal accuracy of our algorithm will then match that of the hydrodynamic scheme. Since we use a third-order Runge-Kutta integrator, the solution is of the third-order accuracy in time.

3. Numerical Examples

3.1. *Polytrope*

First we consider a polytrope in which the density and pressure are expressed as

$$
\rho = \rho_0 \left(1 + \frac{r^2}{3R^2}\right)^{-5/2}, \quad P = \frac{2\pi G {\rho_0}^2}{3}\left(1 + \frac{r^2}{3R^2}\right)^{-3}, \qquad (41)
$$

respectively, where r denotes the spherical radius from the center. Equation (41) denotes a hydrostatic equilibrium for any central density (ρ_0) and size scale (R). In the following numerical test, we assume $\rho_0 = 3$, $R = 1$, $4\pi G = 1$, and $\gamma = 5/3$. We set the initial density and pressure at $t = 0$ to be the cell average of the analytic solution given in Equation (41). The initial velocities are set to zero. We set the mass flux on the outer boundary to zero and fix the hydrodynamic variables and the gravitational potential outside the boundary at their initial values.

We compare two models having different cell widths $h = 7.5 \times 10^{-2}$ and 3.75×10^{-2} to assess the accuracy of our proposed algorithm. Figure 1 shows the relative error in the density in the $z = h/2$ plane of at $t = 8.2$. The number of numerical cells is 64^3 and 128^3 in the left and right panels, respectively. The colorbar ranges differ by a factor 16 so that the colors match when the numerical solutions are of the fourth-order accuracy. As shown in the supplementary movie, the error is due to sound waves and remains of the fourth-order accuracy during the evolution. Relatively large errors near the corner are due to the reflection of the sound waves near the boundary

3.2. *Self-gravitating Sheet in Equilibrium*

To quantify the numerical error, we consider the density and pressure distributions,

$$
\rho(\boldsymbol{x}) = \rho_0 \left[1 + \alpha \cos(\boldsymbol{k}\cdot\boldsymbol{x}) + \frac{\alpha^2}{3}\cos(2\boldsymbol{k}\cdot\boldsymbol{x})\right], \qquad (42)
$$

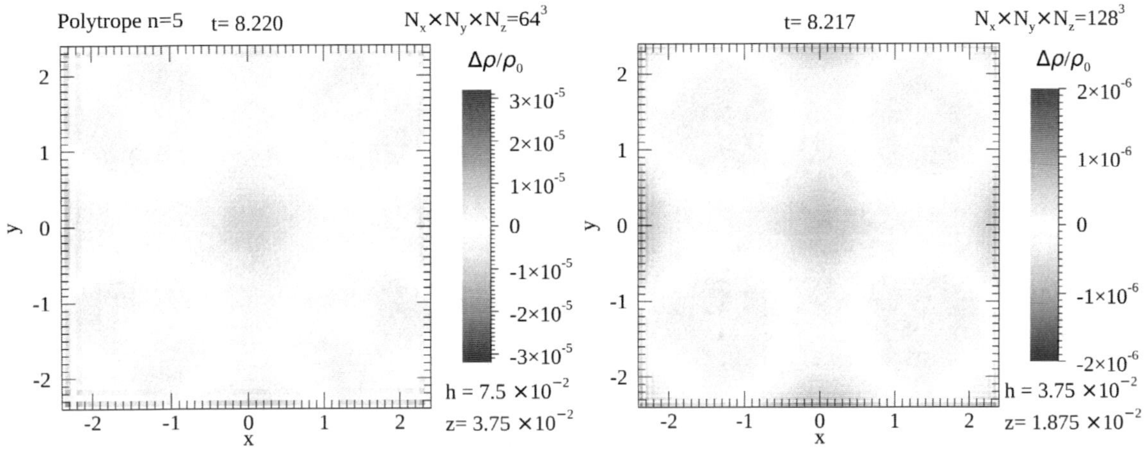

Figure 1. Relative error profiles of the density field ($\Delta\rho/\rho$), after evolving a polytrope for $t \simeq 8.2 \simeq 14/\sqrt{4\pi G\rho_0}$ The spatial resolution improves by a factor of two moving from the left to right panel. The colorbar ranges differ by a factor of 16.

$$P(\boldsymbol{x}) = P_0 + \frac{4\pi G\alpha{\rho_0}^2}{|\boldsymbol{k}|^2}\left[\left(1-\frac{\alpha^2}{12}\right)\cos(\boldsymbol{k}\cdot\boldsymbol{x}) + \frac{\alpha}{3}\cos(2\boldsymbol{k}\cdot\boldsymbol{x}) + \frac{\alpha^2}{12}\cos(3\boldsymbol{k}\cdot\boldsymbol{x}) + \frac{\alpha^3}{144}\cos(4\boldsymbol{k}\cdot\boldsymbol{x})\right], \tag{43}$$

where $\boldsymbol{x} = (x, y, z)$. The symbols $\boldsymbol{k}$ and α denote the wavenumber and amplitude. Since these profiles are periodic in all directions, we set the computational box to cover the rectangular region $0 \le x \le 2\pi/k_x$, $0 \le y \le 2\pi/k_y$, and $0 \le z \le 2\pi/k_z$. As noted in §2.2, we subtract the average density on the right hand side of Equation (7) since we apply periodic boundary conditions. Equations (42) and (43) denote a self-gravitating gas sheet in equilibrium. At the initial stage $t = 0$, the density and pressure are the cell average of the analytic solution. The initial velocity is zero. We use a unit system in which $4\pi G = 1$. Any change from this state is due to numerical errors.

Figure 2 shows the L1 norm of the error in the density as a function of time t for $\rho_0 = 1$, $P_0 = 6$, $\alpha = 0.3$, $\gamma = 5/3$, and $\boldsymbol{k} = \left(\sqrt{6}/6, \sqrt{6}/6, \sqrt{6}/3\right)$. The three black curves denote the L1 norm for three numerical simulations computed with $32\times32\times16$, $64\times64\times32$, and $128\times128\times64$ numerical cells. The red curve shows $1/16$ of the L1 norm for $64 \times 64 \times 32$ cells for comparison with that for $128 \times 128 \times 64$ cells. The L1 norm fluctuates at early times then increases steadily at later times. The time at which the transition from fluctuations to steady increase in the error is increasingly delayed as the spatial resolution is bettered. The errors are mainly due to oscillations at early times and due to numerical diffusion in the late time.

Figure 3 is the same as Figure 2 but for the model with a initial velocity $\boldsymbol{v} = (0.8, 0.6, 0.0)$. The L1 norm fluctuates only at early times $t \le 5$. The error is larger than in the model prescribed with zero velocity. The error should be dominated by the numerical diffusion associated with advection. This model verifies not only the gravitational acceleration but also the gravitational energy release, since the velocity is large (though uniform).

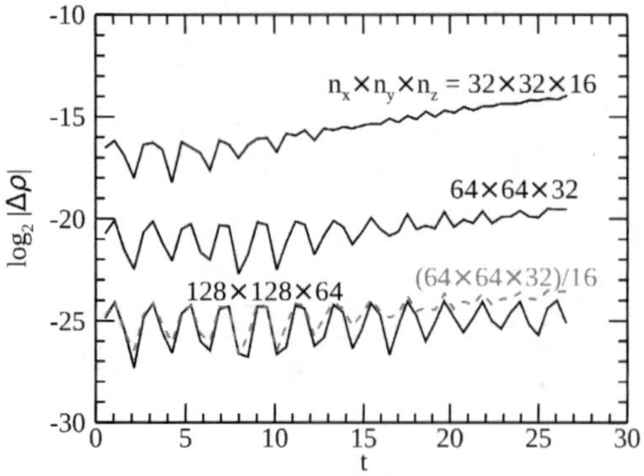

Figure 2. The L1 norm of the error in the density as a function of time, t, for a self-gravitating gas sheet.

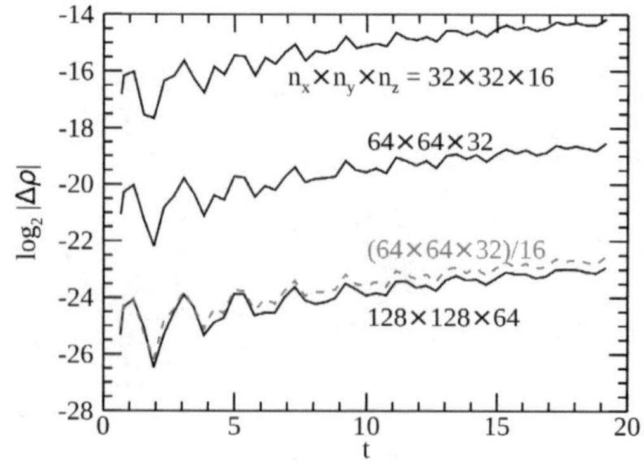

Figure 3. The same as Figure 2 but for the model with uniform velocity, $\boldsymbol{v} = (0.8, 0.6, 0.0)$.

3.3. Standing Gravito-sonic Wave

In this subsection, we consider a standing sonic wave modified by self-gravity. The initial density and pressure are assumed to be uniform at $\rho_0 = 1$ and $P_0 = 1$. We set the initial velocity to be

$$\boldsymbol{v}_0(\boldsymbol{k}) = \frac{\alpha \omega \boldsymbol{k}}{|\boldsymbol{k}|^2} \sin\left(\boldsymbol{k} \cdot \boldsymbol{k}\right), \quad \omega = \sqrt{\gamma \frac{P_0}{\rho_0} |\boldsymbol{k}|^2 - 4\pi G \rho_0}. \tag{44}$$

We consider a small amplitude perturbation with $\alpha = 1.0 \times 10^{-3}$; we set $4\pi G = 1$ and $\gamma = 5/3$. Equation (44) denotes a standing wave when α is infinitesimal. When α is small but finite, non-linearity of the hydrodynamics produces overtones with amplitude proportional to α^2. The overtone modifies the density and pressure profile. However, the specific entropy can change only through numerical errors. Thus, we can measure the accuracy of the gravitational energy release by measuring the change in the specific entropy, $s = \ln P - \gamma \ln \rho$.

We set the initial density, momentum density ($\rho \boldsymbol{v}$), and the total energy density to be cell averaged represenations. We evaluate the specific entropy at the cell center (not the cell average) according to PPM[2]. Remember that the specific entropy derived from the cell average density and pressure are not the cell average specific entropy.

Figure 4 shows the maximum specific entropy (the solid curves) and minimum (the dashed curves) for the solutions solved with $32 \times 32 \times 16$ and $64 \times 64 \times 32$ numerical cells. The red curve denotes 1/16 of the maximum entropy obtained with $32 \times 32 \times 16$ cells. Figure 5 is the same as Figure 4 but for the solutions obtained with $64 \times 64 \times 32$ and $128 \times 128 \times 64$ cells. Since the specific entropy vanishes at the initial state, the deviation from $s = 0$ is of numerical origin.

Although the entropy increases with time, the absolute value is small ($< 2 \times 10^{-8}$). Both the maximum and minimum specific entropy increase linearly as the time passes. The rate of increase is common for the maximum and minimum and proportional to cube of the cell width ($ds_{\rm max}/dt \propto h^3$ and $ds_{\rm min}/dt \propto h^3$). This numerical error is attributed to the hydrodynamic sector. The third-order convergence verifies our method to evaluate the gravitational acceleration and gravitational energy release.

4. Summary and Discussions

As shown in the previous sections, we have devised a numerical algorithm capable of achieving fourth-order accuracy in space in the evaluation of the gravitational acceleration $\rho \boldsymbol{g}$ and the

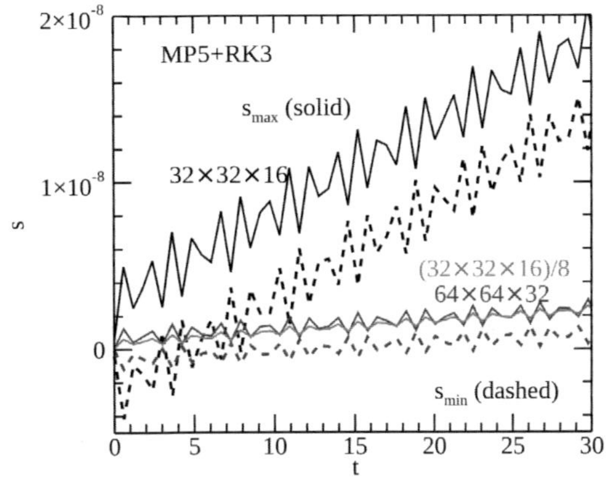

Figure 4. The maximum and minimum specific entropy for standing sonic waves for the solutions obtained with $32 \times 32 \times 16$ and $64 \times 64 \times 32$ numerical cells.

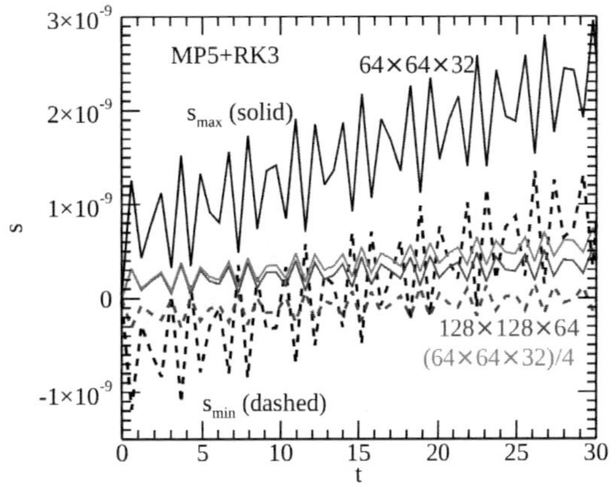

Figure 5. The same as Figure 4 but for the solutions obtained with $64 \times 64 \times 32$ and $128 \times 128 \times 64$ numerical cells.

gravitational energy release $\rho \boldsymbol{v} \cdot \boldsymbol{g}$. When coupling these source terms to the hydrodynamic equations, we can obtain solutions with overall fourth-order accuracy in space. The four examples shown here ensure the fourth-order accuracy in space. The fourth-order accuracy in space will improve the modeling of a self-gravitating astronomical object.

Though our scheme provides fourth-order accuracy in space, there still exist unresolved issues. First, Equations (35), (36), and (37) leave some freedoms in the defining of discretized expressions for the higher order derivatives. They might produce small (at fourth order truncation level), yet significant differences in the results (c.f., Mullen et al.[8] found that particular discretizations of the gravitational stress tensor could produce anomalous accelerations associated with $\nabla \times \boldsymbol{g} \neq 0$). Also, we are still investigating the robustness of the algorithm against various failure modes (e.g., negative pressures); preliminary results demonstrate that we may need to reduce the spatial accuracy around shocks or introduce artificial dissipation. These areas are still under active investigation.

Acknowledgments

This work was supported by JSPS KAKENHI Grant Numbers JP19K03906, JP20H05847. This work has been assigned a document release number LA-UR-23-32412.

References

[1] Suresh A and Hyunh H T 1997 *J. Comp. Phys.* **136** 83
[2] Colella P and Woodward P R 1984 *J. Comp. Phys.* **54** 174
[3] Felker K and Stone J.M. *J. Comp. Phys.* **375** 1365
[4] Liu X-D, Osher S, and Chan T 1994, *J. Comp. Phys.* **115** 200
[5] Trottenberg U, Oosterlee C W, and Schüller A 2001 Multigrid (London: Aacdemic Press)
[6] Shu F H 1992 The Physics of Astrophysics Volume II Gas Dynamics (Mill Valley: University Science Books) p. 46 Gravitational Stress
[7] Jiang Y-F, Belyaev M, Goodman J, and Stone J M 2013 *New Astron.* **19** 48
[8] Mullen P D, Hanawa T, and Gammie, C F. 2021 *Astrophys. J. Suppl.* **252** 30

Journal of Physics: Conference Series **2742** (2024) 012024 doi:10.1088/1742-6596/2742/1/012024

Type Ia supernovae in the age of JWST: Finding the 'right' questions and the path to answers

Peter Hoeflich, Elham Fereidouni, Tyco Brahe Mera

Department of Physics, Florida State University, Keen Bld, Tallahassee, 32306, USA

E-mail: phoeflich@fsu.edu

Abstract. Understanding the Physics of thermonuclear explosions of a White Dwarf star (WD), so called Type Ia Supernovae (SNe Ia), provide a playground for modern physics, computational methods, and are a key to modern cosmology. We identify new and investigate a variety of observational signatures of underlying physical processes related to the thermonuclear runaway, the flame propagation and the environment. Being intrinsically multi-dimensional phenomena, probing the physics requires multi-dimensional radiation-hydrodynamics and MHD simulations. For this task, we developed and employed methods for photon transport for the X-, gamma- and of low energy and of positrons under non-LTE conditions. We identify signatures in the light curves and spectra, in particular, line profiles and polarization spectra. Consistent treatment of high energy processes is critical. Therefore, our framework and results can be used directly a variety of scenarios for SNe Ia including merging WDs and explosions of sub-Chandrasekhar mass WDs. Current simulations have limitations but, nevertheless, when combined with recent JWST and VLT observations solutions emerge to many of decade old problems on the ignition process, flame physics and thermonuclear explosion.

1. Introduction

Type Ia supernovae (SNe Ia) explosively destroy C/O white dwarfs (WDs) in binary or triple systems and produce 50 % of the iron-group elements in the Universe. They have gained wide attention because the luminosity decline rate relation led to the discovery of Dark Energy. But understanding the directional dependence of the luminosity is needed for precision cosmology and to test the underlying fundamental physics. The brilliance of the top-level discoveries outshines the enormous diversity of SNe Ia that has emerged in recent years and led to an even larger diversity of three main explosion scenarios [1]: a) One of the leading explosion scenario considers a surface He detonation (HeD) that triggers a detonation of a sub-$M_{\rm Ch}$ with a low-density core C/O core[2, 3, 1]. b) A second scenario considers delayed detonation (DD) [4], in which the explosion of a near- M_{Ch}-mass WD is triggered by compressional heating in high densities resulting in the production of electron-capture elements (EC). Here, the flame propagation starts as a deflagration and transitions to a detonation. c) In the third leading scenario, two WDs merge or collide, possibly head-on, in a triple system [5], causing large-scale asymmetries and low-density burning. Despite all efforts, none of the current modeling efforts is 'first-principle', and most simulations of light curves and spectra assume spherical geometry, often on angle averaged quantities. For a collection of reviews by all groups, see [6]. With tuning, all models can reproduce overall light curves and total-flux spectra because they depend on the same network of nuclear reactions and similar WD structures, abundances, and explosion energies, a fact described as "stellar amnesia". By contrast, geometry encapsulates explosion- and progenitor-specific information (Fig.1). Geometry is sensitive to the companion star and the

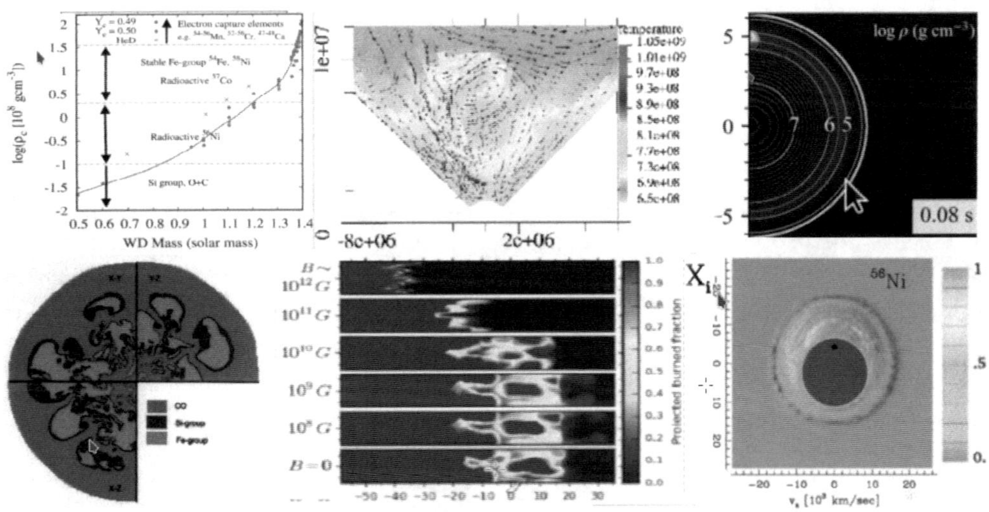

Figure 1: Progenitor structure and multi-dimensional imprints. **Upper** *left:* Central WD density as a function of mass and the influence of element production [7]. *Middle:* Temperature and turbulent velocity structure at the onset of thermonuclear runaway of the inner 100 km of a WD [8]. *Right:* Helium-triggert detonation in a sub-M_{Ch} WD [9]. **Lower** *left:* Developing RT-instabilities in deflagration fronts result in typical scales of $\sim$ 1000 km/sec during the homologous expansion phase [10]. *Middle:* Influence of the magnetic field on the nuclear burning front in a flux tube of 240 km at 0.6 sec for WD-typical conditions. We show the burning rate in arbitrary units in 6 panels for B between 0 and 10^{12} G (bottom to top); all other parameters are identical. B-fields $\leq 10^6 G$ are inferred from late time light curves [11].*Right:*^{56}Ni abundance as a result of an off-center deflagration-detonation transition (black dot) [12].

circumstellar environment(CSM), the physics of the ignition and flame propagation, and a new generation of models make clear and new qualitative and quantitative predictions. Moreover, features of EC-elements are highly blended in the optical and near IR, making JWST with MIR a game-changer. For the first time, it is possible to obtain the full 3-D image of SNe Ia by combining multi-messenger and time-domain observations from days to years after the explosion: optical to NIR total-flux and polarization spectra probe C/O, O/Ne/Mg, and Si/S, nebular optical/NIR spectra cover Si/S/Ca/Co/Fe, and mid-IR spectra Mn and Cr/Ar/Fe/Co/Ni. However, the interpretation requires a new generation of multi-dimensional non-LTE radiation-hydro simulations which, still are challenging and limited by resolution as described below.

2. Numerical Methods

Simulations for transients requires finding the simultaneous solutions of equations for nuclear networks, the statistical equations needed to determine the atomic level population, treat the flame physics, the equations of state, the opacities, and the hydro and radiation problems. Our HYDrodynamical RAdiation *HYDRA* code consists of physics-based modules to provide a solution (Fig. 2). The basic set of equations for solving the individual physical problems can be found in [13, 14], and references therein. Consistency between the solutions is achieved iteratively [15, 16]. The second approach employed is the extended use of hybrid-schemes of grid-based and Monte-Carlo methods with detailed physics problems solved by MC.

Hydrodynamics: The hydro modules use an explicit Piecewise Parabolic Method (PPM) by [21] to solve the compressible reactive flow equations with variable adiabatic gradients. PPM is implemented as a step followed by separate remaps of the thermal and kinetic energy to avoid numerical generation of spurious pressure disturbances. Subsonic reactive fronts are treated [22], based on 3D hydrodynamical simulation. For 3D geometries, [10] in case of spherical symmetry or, in 3D,the burning operator is used as given in equations 8 & 9 of [23].

High-density equation of state and nuclear reactions: The EOS is for a partially degenerate and partially relativistic Bose- and Fermi-gas plus interactions due to effects Coulomb corrections, quantum-relativistic effects on the electron component, and electron-positron pair production and electron screening. For nuclear reactions, we use an extended network of 218 isotopes based on the implementation by [24] but with modified matrix solvers and updated, parameterized reaction rates , and employ nuclear statistical equilibrium for temperatures in excess of $6.5\times10^9 g/cm^2$. All relevant strong, weak and photon interaction have been taken . The

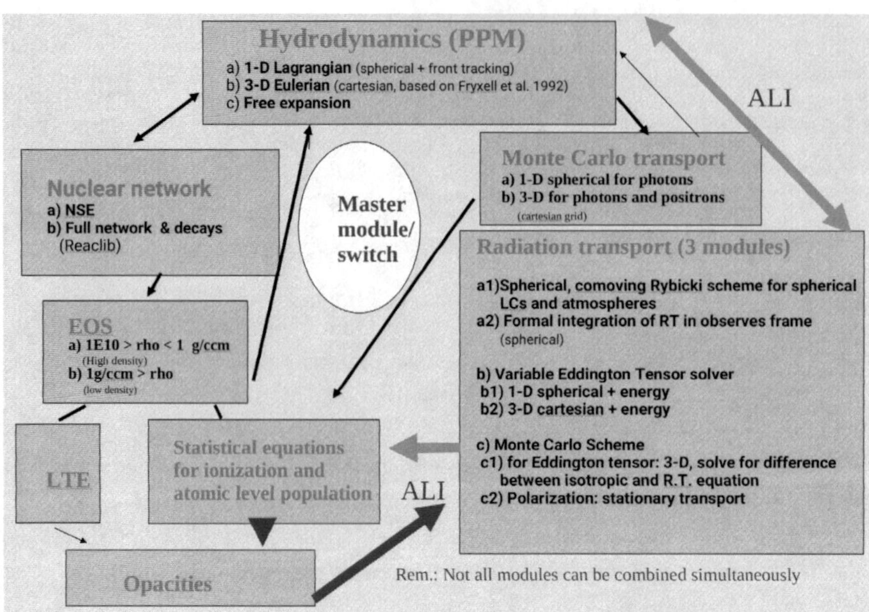

Figure 2: Block diagram of our numerical scheme to solve radiation hydrodynamical problem.

Journal of Physics: Conference Series **2742** (2024) 012024 doi:10.1088/1742-6596/2742/1/012024

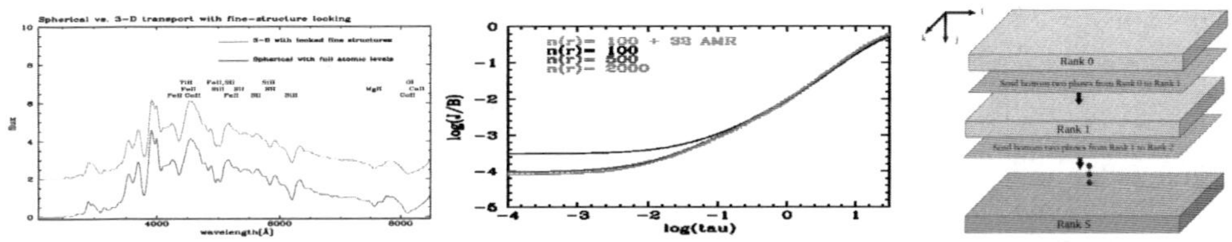

Figure 3: The need for high-resolution multi-D simulations. *Left:* Influence of the discretization on frequency, dimensions, and the complexity of the atomic models based on a spherical sub-luminous DDT model as an example to demonstrate the importance of resolution (R) and complex non-LTE models, reflecting our status of 2002 [15]. The spectra are based on our spherical and full 3-D radiation transport modules which 910 (67/67/67) depth points, 20,000 (2,000) frequency points, and 520 (50) non-LTE-levels, respectively. Even today, many simulations are spherical models with low R ($\sim$50 to 100) [17]. The differences were $\approx 20\%$ in flux at 4000 Å and $B-V$ color, which is comparable to the directional dependence of SN2004dt derived with a spatial grid of 300^2 points ([18], Fig.4). Multi-D needs high resolution explaining why multi-D non-LTE comes to age only now. *Middle:* The effect of the spatial R on the mean intensity relative to a black body is shown using $n(r)$ depth points on an equidistant mesh for conditions resembling an SN Ia density structure at maximum light, namely a scattering optical depth of $\tau = 30$ with a thermalization fraction of 10^{-3} [19, 16]. $J/B = 1$ is required to force LTE-population numbers, and errors of a factor of 3 are found for for n(r)=100 while $n(r) \approx 300$ is actually required [12]. *Right:* As part of an ongoing PhD project for the VET solver, the domain deposition is currently being implemented by Mera Evans using 'pancake slicing' in the spatial domain and cycling between the three Cartesian coordinates. Domain splitting and initialization are shown before bottom-to-top directional sweep [20].

individual rates between isotopes are fitted by expansions in ρ and T with factors tabulated in the reaction-network library REACLIB by Thielemann, and modified weak reactions published by [25]. The expansion of the rates assume a Boltzman distribution for the energies of the particles. Alternatively, the rates are calculated by integration of the cross-sections using a partial degenerate Fermi-gas.

Low density equation of state, opacities and source functions: For the atomic models, we reduce the number of energy levels by use of level-merging/super-levels, with the assumption that coupling between the merged levels being in thermodynamical equilibrium. This implies full frequency redistribution of individual transitions (see below). For the radiation transport. For Monte Carlo transport, in the simulations shown here, we use narrow line limit [26] and take into account the probability for absorption along the way by lines of higher frequency. The module for molecular kinetics use rates [27]. For the atomic level population, we solve the full set of statistical equations, namely for the rate, charge and particle conservation. The time-dependence in the rate equations

We solve the full set of statistical equations including particle and charge conservation to determine the population. The non-thermal excitation by hard γ-rays and electrons are added into the radiative rates R and collisions with fractions according to the individual level densities following the cascading in energy by MC [28, 12]. Time scales in the rate equations are dominated by the slow bound-free transitions. Therefore, time-dependent solution for the ionization balance can be obtained by solving an inhomogeneous ODE analytically or as a simple system of linear equations [29].

Coupling of radiation transport, statistical and hydro equations: We employ to acceleration schemes: a) Accelerated lambda iteration (ALI, e.g. [30]); b) equivalent-2-level approach [31, 32] which drive non-locality in the spatial and in the frequency space via levels to capture both the high-optical depth individual lines and global frequency space needed, e.g. in nebular spectra [23].

Radiation transport for low energy photons: For the time-dependent transport, we use variable Eddington Tensor solvers with full Monte-Carlo simulation in space and frequency [33, 15, 34, 16]. For the 3D case, the variable Eddington-Tensor method is used which is accurate to the order O(u/v) and we neglect acceleration terms [35, 36, 33, 16, 37]. The system of equations is given by

$$\frac{1}{c}\frac{\delta F_\nu}{\delta t} + \frac{1}{c}\nabla(uF_\nu + c\nabla(T_\nu E_\nu)) = -k_\nu F_\nu \tag{1}$$

$$\rho\frac{D(E_\nu/\rho)}{Dt} + E_\nu T_\nu \nabla u - \nabla[\frac{cT_\nu}{\kappa_n u\rho q_\nu}\nabla(q_\nu E_\nu)] = 4\pi j_\nu - \kappa_\nu \rho c E_\nu \tag{2}$$

with the tensor T and, following Auer (1971), the variable q_ν as defined by:

$$\nabla T_\nu \nabla log\ q_\nu = \nabla(\nabla T_\nu) \tag{3}$$

$$\frac{\partial ln(r^2 q_\nu)}{\partial r} = \frac{3f_\nu - 1}{f_\nu r} \tag{4}$$

and

$$E_\nu(\vec{x}) = 1/c\int \vec{I_\nu}(\vec{x},\vec{\Omega})d\vec{\Omega},\ \ \vec{F_\nu}(\vec{x}) = 1/c\int \vec{I_\nu}(\vec{x},\vec{\Omega})\vec{\Omega}d\vec{\Omega},\ \text{and}\ \underline{P}_\nu(\vec{x}) = 1/c\int \vec{I_\nu}(\vec{x},\vec{\Omega})\vec{\Omega}\otimes\vec{\Omega}d\vec{\Omega}. \tag{5}$$

with the integrals over ν denote the corresponding radiative components in the hydro-equations. For 3D, the advection frequency redistribution terms enter via the MC solution. For some tests of 3D vs. spherical solutions (Fig. 2). Again, we use stationary transport for the closure relation and tested the result against the spherical transport module. Monte Carlo (MC) methods are used including advection and aberration for the closure relation of low energy photons for flux and polarization spectra, and similar to hard γ and positron transport [38, 39, 40, 41], and the cascading of high-energy thermal leptons [12]. Beyond velocity gradients, the redistribution of energy in frequency space is caused by a large number of overlapping lines are important. E.g., in the optical, a typical of 10^6 transitions are taking into account over $\approx 10^4$ $\mathring{A}$. This leads to a coupling of photons in the frequency space via the rate-equations. The same MC transport used for the closure allows to calculate the redistribution functions ψ_ν in the source terms numerically by weighting the neighboring frequencies with the overlap of a specific transition so that the frequency integral over S_ν is conserved [29].

3. Some applications and first results based on JWST and SPECPOL

Up to now, all 5 'classical' normal thermonuclear SNe with MIR spectra show nebular spectra dominated by forbidden iron-group elements and multiple stable ^{58}Ni, indicative for high-density burning (Fig. 1) found in M_{Ch} [50, 51, 52, 46] including one underluminous SN2022xkq (Fig. 5). The small Doppler shift of Ni suggests close to center, single spot ignition. Pre-existing

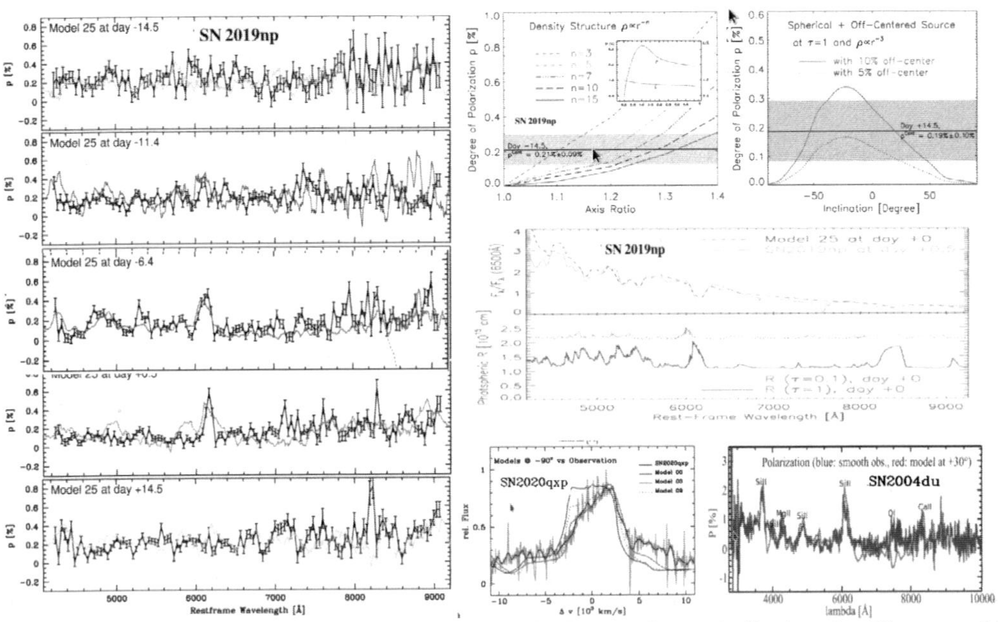

Figure 4: An off-center DDT model for a normal-bright SNe Ia viewed at an inclination of +45° compared to SN2019np observed by our SPECPOL VLT/FORS program on days −14.5, −11.4, −6.4, +5 and +14.5 w.r.t. maximum light [42]. *Left:* the evolution of the polarization spectra. *Right, upper:* Continuum polarization at days −14.5 (middle) and +14.5 (right) with the colored areas indicating the value consistent with the observations. *Right, middle:* Total-flux spectrum and photospheric radius at maximum. The observed p is reproduced both across SiII λ6360 and in the continuum. Many of the weak features in the polarization spectra (left) can be attributed to line transitions. SN 2019np was an overall spherical explosion with an asymmetric surface layer of $\sim 5\ 10^{-3} M_\odot$ and an aspherical ^{56}Ni core. Polarization and total-flux spectra form in the same part of thev extended atmosphere (right middle vs. insert in upper middle panel). A comparison between an off-center DDT model for SN2004du (lower right) shows similar features but much higher p [43] in lines though, still, a low continuum p. Within this framework, this can be understood such that in SN 2004du the DDT was triggered at $M_{DDT} = 0.25 M_\odot$ (vs. $M_{DDT} = 0.5 M_\odot$ in SN 2019np) and at an optimal viewing angle, namely +30°. *Lower, middle:* The unblended nebular [Fe II] line profile at 1.644 μm on day +210 viewed from -90° with $M_{\rm DDT} = 0.$, 0.3 and 0.9 $M_{\rm WD}$ in comparison to the underluminous SN2020qxp (black) observed at Keck (PI: C. Ashall) shows that nebular spectra offer complementary means to determine $M_{\rm DDT}$ and the viewing angle [44]. Discrepancies in the results would hint towards problems with the explosion scenario employed.

turbulent velocities are consistent with the line-width [53, 48] whereas off-center, multi-spot off-center ignitions would produce extended mixing by RT-instabilities and, thus, broad lines [49]. The lack of microscopic mixing of Ni possibly due to high B-fields [11] is evident in SN2022aefx and SN2022xkq based on our simulations. Line-asymmetries in Ar and Co are consistent with off-center delayed-detonation transitions in the regime of distributed burning from both spectropolarimetry (Fig. 4) and line profiles [23, 54]. Our current simulations have resolutions of $R \approx 300$ posing a severe restriction for probing small scales. Detailed results of the JWST MIR-spectra and spectropolrimetry programs can be found in [12, 52, 54, 46].

Acknowledgements: P.H. acknowledge support by NASA grants 80NSSC20K0538, JWST-GO-02114 and JWST-GO-02122 NASA grants, and NSF grant AST-1715133 and AST-2306395, and thanks to all members of our JWST and SPECPOL/VLT teams.

References

[1] Hoeflich P and Khokhlov A 1996 *ApJ* **457** 500 (*Preprint* arXiv:astro-ph/9602025)

Journal of Physics: Conference Series **2742** (2024) 012024 doi:10.1088/1742-6596/2742/1/012024

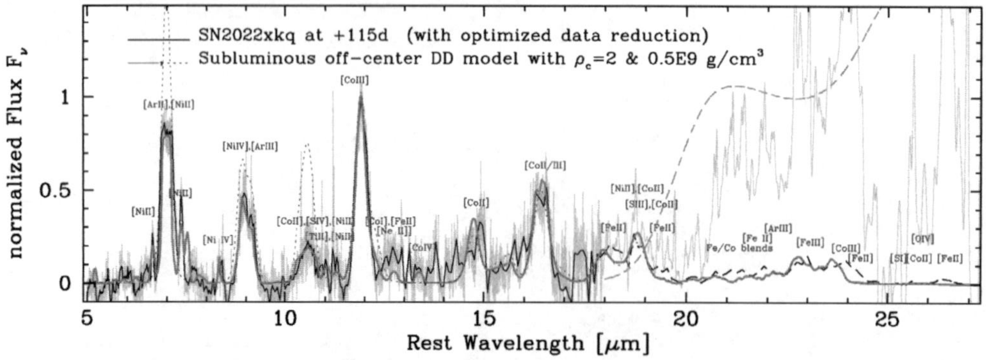

Figure 5: Comparison between the MIR spectra of the subluminous SN2022xkq [45, 46] obtained with MIRI/MRS at JWST [47] and models with $\rho_c = 2 \times 10^9$ g cm^{-3}(solid red) and 5×10^8 g cm^{-3}(dotted blue) including line identification based on the simulations. All spectra have been normalized using a distance module of 32.52 mag. The JWST pipeline provides excellent calibration and background reduction, a combination of observations and synthetic spectra is needed to properly reduce the background if it dominates the flux as in Channel 4 of JWST. The correction log of the calibration factor is given in cyan. The simulations allow to analyze the data on a 10% level. The low central density model strongly over-predicts the flux in the 7 and 9 μm [Ar] providing tight limit on the central density of the exploding WD and the presence of narrow Ni lines, somewhat surprisingly, identify the as a M_{Ch} type rather than a sub-M_{Ch}, helium triggered explosion. The narrow line strongly suggest single-point and mostly central ignition [8, 48] rather than a strong off-center multi-spot ignition [49].

[2] Woosley S E, Weaver T A and Taam R E 1980 Models for Type I supernovae *Texas Workshop on Type I Supernovae* ed Wheeler J C pp 96–112

[3] Nomoto K 1982 *ApJ* **257** 780–792

[4] Khokhlov A M 1989 *MNRAS* **239** 785–808

[5] Webbink R F 1984 *ApJ* **277** 355

[6] Alsabti A W and Murdin P 2017 *Handbook of Supernovae*

[7] Hoeflich P, Ashall C, Fisher A, Hristov B, Collins D, Hsiao E, Wiedenhoever I, Chakraborty S and Diamond T 2019 Thermonuclear Supernovae: Prospecting in the Age of Time-Domain and Multi-Wavelength Astronomy *Nuclei in the Cosmos XV* vol 219 pp 187–194 (*Preprint* 1810.05910)

[8] Höflich P and Stein J 2002 *ApJ* **568** 779–790 (*Preprint* astro-ph/0104226)

[9] Boos S J, Townsley D M, Shen K J, Caldwell S and Miles B J 2021 *ApJ* **919** 126 (*Preprint* 2101.12330)

[10] Gamezo V N, Khokhlov A M, Oran E S, Chtchelkanova A Y and Rosenberg R O 2003 *Science* **299** 77–81 (*Preprint* astro-ph/0212054)

[11] Hristov B, Collins D C, Hoeflich P, Weatherford C A and Diamond T R 2018 *ApJ* **858** 13 (*Preprint* 1711.11103)

[12] Hoeflich P, Ashall C, Bose S, Baron E, Stritzinger M D, Davis S, Shahbandeh M, Anand G S, Baade D, Burns C R, Collins D C, Diamond T R, Fisher A, Galbany L, Hristov B A, Hsiao E Y, Phillips M M, Shappee B, Suntzeff N B and Tucker M 2021 *ApJ* **922** 186 (*Preprint* 2109.03359)

Journal of Physics: Conference Series **2742** (2024) 012024 doi:10.1088/1742-6596/2742/1/012024

[13] Hristov B, Hoeflich P and Collins D 2021 *ApJ* **submitted**

[14] Hoeflich P, Hsiao E Y, Ashall C, Burns C R, Diamond T R, Phillips M M, Sand D, Stritzinger M D, Suntzeff N, Contreras C, Krisciunas K, Morrell N and Wang L 2017 *ApJ* **846** 58 (*Preprint* 1707.05350)

[15] Höflich P 2003 ALI in Rapidly Expanding Envelopes *Stellar Atmosphere Modeling* (*Astronomical Society of the Pacific Conference Series* vol 288) ed Hubeny I, Mihalas D and Werner K p 185

[16] Höflich P 2009 Multi-dimensional Radiation Transport in Rapidly Expanding Envelopes *American Institute of Physics Conference Series* (*American Institute of Physics Conference Series* vol 1171) ed Hubeny I, Stone J M, MacGregor K and Werner K pp 161–172

[17] Blondin S, Dessart L and Hillier D J 2018 *Mon. Not. R. Astron. Soc.* **474** 3931–3953 (*Preprint* 1711.09107)

[18] Hoeflich P, Gerardy C L, Marion H and Quimby R 2006 *New Astronomy* **50** 470–473

[19] Hubeny I 2003 Accelerated Lambda Iteration: An Overview *Stellar Atmosphere Modeling* (*Astronomical Society of the Pacific Conference Series* vol 288) ed Hubeny I, Mihalas D and Werner K p 17

[20] Tro S, Mera Evans T, Aslam T D, Lozano E and Culp D B 2023 *Journal of Computational Physics* **474** 111785

[21] Colella P and Woodward P R 1984 *Journal of Computational Physics* **54** 174–201

[22] Domínguez I and Höflich P 2000 *ApJ* **528** 854–860 (*Preprint* astro-ph/9908204)

[23] Hristov B, Hoeflich P and Collins D C 2021 *ApJ* **923** 210 (*Preprint* 2106.14589)

[24] Thielemann F K, Nomoto K and Hashimoto M 1994 Nucleosynthesis in supernovae *Supernovae* ed Bludman S A, Mochkovitch R and Zinn-Justin J p 629

[25] Cyburt R H, Amthor A M, Ferguson R, Meisel Z, Smith K, Warren S, Heger A, Hoffman R D, Rauscher T, Sakharuk A, Schatz H, Thielemann F K and Wiescher M 2010 *ApJS* **189** 240–252

[26] Castor J I 2004 *Radiation Hydrodynamics, Springer Press*

[27] Dominik C 2009 Physical Processes: Dust Coagulation and Fragmentation *Cosmic Dust - Near and Far* (*Astronomical Society of the Pacific Conference Series* vol 414) ed Henning T, Grün E and Steinacker J p 494

[28] Fransson C and Kozma C 1993 *ApJ Let.* **408** L25

[29] Hoeflich P 1995 *ApJ* **443** 89–108

[30] Scharmer G B 1984 Accurate solutions to non-LTE problems using approximate Lambda operators. *Methods in Radiative Transfer* pp 173–210

[31] Athay R G 1972 *Radiation Transport in Spectral Lines*

[32] Hoeflich P 1990 *A&A* **229** 191–195

[33] Stone J M, Mihalas D and Norman M L 1992 *ApJS* **80** 819–845

[34] Castor J I Toward a fully consistent radiation hydrodynamics

[35] Buchler J R 1979 *JQSRT* **22** 293

[36] Buchler J R 1983 *JQSRT* **30** 395–407

[37] Hubeny I and Mihalas D 2014 *Theory of Stellar Atmospheres, Princeton University Press*

[38] Hoeflich P 1991 *A&A* **246** 481

[39] Hoeflich P, Khokhlov A and Müller E 1992 *A&A* **259** 549–566

[40] Hoeflich P 1995 *ApJ* **440** 821

[41] Penney R and Hoeflich P 2014 *ApJ* **795** 84 (*Preprint* 1409.2159)

[42] Hoeflich P, Yang Y, Baade D, Cikota A, Maund J R, Mishra D, Patat F, Patra K C, Wang L, Wheeler J C, Filippenko A V, Gal-Yam A and Schulze S 2023 *Mon. Not. R. Astron. Soc.* **520** 560–582 (*Preprint* 2301.04721)

[43] Höflich P, Gerardy C L, Marion H and Quimby R 2006 *New Astronomy Review* **50** 470–473

[44] Hoeflich P, Ashall C, Bose S, Baron E, Stritzinger M D, Davis S, Shahbandeh M, Anand G S, Baade D, Burns C R, Collins D C, Diamond T R, Fisher A, Galbany L, Hristov B A, Hsiao E Y, Phillips M M, Shappee B, Suntzeff N B and Tucker M 2021 *arXiv e-prints* arXiv:2109.03359 (*Preprint* 2109.03359)

[45] Pearson J, Sand D J, Lundqvist P, Galbany L, Andrews J E, Bostroem K A, Dong Y, Hoang E, Hosseinzadeh G, Janzen D, Jencson J E, Lundquist M J, Mehta D, Meza Retamal N, Shrestha M, Valenti S, Wyatt S, Anderson J P, Ashall C, Auchettl K, Baron E, Blondin S, Burns C R, Cai Y, Chen T W, Chomiuk L, Coulter D A, Cross D, Davis K W, de Jaeger T, DerKacy J M, Desai D D, Dimitriadis G, Do A, Farah J R, Foley R J, Gromadzki M, Gutiérrez C P, Haislip J, González Hernández J I, Hinkle J T, Hoogendam W B, Howell D A, Hoeflich P, Hsiao E, Huber M E, Jha S W, Jiménez Palau C, Kilpatrick C D, Kouprianov V, Kumar S, Kwok L A, Larison C, LeBaron N, Le Saux X, Lu J, McCully C, Mera Evans T, Milne P, Modjaz M, Morrell N, Müller-Bravo T E, Newsome M, Nicholl M, Padilla Gonzalez E, Payne A V, Pellegrino C, Phan K, Pineda-García J, Piro A L, Piscarreta L, Polin A, Reichart D E, Rojas-Bravo C, Ryder S D, Salmaso I, Schwab M, Shahbandeh M, Shappee B J, Siebert M R, Smith N, Strader J, Taggart K, Terreran G, Tinyanont S, Tucker M A, Valerin G and Young D R 2023 *arXiv e-prints* arXiv:2309.10054 (*Preprint* 2309.10054)

[46] DerKacy J M, Ashall C, Hoeflich P, Baron E, Shahbandeh M, Shappee B J, Andrews J, Baade D, Balangan E F, Bostroem K A, Brown P J, Burns C R, Burrow A, Cikota A, de Jaeger T, Do A, Dong Y, Dominguez I, Fox O, Galbany L, Hoang E T, Hsiao E Y, Janzen D, Jencson J E, Krisciunas K, Kumar S, Lu J, Lundquist M, Mera Evans T B, Maund J R, Mazzali P, Medler K, Meza Retamal N E, Morrell N, Patat F, Pearson J, Phillips M M, Shrestha M, Stangl S, Stevens C P, Stritzinger M D, Suntzeff N B, Telesco C M, Tucker M A, Valenti S, Wang L and Yang Y 2023 *arXiv e-prints* arXiv:2310.09153 (*Preprint* 2310.09153)

[47] Ashall C, Baron E, Hoeflich P A, Baade D, Brown P J, Burns C, Burrow A, Cikota A, Davis S, DerKacy J M, Do A J M, Fisher A J, Galbany L, Hsiao E, Karamehmetoglu E, Krisciunas K, Kumar S, LU J, Mera Evans T B, Morrell N, Patat F, Phillips M M, Shahbandeh M, Shappee B J, Stritzinger M, Suntzeff N B, Telesco C, Tucker M, Wang

L, Yang Y and de Jaeger T 2021 MIR Spectroscopy of Type Ia Supernovae: The Key to Unlocking their Explosions and Element Production JWST Proposal. Cycle 1

[48] Zingale M, Nonaka A, Almgren A S, Bell J B, Malone C M and Woosley S E 2011 *ApJ* **740** 8

[49] Seitenzahl I R, Ciaraldi-Schoolmann F, Röpke F K, Fink M, Hillebrandt W, Kromer M, Pakmor R, Ruiter A J, Sim S A and Taubenberger S 2013 *Mon. Not. R. Astron. Soc.* **429** 1156–1172 (*Preprint* 1211.3015)

[50] Gerardy C L, Meikle W P S, Kotak R, Hoeflich P, Farrah D, Filippenko A V, Foley R J, Lundqvist P, Mattila S, Pozzo M, Sollerman J, Van Dyk S D and Wheeler J C 2007 *ApJ* **661** 995–1012 (*Preprint* arXiv:astro-ph/0702117)

[51] Telesco C M, Hoeflich P, Li D and etal 2015 *ApJ* **798** 93 (*Preprint* 1409.2125)

[52] Kwok L A and Jha S W e a 2023 *ApJ Let.* **944** L3 (*Preprint* 2211.00038)

[53] Hoeflich P and Stein J 2002 *ApJ* **568** 779–790 (*Preprint* arXiv:astro-ph/0104226)

[54] DerKacy J M, Ashall C, Hoeflich P, Baron E, Shappee B J, Baade D, Andrews J, Bostroem K A, Brown P J, Burns C R, Burrow A, Cikota A, de Jaeger T, Do A, Dong Y, Dominguez I, Galbany L, Hsiao E Y, Karamehmetoglu E, Krisciunas K, Kumar S, Lu J, Evans T B M, Maund J R, Mazzali P, Medler K, Morrell N, Patat F, Phillips M M, Shahbandeh M, Stangl S, Stevens C P, Stritzinger M D, Suntzeff N B, Telesco C M, Tucker M A, Valenti S, Wang L, Yang Y, Jha S W and Kwok L A 2023 *ApJ Let.* **945** L2 (*Preprint* 2301.03647)

High-order genuinely multidimensional finite volume methods via kernel-based WENO

I C T May[1] **and D Lee**[1]

[1]Applied Mathematics, University of California Santa Cruz,
1156 High St, Santa Cruz, CA 95064, USA

E-mail: imay1@ucsc.edu

E-mail: dlee79@ucsc.edu

Abstract. In this paper a family of fully multidimensional kernel-based reconstruction schemes for use in finite volume methods (FVMs) will be presented. These methods are intended for use in shock dominated problems, and stability is achieved through a suitable adaptation of the Adaptive Order Weighted Essentially Non-Oscillatory (WENO-AO) method to the proposed kernel-based reconstruction schemes. There are a number of key difficulties in the design of high-order finite volume schemes which will be discussed and addressed. High (4th and 6th) order convergence will be demonstrated on smooth exact solutions of the ideal MHD equations. The very same scheme will then be applied to extremely stringent astrophysical benchmark problems.

1. Motivation

The method developed herein is concerned with the solution of systems of hyperbolic conservation laws. In divergence form these systems can be written as

$$\frac{\partial \mathbf{U}}{\partial t} + \nabla \cdot \mathbf{F}(\mathbf{U}) = \mathbf{S}(\mathbf{U}), \tag{1}$$

where $\mathbf{U} \in \mathbb{R}^n$ is a vector of conserved quantities, $\mathbf{F} \in \mathbb{R}^{n\times d}$ is a flux tensor for the n conserved quantities in d space dimensions, and $\mathbf{S} \in \mathbb{R}^n$ contains any source terms present. Let Ω be the domain where a solution is sought, and $\{\Omega_h\}$ be a non-overlapping partitioning of Ω into distinct sub-volumes indexed by h. Integrating Eq. (1) over Ω_h and applying the divergence theorem yields

$$\frac{\partial \overline{\mathbf{U}}_h}{\partial t} = -\frac{1}{||\Omega_h||} \oint_{\partial\Omega_h} \hat{\mathbf{F}}\left(\mathbf{U}^-, \mathbf{U}^+\right) \cdot \mathbf{n} ds + \overline{\mathbf{S}(\mathbf{U})}_h, \tag{2}$$

where an overline denotes the cell-average

$$\overline{\mathbf{U}}_h = \frac{1}{||\Omega_h||} \int_{\Omega_h} \mathbf{U}(\boldsymbol{x}) d\mathbf{x}, \tag{3}$$

and $\hat{\mathbf{F}}$ denotes a consistent (numerical) flux as a function of the inner and outer traces of the solution $\mathbf{U}^-$ and $\mathbf{U}^+$ respectively. Time integration will be done using standard Runge-Kutta methods, specifically those from [1, 2], though discussion of them will be suppressed to focus on the spatial semi-discretization. The boundary and volume integrals in Eq. (2) will be discretized by standard quadrature rules. The flux $\hat{\mathbf{F}}$ will be evaluated using a standard approximate Riemann solver, and hence the input states $\mathbf{U}^{\pm}$ will be referred to as Riemann states. A core operation of the FVM, which will be the focus of the present article, is the formulation of a stable and accurate procedure for reconstructing these Riemann states, as well interior states if needed, from available cell-average states.

This reconstruction scheme needs to satisfy several conflicting objectives, and hence the design of such a scheme is quite involved. We seek accurate solutions to Eq. (1), and modern computing hardware strongly favors high-order methods that provide better truncation orders at the cost of potentially more operations to reduce loads of data from main memory. On the other hand, Eq. (1) generally supports discontinuous solutions, which may arise even from initially smooth solutions. Naively designed high-order methods will fail due to the onset of Gibb's oscillations near these discontinuities. Indeed, by Godunov's theorem [3] any method beyond first order accuracy must be nonlinear with respect to the input data in order to be stable. Furthermore, in the consideration of problems of astrophysical significance these discontinuities can be extremely strong, and even otherwise useful schemes may fail to preserve the positivity of quantities like fluid density and internal energy.

Beyond these fundamental difficulties there are also difficulties inherent to FVMs in multiple space dimensions. The FVM evolves cell-averages of the solution through time, but in order to evaluate the integrals in Eq. (2) one needs point values of the solution. In one space dimension it is straightforward to obtain point values of the solution from these cell-averages, and the underlying schemes work nearly identically to those present in finite difference methods (FDMs). This simplicity and the relation to FDMs has lead to a substantial number of developed schemes. See [4] and references therein for more information. In higher dimensions FDMs can trivially be applied dimension-by-dimension. Unfortunately, naively applying one-dimensional FVM schemes in a similar fashion will lead to an immediate drop to second order accuracy [5, 6]. A variety of workarounds have been presented over time. Here, a genuinely multidimensional reconstruction scheme will be developed which will produce multiple point values on all faces of a cell in a single step. This method is multidimensional insofar as it utilizes cell-averages scattered in multiple dimensions simultaneously, rather than building up a reconstruction through multiple one dimensional processes. The presented scheme utilizes a kernel-based approach which sidesteps some common issues that arise in multidimensional polynomial schemes [7]. For simplicity, only uniform Cartesian grids with spacing Δ will be considered, but the reconstruction scheme is designed with more general, and even fully unstructured, grids in mind.

The following subsection will introduce the systems of conservation laws that the developed scheme will be applied to. Section 2 will introduce a high-order kernel-based reconstruction scheme that is linear with respect to the data, and apply this scheme to the iso-density vortex problems of [8] and [9]. Section 3 will then adapt this linear scheme into a nonlinear WENO scheme appropriate for use in problems with discontinuities. Choices of reconstruction variables and a particular local smoothness indicator allowing WENO to be skipped will also be deliberated on. In closing a set of high-Mach number astrophysical jet problems will be presented, and finally Section 4 will draw final conclusions.

1.1. Systems of interest

The present work considers two systems of hyperbolic conservation laws written in the form of Eq. (1). First are the compressible Euler equations where the vector of conserved quantities and fluxes are

$$\mathbf{U} = \begin{pmatrix} \rho \\ \rho u_i \\ E \end{pmatrix} \quad \mathbf{F}_j = \begin{pmatrix} \rho u_j \\ \rho u_i u_j + p\delta_{ij} \\ u_j(E+p) \end{pmatrix}, \tag{4}$$

where ρ is the fluid density, u_i are the velocities with $i = 1, 2, 3$, E is the total energy density, p is the pressure, and δ_{ij} is the Kronecker delta. The second system of interest is the ideal magnetohydrodynamic equations with generalized Lagrange multipliers (ideal MHD-GLM) as given in [10]. Here the conserved quantities, fluxes, and source terms are given by

$$\mathbf{U} = \begin{pmatrix} \rho \\ \rho u_i \\ E \\ B_i \\ \psi \end{pmatrix} \quad \mathbf{F}_j = \begin{pmatrix} \rho u_j \\ \rho u_i u_j + p_* \delta_{ij} - B_i B_j \\ F_j^E \\ B_i u_j - B_j u_i + C_h \psi \delta_{ij} \\ C_h B_j \end{pmatrix} \quad \mathbf{S} = \begin{pmatrix} 0 \\ -B_i \partial_k B_k \\ -(u_k B_k)\partial_k B_k - \psi u_k \partial_k \psi \\ -u_i \partial_k B_k \\ -u_k \partial_k \psi \end{pmatrix}, \tag{5}$$

$$F_j^E = u_j(E + p_* - \frac{1}{2}\psi^2) + B_j(C_h \psi - u_k B_k)$$

where B_i are the magnetic field components, $p_* = p + B_k B_k/2$ is the total pressure, C_h is the hyperbolic cleaning speed, ψ is the GLM field, and $\partial_k = \partial/\partial x_k$. Einstein summation convention has been used regarding any repeated dummy indices. Both systems are closed by an ideal gas equation of state giving the thermal pressure as

$$p = (\gamma - 1)\left(E - \frac{1}{2}\rho u_k u_k - \frac{1}{2}B_k B_k - \frac{1}{2}\psi^2\right), \tag{6}$$

where γ is the ratio of specific heats, and the magnetic terms are zero in the Euler case. Finally, the solenoidal magnetic field constraint in the ideal MHD equations is enforced weakly through the generalized Lagrange multiplier (GLM) field ψ following the Galilean invariant formulation of [10].

2. Kernel based reconstruction

The first task in evaluating the right hand side of Eq. (2) is to reconstruct point values of the solution at all quadrature points within a cell given nearby cell-average values. Let Ω_1 be the cell where reconstruction is being performed and $\mathcal{S}_0 = \{\Omega_h\}_{h=1}^N$ be a set of cells making up the stencil where data is to be drawn from. These stencils will be formed by considering all cells whose center lies within a given radius of the center of Ω_1. This radius R is scaled to the grid spacing and hence given simply as an integer. In two dimensions the roughly circular stencils of radii $R = 2$ and $R = 3$ contain 13 and 29 cells respectively. Similarly, in three dimensions the roughly spherical stencil of radius $R = 2$ contains 33 cells. The two dimensional radius $R = 2$ stencil can be seen in Figure 1.

In the kernel-based approach to reconstruction we seek functions of the form

$$\widetilde{f}(\boldsymbol{x}) = \sum_{l=1}^{N} a_c K(\boldsymbol{x}, \boldsymbol{x}_l) + \sum_{|\boldsymbol{\alpha}_{(v)}| \le D} b_v \boldsymbol{x}^{\boldsymbol{\alpha}_{(v)}}, \tag{7}$$

where the kernel $K : \mathbb{R}^d \times \mathbb{R}^d \to \mathbb{R}$ is a symmetric positive definite function [11, 12, 13], N is the total number of cells in the stencil, and $\boldsymbol{x}_l$ is the center point of the lth cell. This work considers the squared exponential kernel, $K(\boldsymbol{x}, \boldsymbol{y}) = e^{-\frac{||\boldsymbol{x}-\boldsymbol{y}||^2}{2L^2}}$, where $L > 0$ is a constant length

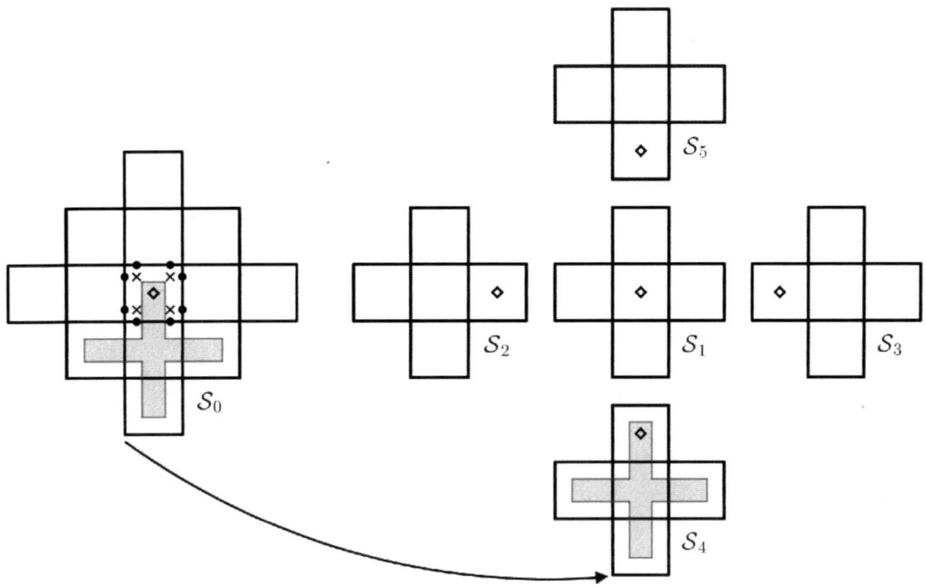

Figure 1. The full radius $R = 2$ stencil, $\mathcal{S}_0$, in $d = 2$ dimensions is shown on the left with the boundary and interior quadrature points indicated by circles and crosses respectively. Shown to the right are the 5 substencils, $\mathcal{S}_q$. In all (sub)stencils the cell marked with an open diamond is the same to indicate how the substencils fit into the full stencil, and substencil $\mathcal{S}_4$ has been annotated with a shaded region to emphasize the relationship.

scale taken here as $L = 5\Delta$. The second term in Eq. (7) is a tail consisting of all monomials up to a total degree of D and each $\boldsymbol{\alpha}_{(v)}$ denotes a multi-index. Parentheses here emphasize that this is the vth such multi-index rather than the vth entry of a multi-index.

For clarity, within this section, let the cell-averaging functional from Eq. (3) be written in a slightly longer form as

$$\lambda_h^{(\boldsymbol{x})} \cdot := \frac{1}{||\Omega_h||} \int\limits_{\Omega_h} \cdot d\boldsymbol{x}, \tag{8}$$

where the subscript indicates which cell is being averaged over, and the superscript indicates the variable being averaged over. This latter notation is necessary since the kernel function is bivariate, and it is important to be able to apply functionals unambiguously to just one argument. Let $\mathbf{g}$ be a vector whose entries are the cell-averages of the (otherwise unknown) function being approximated, that is $\mathbf{g}_h = \lambda_h^{(\boldsymbol{x})} f(\boldsymbol{x})$. The reconstructed function is forced to match the known data by averaging Eq. (7) over each cell. Gathering the unknown coefficients in Eq. (7) into the vectors $\mathbf{a}$ and $\mathbf{b}$, then yields

$$\mathbf{Qa} + \mathbf{Pb} = \mathbf{g}, \tag{9}$$

$$\mathbf{Q}_{hl} = \lambda_h^{(\boldsymbol{x})} K(\boldsymbol{x}, \boldsymbol{x}_l), \tag{10}$$

$$\mathbf{P}_{hv} = \lambda_h^{(\boldsymbol{x})} \boldsymbol{x}^{\boldsymbol{\alpha}_{(v)}}. \tag{11}$$

The entries of the kernel matrix $\mathbf{Q}$ derive from the kernel function by averaging over the hth cell in one argument and point evaluation at the center of the lth cell in the other argument. Similarly, the polynomial matrix $\mathbf{P}$ consists of cell averages of the various monomials up to total degree D.

Journal of Physics: Conference Series **2742** (2024) 012025 doi:10.1088/1742-6596/2742/1/012025

At present Eq. (9) is under-determined as there is too much freedom to fit the given data using both the kernel and polynomial parts of the expansion. Note however that if $\mathbf{g}$ happens to lie in the span of the columns of $\mathbf{P}$ then the kernel part is entirely unnecessary, and the given data can be completely described by the polynomial part. To enforce this polynomial-exactness the additional requirement $\mathbf{P}^T\mathbf{a} = 0$ is added, and the system can be solved. Additionally, by having this exactness property the accuracy of the reconstruction is guaranteed regardless of how the length scale L is chosen. A deeper discussion on the inclusion of polynomials, and on this construction as a whole, can be found in [14]. With the stencils used in this work the polynomial tail can handle terms to total degree $D = 2R - 1$, and hence the reconstructions will be accurate to $\mathcal{O}(\Delta^{2R})$.

Notably, the approximation $\widetilde{f}$ is defined everywhere, but is only actually needed at the quadrature points. Let $\boldsymbol{x}_*$ be one such point, and let the vectors $\mathbf{T}$ and $\mathbf{S}$ be evaluations of the kernel and polynomial parts of Eq. (7). The entries of these vectors are thus

$$\mathbf{T}_l = K(\boldsymbol{x}_*, \boldsymbol{x}_j), \quad \mathbf{S}_v = \boldsymbol{x}_*^{\boldsymbol{\alpha}(v)}. \tag{12}$$

Then reconstruction specifically to the point $\boldsymbol{x}_*$ can instead be found by solving the block system

$$\begin{bmatrix} \mathbf{Q}^T & \mathbf{P} \\ \mathbf{P}^T & 0 \end{bmatrix} \begin{pmatrix} \mathbf{r} \\ \mathbf{w} \end{pmatrix} = \begin{pmatrix} \mathbf{T} \\ \mathbf{S} \end{pmatrix}, \tag{13}$$

for the reconstruction vector $\mathbf{r}$ yielding $\widetilde{f}(\boldsymbol{x}_*) = \mathbf{T}^T\mathbf{a} + \mathbf{S}^T\mathbf{b} = \mathbf{r}^T\mathbf{g}$. This rearrangement is crucial as now only the reconstruction vectors are needed and all linear solves coming from Eq. (13) can be handled in a single set up step ahead of time. As a final note, the weights $\mathbf{w}$ are essentially just Lagrange multipliers from the polynomial-exactness constraint and can simply be discarded.

2.1. Evaluation of GLM source terms

The previously discussed reconstruction method is designed to handle arbitrary scalar fields, and is hence a useful building block for handling any system of equations. The somewhat complicated source term present in the ideal MHD-GLM equations in particular deserves more attention. After all reconstructions are performed, using either the previous linear scheme or the nonlinear WENO scheme described below, one will have solution values at both the boundary quadrature points (circles in Figure 1) and the interior quadrature points (crosses in Figure 1). These can then be supplied to numerical flux functions and source terms respectively.

However, annotating the source in Eq. (1) as a function of $\mathbf{U}$ is a slight misnomer when considering the GLM terms from Eq. (5). Here the source term depends also on derivatives of $\mathbf{U}$, in particular those of the magnetic field $\mathbf{B}$ and the scalar field ψ. The reconstructions generated in the previous section are well-defined functions of space, and can be easily differentiated simply by modifying the $\mathbf{T}$ and $\mathbf{S}$ vectors in Eq. (13) to hold samples of the derivatives of their respective terms. This is indeed how the smoothness indicators for WENO are defined, which will be discussed more below.

On the other hand, most source terms in general will not have a dependence on derivatives of the solution. The GLM sources are inherent to this particular system of equations. To support both user supplied sources (e.g. gravitational forces in some background) and the GLM sources a different tact is needed. Reporting all solution values and all derivatives at the interior quadrature points is unnecessary for most cases. Instead, for the GLM sources internally we can get the needed derivatives by slicing across lines of quadrature points (including the Riemann states) and applying standard univariate polynomial differentiation matrices [15]. In this way the GLM sources are treated to full accuracy, and a more useful standard interface for source

Table 1. The L_1 errors and experimental orders of accuracy as obtained from the iso-density vortex problems are shown.

Grid	L_1 Error	L_1 Order	Grid	L_1 Error	L_1 Order	Grid	L_1 Error	L_1 Order
	2D, $R = 2$			2D, $R = 3$			3D, $R = 2$	
64	1.64×10^{-3}	–	64	2.75×10^{-4}	–	32	5.68×10^{-4}	–
128	7.73×10^{-5}	4.41	128	3.10×10^{-6}	6.47	64	3.50×10^{-5}	4.02
256	4.17×10^{-6}	4.21	256	5.00×10^{-8}	5.95	128	1.76×10^{-6}	4.32
512	2.21×10^{-7}	4.24	512	7.67×10^{-10}	6.03			

terms can be created. As a final note, it should be mentioned that this construction does not rely on having a uniform grid, and would apply equally well to any quadrilateral or hexahedral grid.

2.2. Iso-density vortex problems

The previous reconstruction scheme is purely linear with respect to the input data, and is hence inappropriate for use in problems with shocks. It is however sufficient for smooth problems and the accuracy of the scheme can be tested. Here we consider two known exact solutions to the ideal MHD equations, and by proxy exact solutions of the ideal MHD-GLM equations with $\psi = 0$. These are the iso-density vortex solutions, where the two dimensional form comes from [8], and the three dimensional form comes from [9].

These problems were solved in two dimensions with radius $R = 2$ and $R = 3$ schemes, and in three dimensions with the radius $R = 2$ scheme, over a variety of grid resolutions. The error in the final state is easily obtained by comparison with the initial state. Time integration was done using the $[4S]^*_+$ integrator from [1] with absolute and relative tolerances of 10^{-10}, and numerical fluxes were obtained using the local Lax-Friedrichs flux from [10]. Table 1 displays the L_1 errors and experimental orders of accuracy for each scheme, and in all cases we observe the expected order of accuracy.

3. Kernel based WENO

The reconstruction procedure developed in Section 2 is linear with respect to the known data on each stencil. To handle problems with discontinuities and shocks this procedure will need to be made manifestly nonlinear. Weighted essentially nonoscillatory (WENO) methods are one approach to introduce the needed nonlinearity. In these schemes reconstruction is done over a set of substencils, and the contributions from each substencil are combined together with weights that are chosen in a data-dependent manner; substencils containing rough data are down-weighted to remove their contribution.

Traditional WENO methods utilize only the substencil reconstructions and seek to define some set of optimal (or linear) weights that recreate the action of the full stencil implicitly when all data is smooth. The existence of such (positive) weights is already tenuous in the simpler case of univariate reconstruction [16], and such weights do not exist in the more complex setting here. Instead, the full stencil can be explicitly included in the reconstruction procedure through an adaptation of the adaptive order WENO (WENO-AO) method [17].

The full stencil $\mathcal{S}_0$ is partitioned into 5(7) substencils in 2(3) space dimensions. First the substencil $\mathcal{S}_1$ is set centrally as a stencil of radius $R - 1$, then 4(6) biased stencils are set conically outwards from the central cell. An example decomposition for the two dimensional radius $R = 2$ stencil can be seen in Figure 1, with similar partitionings for other stencil sizes.

Journal of Physics: Conference Series **2742** (2024) 012025 doi:10.1088/1742-6596/2742/1/012025

The approximated point value is now given as

$$\widetilde{f}(\boldsymbol{x}_*) = \frac{\omega_0}{\gamma_0}\mathbf{r}_{(0)}^T\mathbf{g}_{(0)} + \sum_{q=1}^{N_S}\left(\omega_q - \omega_0\frac{\gamma_q}{\gamma_0}\right)\mathbf{r}_{(q)}^T\mathbf{g}_{(q)}, \tag{14}$$

where each γ_q is a fixed linear weight associated to the qth (sub)stencil for smooth data, ω_q is the associated nonlinear weight to be described next, $\mathbf{r}_{(q)}$ is the reconstruction vector defined on the qth (sub)stencil, and finally $\mathbf{g}_{(q)}$ is the restriction of the known data to q^{th} (sub)stencil. As above, the parenthesized subscripts are here to indicate that $\mathbf{r}_{(q)}$ and $\mathbf{g}_{(q)}$ are full vectors rather than entries thereof.

3.1. Linear and nonlinear weights

As the full stencil reconstruction is directly present in Eq. (14) the linear weights γ_q can be chosen purely for stability, with the only requirement being that they sum to one. Note also that if the nonlinear weights reduce to their linear counterparts, $\omega_q \to \gamma_q$ throughout Eq. (14), the only surviving term is the full stencil reconstruction. Alternatively, if any ω_q is driven to zero then that substencil will not participate, which crucially relies on the fact that ω_0 will naturally also be driven to zero as the full stencil contains all substencil data in itself. The nonlinear weights are adapted from the WENO-Z method [18, 19], and their particular construction can be found in [20].

3.2. Reconstruction variables

The reconstruction methods described only apply to scalar fields. The linear reconstruction defined in Section 2 can simply be applied componentwise to whatever variables are present, as was done in solving the iso-density vortex problems in Section 2.2. The situation is more complicated for nonoscillatory reconstruction. As is well known, performing reconstruction componentwise over the conservative variables leads to poor results near discontinuities.

Generically we can map the input cell-averages $\overline{\mathbf{U}}_h$ to another set of variables denoted by $\mathbf{W}$ via

$$\overline{\mathbf{W}}_h = \mathbf{\Phi}\overline{\mathbf{U}}_h, \quad \forall\Omega_h \in \mathcal{S}_0, \quad \mathbf{\Phi} \in \mathbb{R}^{n\times n}. \tag{15}$$

It is crucial that $\mathbf{\Phi}$ be a fixed (invertible) linear map over the whole stencil in order to maintain accuracy. This means that the new system of variables $\overline{\mathbf{W}}_h$ will still correspond to some set of cell-averaged quantities, and hence will be compatible with the developed reconstruction schemes.

The most classical approach is to consider the characteristic variables given by

$$\left.\frac{\partial \mathbf{F}_\eta}{\partial \mathbf{U}}\right|_{\widetilde{\mathbf{U}}} = \mathbf{R}\mathbf{\Lambda}\mathbf{L}, \quad \mathbf{\Phi} = \mathbf{L}, \quad \mathbf{\Phi}^{-1} = \mathbf{R}, \tag{16}$$

where $\mathbf{F}_{\boldsymbol{\eta}}$ is the flux in the $\boldsymbol{\eta}$ direction, and the Jacobian is evaluated with respect to some reference state $\widetilde{\mathbf{U}}$ usually taken as the cell-average of the central cell. There are several observations worth making. First, the characteristic variables are direction-dependent and need to be re-evaluated when doing reconstruction on different faces of a cell (not to mention the ambiguity in selecting $\boldsymbol{\eta}$ when reconstructing interior values). Second, it does not matter that the eigenstructure is only evaluated to second order accuracy, which naturally happens when using a cell-average state as the reference. Accuracy of the reconstruction is maintained for arbitrary (invertible) $\mathbf{\Phi}$. Finally, this approach is most well suited to problems where the full eigenstructure is known a-priori, as otherwise one will need to embed an eigenvalue solver inside the reconstruction procedure.

Journal of Physics: Conference Series **2742** (2024) 012025 doi:10.1088/1742-6596/2742/1/012025

The direction-dependence of characteristic variables make them an unsuitable choice for multidimensional reconstruction, where a single function represents all states in the cell. The next most natural choice would be to use the primitive variables $\mathbf{V}$, but these are unfortunately nonlinear functions of the conservative variables, and hence their use immediately drops the reconstruction to second order accuracy. Alternatively one can consider a linearization of the mapping from conservative to primitive variables. This approach is discussed more fully in [20], but one ultimately finds the transformation matrices

$$\mathbf{\Phi} = \left.\frac{\partial \mathbf{V}}{\partial \mathbf{U}}\right|_{\tilde{\mathbf{U}}}, \quad \mathbf{\Phi}^{-1} = \left.\frac{\partial \mathbf{U}}{\partial \mathbf{V}}\right|_{\mathbf{V}(\tilde{\mathbf{U}})}. \tag{17}$$

This approach is used in the astrophysical jet problems considered in Section 3.4.

3.3. Complimentary techniques

In addition to the techniques described herein there are two additional techniques that feature prominently in the subsequent astrophysical jet benchmark.

The described kernel-based WENO-AO reconstruction has been designed with efficiency in mind, but nevertheless remains the most expensive part of any calculation. This balance of effort is not unique to this particular reconstruction scheme, and a variety of local smoothness indicators (or troubled cell indicators) have arisen. In this work we consider the indicator described in [21], slightly modified to match the time integrators used here. In this case WENO is only applied in cells where the change in the internal energy e_h from one time step to the next is sufficiently large. Concretely, WENO is applied whenever

$$\frac{|e_h^{n+1} - e_h^n|}{\min(e_h^{n+1}, e_h^n)} > C\Delta t, \tag{18}$$

holds. Here $C = \mathcal{O}(10)$ is a fixed constant influencing how aggressively WENO is turned on or off.

On the other hand, even when WENO is applied one may find unphysical solution values near very strong discontinuities. This again is not a feature specific to the proposed scheme, and positivity preservation has been a subject of much research. In this work we incorporate the self-adjusting positivity preserving limiter from [22]. This limiter is applied after reconstruction by mixing in small amounts of first order corrections as needed to bring the solution within physical limits. The novel development in this limiter is the selection of physical bounds on fluid density and internal energy in a solution dependent fashion. That is, no a-priori floors on these values need to be supplied.

3.4. High-Mach astrophysical jets

High-mach number astrophysical jets have become a useful benchmark problem for the assessment of schemes meant to be used in extreme settings. These problems are very stringent tests as maintaining positivity of both fluid density and internal energy is challenging. Here we consider three dimensional variants of the high-density unmagnetized [22] and magnetized [23] jets. In all cases the jet propagates into a quiescent background through a circular region on the lower face of the domain. The jet density is 10 times larger than the background, the pressure is matched, and the inflow velocity is set to give the jet a Mach number of 800. The magnetized jet additionally has a uniform magnetic field of strength $B = \sqrt{20000}$ aligned with the jet direction, yielding a plasma β value of just 10^{-4}.

These jets are evolved to a final time of $T = 0.002$ using the $R = 2$ scheme on grids with resolution $160 \times 480 \times 160$. Symmetry conditions are applied on the lower x and z boundaries,

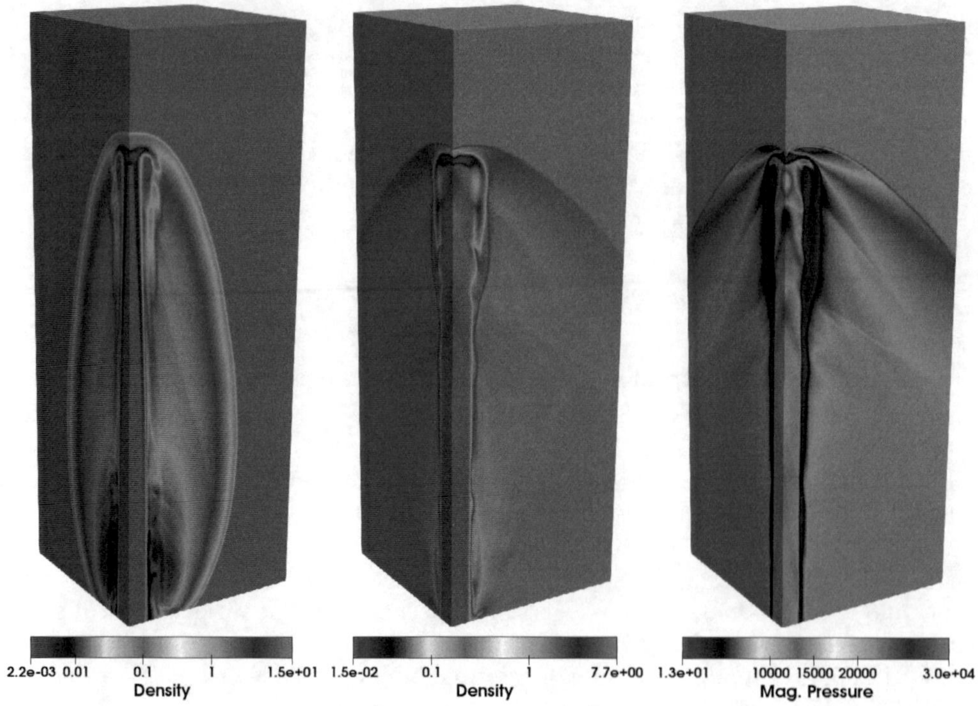

Figure 2. The left panel shows the logarithmic density for the unmagnetized jet, while the center and right panels show the logarithmic density and magnetic pressure for the magnetized jet.

mixed inflow/outflow conditions are applied on the lower y boundary, and outflow conditions are applied elsewhere. Time integration is performed using the $SSP(4,3)$ method from [2] with an embedded second order scheme from [24] where the relative and absolute tolerances are each set to 10^{-3}. The numerical fluxes are found using the HLL [25] for the unmagnetized case, and the local Lax-Friedrichs flux from [10] for the magnetized case. The resulting logarithmic densities and the magnetic pressure can be seen in Figure 2. Note that the magnetic pressure varies over 5 orders of magnitude.

4. Conclusion

In this short article a genuinely multidimensional finite volume method has been presented for the solution of systems of hyperbolic conservation laws, and applied in particular to the compressible Euler equations and ideal MHD-GLM equations. At the heart of the proposed method is a kernel-based reconstruction scheme that sidesteps some of the inherent difficulties in the use of multivariate polynomials for reconstruction. The WENO-AO method was adapted to this new reconstruction scheme to allow the simultaneous treatment of problems with smooth solutions and those containing strong discontinuities. This baseline scheme was combined with a local smoothness indicator to improve efficiency by avoiding WENO when possible, and a positivity preserving limiter to allow the treatment of extreme problems. Additionally, a set of direction-independent reconstruction variables suitable for high-order schemes were presented. The efficacy of the proposed scheme was demonstrated on the smooth iso-density vortex problems in two and three dimensions as well as two highly stringent astrophysical jet problems in three dimensions.

Journal of Physics: Conference Series **2742** (2024) 012025 doi:10.1088/1742-6596/2742/1/012025

References

[1] Ranocha H, Dalcin L, Parsani M and Ketcheson D I 2022 *Communications on Applied Mathematics and Computation* **4** 1191–1228 ISSN 2661-8893 URL https://doi.org/10.1007/s42967-021-00159-w
[2] Kraaijevanger J F B M 1991 *BIT Numerical Mathematics* **31** 482–528 ISSN 1572-9125 URL https://doi.org/10.1007/BF01933264
[3] Godunov S K 1959 *Matematicheskii Sbornik* **47(89)** 271–306
[4] Shu C W 2009 *SIAM review* **51** 82–126
[5] Buchmüller P and Helzel C 2014 *Journal of Scientific Computing* **61** 343–368 ISSN 1573-7691 URL https://doi.org/10.1007/s10915-014-9825-1
[6] Buchmüller P, Dreher J and Helzel C 2016 *Applied Mathematics and Computation* **272** 460–478 ISSN 0096-3003 recent Advances in Numerical Methods for Hyperbolic Partial Differential Equations URL https://www.sciencedirect.com/science/article/pii/S0096300315003926
[7] Bourgeois R and Lee D 2022 *Journal of Computational Physics* **471** 111603
[8] Balsara D S 2004 *The Astrophysical Journal Supplement Series* **151** 149 URL http://stacks.iop.org/0067-0049/151/i=1/a=149
[9] Mignone A, Tzeferacos P and Bodo G 2010 *Journal of Computational Physics* **229** 5896 – 5920 ISSN 0021-9991 URL http://www.sciencedirect.com/science/article/pii/S0021999110001890
[10] Derigs D, Winters A R, Gassner G J, Walch S and Bohm M 2018 *Journal of Computational Physics* **364** 420–467 ISSN 0021-9991 URL https://www.sciencedirect.com/science/article/pii/S0021999118301463
[11] Fasshauer G E and McCourt M J 2015 *Kernel-based approximation methods using Matlab* vol 19 (World Scientific Publishing Company)
[12] Schaback R and Wendland H 2006 *Acta Numerica* **15** 543–639
[13] Wendland H 2010 *Scattered Data Approximation* Cambridge Monographs on Applied and Computational Mathematics (Cambridge University Press) ISBN 9780521131018 URL http://books.google.com/books?id=Q4ByQgAACAAJ
[14] Flyer N, Fornberg B, Bayona V and Barnett G A 2016 *Journal of Computational Physics* **321** 21–38 ISSN 0021-9991 URL https://www.sciencedirect.com/science/article/pii/S0021999116301632
[15] Giraldo F X 2020 *An Introduction to Element-Based Galerkin Methods on Tensor-Product Bases* (Springer Cham) URL https://link.springer.com/book/10.1007/978-3-030-55069-1
[16] Shi J, Hu C and Shu C W 2002 *Journal of Computational Physics* **175** 108–127
[17] Balsara D S, Garain S and Shu C W 2016 *Journal of Computational Physics* **326** 780–804
[18] Borges R, Carmona M, Costa B and Don W S 2008 *Journal of Computational Physics* **227** 3191–3211 ISSN 0021-9991 URL https://www.sciencedirect.com/science/article/pii/S0021999107005232
[19] Castro M, Costa B and Don W S 2011 *Journal of Computational Physics* **230** 1766–1792 ISSN 0021-9991 URL https://www.sciencedirect.com/science/article/pii/S0021999110006431
[20] May I and Lee D 2023 Multidimensional adaptive order gp-weno via kernel-based reconstruction (*Preprint* 2304.03823)
[21] Chertock Alina Chu S and Kurganov A 2023 *East Asian Journal on Applied Mathematics* **13** 576–609 ISSN 2079-7370 URL http://global-sci.org/intro/article_detail/eajam/21723.html
[22] Balsara D S 2012 *Journal of Computational Physics* **231** 7504–7517
[23] Wu K and Shu C W 2019 *Numerische Mathematik* **142** 995–1047
[24] Fekete I, Conde S and Shadid J N 2022 *Journal of Computational and Applied Mathematics* **412** 114325 ISSN 0377-0427 URL https://www.sciencedirect.com/science/article/pii/S0377042722001388
[25] Harten A, Lax P D and Leer B v 1983 *SIAM Review* **25** 35–61 (*Preprint* https://doi.org/10.1137/1025002) URL https://doi.org/10.1137/1025002

Journal of Physics: Conference Series **2742** (2024) 012026 doi:10.1088/1742-6596/2742/1/012026

Quasi all-speed schemes for magnetohydrodynamic flows in a wide range of Mach numbers

Takashi Minoshima

Center for Mathematical Science and Advanced Technology, Japan Agency for Marine-Earth Science and Technology, 3173-25, Showa-machi, Kanazawaku, Yokohama 236-0001, Japan

E-mail: minoshim@jamstec.go.jp

Abstract. We present novel numerical schemes for ideal magnetohydrodynamic (MHD) simulations aimed at enhancing stability against numerical shock instability and improving the accuracy of low-speed flows in multidimensions. Stringent benchmark tests confirm that our scheme is more robust against numerical shock instability and is more accurate for low-speed, nearly incompressible flows than conventional shock-capturing schemes. Our scheme is a promising tool for tackling MHD systems, including both high and low Mach number flows.

1. Introduction

A magnetohydrodynamic (MHD) simulation is an indispensable tool for investigating macroscopic dynamics of laboratory, space, and astrophysical plasmas. In the field of compressible MHD simulations, a family of upwind, shock-capturing schemes have been developed based on the solution to the Riemann problem in one-dimensional hyperbolic conservation laws. These schemes enable us to effectively resolve shocks and discontinuities in supersonic flows. Numerous studies have been dedicated to the development of numerical schemes to ensure the robustness and accuracy of MHD simulations. Notably, the Roe [1, 2] and the Harten-Lax-van Leer discontinuities (HLLD) [3] approximate Riemann solvers are extensively implemented in modern MHD simulation codes.

In practical MHD simulations, however, conventional shock-capturing schemes may encounter numerical difficulties. For high Mach number flows, these schemes tend to be vulnerable to numerical shock instability when a multidimensional shock closely aligns with the grid spacing [4]. Conversely, in the case of low Mach number flows, these schemes lead to a degradation in solution accuracy as the Mach number decreases. In the field of computational aerodynamics, a family of the advection upstream splitting method (AUSM) [5, 6] has been extensively employed due to its high flexibility, and has been extended to the "all-speed" regime to permit reliable simulations in a wide range of Mach numbers, including nearly incompressible flows and high Mach number shocks [7, 8, 9].

In this paper, we present novel numerical schemes designed for MHD simulations including both low and high Mach number flows. The first scheme is the multistate low-dissipation advection upstream splitting method (MLAU) [10], an MHD extension of the all-speed AUSM scheme. The second scheme is the low-dissipation HLLD scheme (LHLLD) [11], which incorporates techniques from the MLAU scheme, allowing users of the HLLD scheme to easily enjoy its capabilities. Both schemes provide accurate solutions for nearly incompressible flows that remain super-Alfvénic, a feature we refer to as *quasi all-speed* capability.

2. Numerical schemes

We consider one-dimensional ideal MHD equations written in the following conservative form:

$$\frac{\partial \boldsymbol{U}}{\partial t} + \frac{\partial \boldsymbol{F}}{\partial x} = 0, \tag{1}$$

$$\boldsymbol{U} = \left[\rho, \rho u, \rho v, \rho w, B_y, B_z, e\right]^T, \tag{2}$$

$$\boldsymbol{F} = \begin{bmatrix} \rho u \\ \rho u^2 + P + |\boldsymbol{B}|^2/2 - B_x^2 \\ \rho v u - B_x B_y \\ \rho w u - B_x B_z \\ B_y u - B_x v \\ B_z u - B_x w \\ \left(e + P + |\boldsymbol{B}|^2/2\right) u - B_x \left(\boldsymbol{u}\cdot\boldsymbol{B}\right) \end{bmatrix}, \tag{3}$$

where $\boldsymbol{U}$ and $\boldsymbol{F}$ are the state vector of conservative variables and the corresponding flux vector, respectively, and $\rho, \boldsymbol{u} = (u, v, w), \boldsymbol{B} = (B_x, B_y, B_z)$, and e are the mass density, velocity, magnetic field, and total energy density. The gas pressure P is determined from the equation of state for the ideal gas,

$$P = (\gamma - 1)\left(e - \frac{\rho|\boldsymbol{u}|^2}{2} - \frac{|\boldsymbol{B}|^2}{2}\right), \tag{4}$$

where γ is the specific heat ratio. The solenoidal condition of the magnetic field gives B_x = constant in one dimension. Equation (1) is discretized on computational cells into a finite volume form as follows:

$$\frac{d\boldsymbol{U}_i^n}{dt} = -\frac{\hat{\boldsymbol{F}}_{i+1/2} - \hat{\boldsymbol{F}}_{i-1/2}}{\Delta x}, \quad \boldsymbol{U}_i^n = \frac{1}{\Delta x}\int_{x_i-\Delta x/2}^{x_i+\Delta x/2} \boldsymbol{U}(x, t_n)dx, \tag{5}$$

where $\hat{\boldsymbol{F}}_{i\pm 1/2}$ is the numerical flux at the interfaces of a cell $[x_i - \Delta x/2, x_i + \Delta x/2]$. The quality of numerical solutions is significantly influenced by the assessment of the numerical flux.

2.1. Multistate low-dissipation advection upstream splitting method (MLAU)

Following the methodology of the AUSM-family schemes, we split the flux (equation (3)) into three parts, namely, advection, pressure, and magnetic tension parts, such that

$$\boldsymbol{F} = \rho u \boldsymbol{\Phi} + P_t \boldsymbol{N} - \boldsymbol{T}, \tag{6}$$

where,

$$\boldsymbol{\Phi} = \left(1, u, v, w, B_y/\rho, B_z/\rho, h\right)^T, \tag{7}$$

$$\boldsymbol{N} = \left(0, 1, 0, 0, 0, 0, 0\right)^T, \tag{8}$$

$$\boldsymbol{T} = B_x\left(0, B_x/2, B_y, B_z, v, w, \boldsymbol{u}_t \cdot \boldsymbol{B_t}\right)^T, \tag{9}$$

$$\boldsymbol{u}_t = (0, v, w), \boldsymbol{B}_t = (0, B_y, B_z), \tag{10}$$

$$h = \gamma P/(\gamma - 1)\rho + |\boldsymbol{u}|^2/2 + |\boldsymbol{B}_t|^2/\rho, \tag{11}$$

and

$$P_t = P + |\boldsymbol{B}_t|^2/2, \tag{12}$$

is the total pressure (excluding the contribution of B_x). Subsequently, the numerical flux at the interface is

$$\hat{\boldsymbol{F}}_{i+1/2} = \dot{m}\left(d_L\boldsymbol{\Phi}_{i+1/2,L} + d_R\boldsymbol{\Phi}_{i+1/2,R}\right) + \hat{P}_{t,i+1/2}\boldsymbol{N} - \hat{\boldsymbol{T}}_{i+1/2}, \tag{13}$$

$$d_L = \frac{1+\mathrm{sgn}(\dot{m})}{2}, d_R = \frac{1-\mathrm{sgn}(\dot{m})}{2}, \dot{m} = (\rho u)_{i+1/2}, \tag{14}$$

where the subscripts L and R refer to the left and right states at the interface. The mass flux $\dot{m}$ and the pressure flux $\hat{P}_t$ are built based on the all-speed AUSM scheme [7], which utilizes a weighted average of the left and right states with respect the Mach number of the fast magnetosonic speed c_f. The magnetic tension flux $\hat{\boldsymbol{T}}$ is built to be consistent with the HLLD scheme.

Liou argued that the scheme becomes vulnerable to numerical shock instability in multidimensions when the mass flux $\dot{m}$ includes the term proportional to the pressure difference $(P_{i+1/2,R} - P_{i+1/2,L})$ and aligns with the shock surface; although, this term is effective for stabilizing one-dimensional shocks [12]. To prevent the instability while maintaining the robustness of one-dimensional shocks, we introduce a shock-detecting factor θ and multiply it by the pressure difference term (see equation (20) in the next subsection),

$$\begin{aligned}
\theta &= \min\left(1, \frac{-\min(\Delta u, 0) + c_f}{-\min(\Delta v, \Delta w, 0) + c_f}\right)^4, && (15)\\
\Delta u &= u_{i+1,j,k} - u_{i,j,k}, && (16)\\
\Delta v &= \min(v_{i,j,k} - v_{i,j-1,k}, v_{i,j+1,k} - v_{i,j,k}, \\
&\quad v_{i+1,j,k} - v_{i+1,j-1,k}, v_{i+1,j+1,k} - v_{i+1,j,k}), && (17)\\
\Delta w &= \min(w_{i,j,k} - w_{i,j,k-1}, w_{i,j,k+1} - w_{i,j,k}, \\
&\quad w_{i+1,j,k} - w_{i+1,j,k-1}, w_{i+1,j,k+1} - w_{i+1,j,k}), && (18)
\end{aligned}$$

which quickly approaches zero when the shock surface is nearly parallel to the $z-x$ or $x-y$ plane (i.e., the shock normal direction is orthogonal to x).

The pressure flux $\hat{P}_t$ is expressed as a combination of the central average, pressure difference, and the velocity difference of the left and right states. While the pressure difference term scales with the fluid velocity and acts as upwinding, the velocity difference term scales with the fast magnetosonic speed and serves as diffusion in one-dimensional compressible flows. However, it causes excessive diffusion for low-speed flows in multidimensions. To correct the scaling of the pressure flux, we introduce a modified fast magnetosonic speed c_u,

$$c_u^2 = \frac{1}{2}\left[\left(c_a^2 + |\boldsymbol{u}|^2\right) + \sqrt{(c_a^2 + |\boldsymbol{u}|^2)^2 - 4|\boldsymbol{u}|^2 c_{ax}^2}\right], \tag{19}$$

where $c_a^2 = |\boldsymbol{B}|^2/\rho$ and $c_{ax}^2 = B_x^2/\rho$ are Alfvén speeds. We then multiples c_u/c_f by the velocity difference term. This correction appropriately scales down the pressure flux to $|\boldsymbol{u}| = c_a$, and as a result, it is expected to provide accurate solutions for super-Alfvénic flows, particularly in high-beta and low-speed plasma. For more comprehensive information about the scheme design, please refer to [10, 11].

2.2. *Low-dissipation HLLD approximate Riemann solver (LHLLD)*

The novel techniques used in the MLAU scheme are implemented to the HLLD scheme to enhance stability against numerical shock instability and improve the accuracy of low-speed flows. The normal velocity S_M and the total pressure P_t^* in the Riemann fan are modified to incorporate the shock detection in equation (15) and the pressure correction using equation (19), such that

$$S_M = \frac{(S_R - u_R)\rho_R u_R - (S_L - u_L)\rho_L u_L - \theta(P_{tR} - P_{tL})}{(S_R - u_R)\rho_R - (S_L - u_L)\rho_L}, \tag{20}$$

$$P_t^* = \frac{(S_R - u_R)\rho_R P_{tL} - (S_L - u_L)\rho_L P_{tR} + \phi\rho_L\rho_R(S_R - u_R)(S_L - u_L)(u_R - u_L)}{(S_R - u_R)\rho_R - (S_L - u_L)\rho_L}, \tag{21}$$

where $S_L = \min(0, \min(u_L, u_R) - \max(c_{f,L}, c_{f,R}))$ and $S_R = \max(0, \max(u_L, u_R) + \max(c_{f,L}, c_{f,R}))$. The factor ϕ multiplied by the third term in equation (21) should satisfy the condition $\propto c_u/c_f$ ($c_u/c_f \ll 1$) to improve the accuracy of low Mach number flows and equal 1 ($c_u/c_f \geq 1$) to reduce to the original scheme at high Mach numbers. We use

$$\phi = \chi(2 - \chi), \quad \chi = \min(1, \max(c_{u,L}, c_{u,R})/\max(c_{f,L}, c_{f,R})). \tag{22}$$

For more comprehensive information about the scheme design, please refer to [11].

3. Numerical experiments

We conduct stringent numerical experiments to assess the capability of the present schemes. Since the numerical results obtained using the MLAU and LHLLD schemes are nearly indistinguishable, we present the LHLLD scheme for clarity. Our numerical code incorporates the second-order MUSCL scheme with a minmod limiter for interpolating physical variables, the third-order strong stability preserving Runge-Kutta method for time integration, and the central upwind constrained transport method to preserve the solenoidal condition of the magnetic field [13]. We employ a CFL number of 0.4.

The first problem is the two-dimensional Kelvin-Helmholtz instability (KHI) in nearly incompressible flows. The initial condition has a velocity shear $\boldsymbol{u} = (V_0/2)\tanh((y - 10\lambda)/\lambda)\boldsymbol{e}_x$, uniform density and pressure $\rho = \rho_0, P = P_0$, and a uniform magnetic field $\boldsymbol{B} = B_0(\cos\theta, 0, \sin\theta)$, with $\rho_0 = V_0 = B_0 = \lambda = 1.0$ and $\theta = 71.565°$. We use $\gamma = 2.0$. To initiate the instability, we impose a small perturbation into the y-component of the velocity along the shear layer with a wavelength of 20λ, corresponding to the fastest growing mode. The computational domain covers the range from $0 \leq x < 20\lambda$ and $0 \leq y < 20\lambda$, and it is resolved by $N \times N$ cells, where $N = 64, 128, 256$, and 1024 for reference. The boundary condition is periodic and symmetric in the x and y directions.

Figure 1(a)-(c) shows the stream lines at $t = 119$ with $P_0 = 500$ (corresponding Mach number of 0.0158), obtained using the HLLD and LHLLD schemes at $N = 256$, and the HLLD scheme at $N = 1024$ as a reference. The in-plane magnetic field is amplified by stretching, and the tension force distorts the flow within the vortex in $5 < x < 15$ and $7 < y < 13$. It is evident that the LHLLD scheme outperforms the HLLD scheme in resolving the flow pattern, exhibiting good agreement with the reference solution that has 4^2 times higher grid resolution. Figure 1(d) shows the growth rate at $N = 64, 128, 256$ with $P_0 = 500$. The linear growth shown in a dashed line, obtained by solving the linearized MHD equations (equation (68) in [10]), is mostly independent of the pressure in nearly incompressible flows. The LHLLD scheme shown in red lines converges to a linear growth rate of $\gamma = 0.051V_0/\lambda$ for all runs, while the HLLD scheme shown in blue lines requires $N = 256$ to achieve convergence. Figure 1(e) shows the growth rate at $N = 64$ with $P_0 = 500$, 5000, 50000, and 500000. Thanks to the pressure correction (equation (21)), the solutions obtained using the LHLLD scheme (red lines) are mostly independent of the

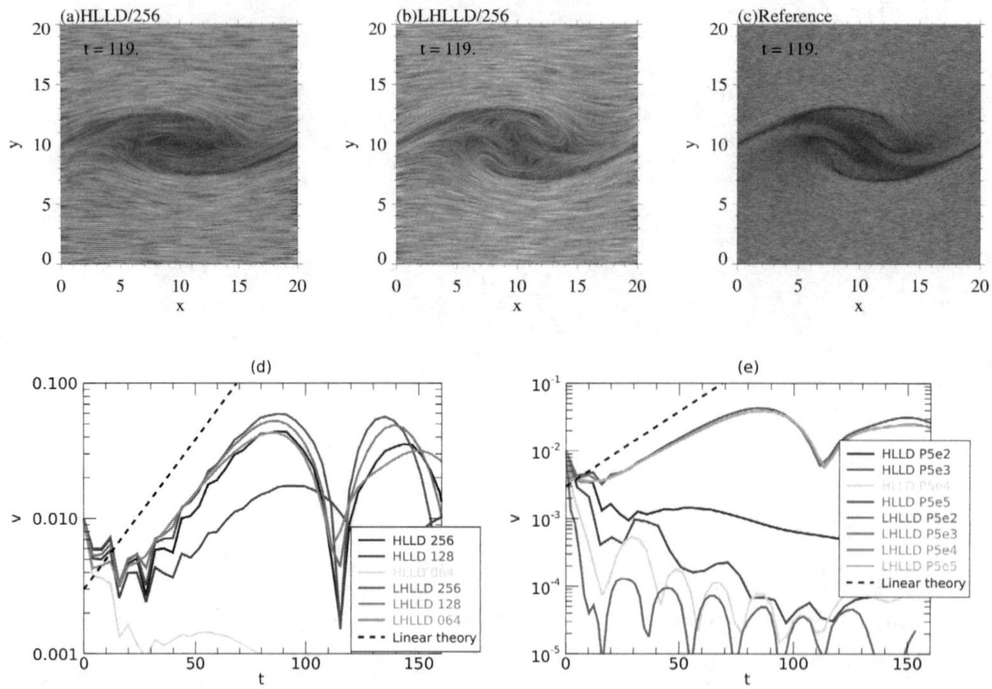

Figure 1. Numerical simulation of the Kelvin-Helmholtz instability: (a-c) Stream lines obtained using the HLLD scheme at $N = 256$, the LHLLD scheme at $N = 256$, and the HLLD scheme at $N = 1024$. (d-e) Time profiles of the fastest growing mode of the y-component of the velocity, demonstrating the dependence on grid resolution and the initial pressure.

pressure, while the HLLD scheme (blue lines) gets worse as the pressure increases. Consequently, we can conclude that the LHLLD scheme possesses the capability to provide accurate solutions for nearly incompressible flows.

The second problem is the two-dimensional Richtmyer-Meshkov instability (RMI) occurring in hypersonic flows with a very weak magnetic field. The initial condition is as follows:

$$(\rho, \boldsymbol{u}, \boldsymbol{B}, P) = \begin{cases} (1.0, 0, -1.0, 0, 0.000034641, 0, 0, 0.00006) & (y > 0) \\ (3.9988, 0, -0.250075, 0, 0.000138523, 0, 0, 0.75) & (y < 0) \end{cases} \tag{23}$$

which satisfies the Rankine-Hugoniot condition for perpendicular MHD shocks. The upstream Mach number is 100 and the plasma beta is $\beta = 2P/|\boldsymbol{B}|^2 = 10^5$. We use $\gamma = 5/3$. A corrugated contact discontinuity is imposed in the upstream region, $y_{\rm cd} = Y_0 + \psi_0 \cos(2\pi x/\lambda)$, where $Y_0 = 1.0$, $\psi_0 = 0.1$ is the corrugation amplitude, and $\lambda = 1.0$ is the wavelength. We shift a frame moving at $v = -0.625$, which corresponds to the interface velocity after the collision with the incident shock, ensuring that the structure of the RMI remains at approximately $y = 0$ throughout the simulation run. The computational domain covers the range from $0 \leq x < \lambda$ and $-40\lambda \leq y < 40\lambda$, and it is resolved by $N \times 80N$ cells, where $N = 128$. The boundary condition is periodic in the x direction and is fixed to the initial state in the y direction.

Figure 2 shows the density profile at $t = 25$ obtained using the Roe, HLLD, and LHLLD schemes. The surface at $y = 4$ corresponds to the transmitted shock, and the distinctive mushroom-shaped structure in $0 < y < 1$ corresponds to the nonlinear evolution of the RMI. In panel (a), the solution obtained using the Roe scheme is severely distorted by numerical shock

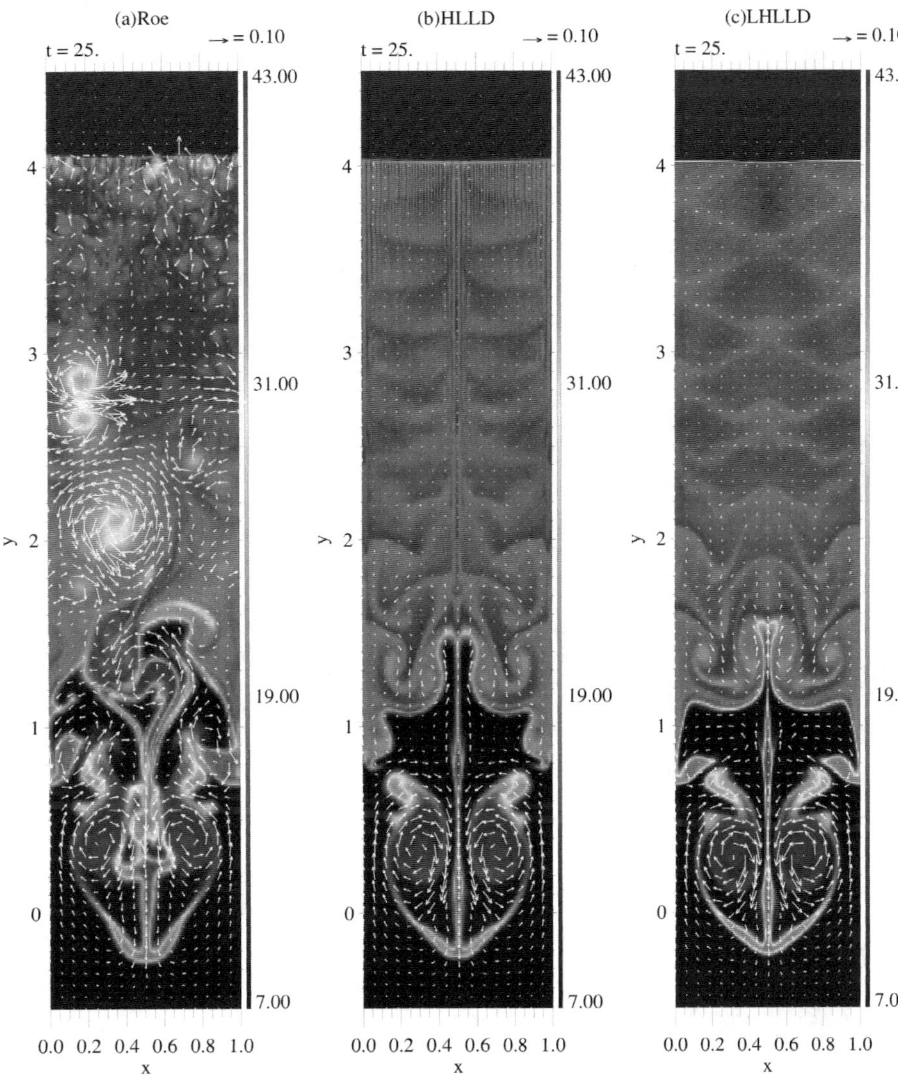

Figure 2. Density profiles in the Richtmyer-Meshkov instability at $t = 25$, obtained using (a) the Roe scheme, (b) the HLLD scheme, and (c) the LHLLD scheme. The arrows represent the rotational velocity component.

instability. The instability generates multiple unphysical voids in the shocked region, disrupts the structure of the region of interest ($0 < y < 1$), and causes a loss of solution symmetry. In panel (b), the HLLD scheme significantly mitigates the catastrophic structures observed in the Roe scheme, but it still exhibits grid-scale oscillation around the shock front. In contrast, the LHLLD scheme in panel (c) succeeds in completely eliminating unphysical structures by utilizing the shock detection, resulting in a smooth solution. Furthermore, the solution obtained using the LHLLD scheme exhibits faster roll-up flows in the region $0 < y < 1$ compared to the HLLD scheme, which is attributed to the pressure correction for subsonic flows in the Mach number range of $0.01 - 0.1$.

The last problem is the three-dimensional MHD turbulence induced by the magnetorotational instability (MRI). Building upon our previous research [14], we solve isothermal MHD equations

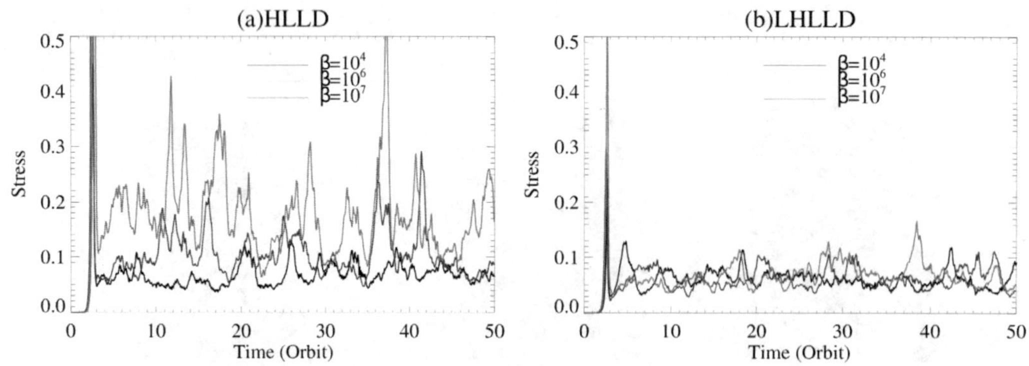

Figure 3. Time profiles of the volume-averaged stress in the MRI-induced turbulence with plasma beta values of 10^4 (black lines), 10^6 (blue lines), and 10^7 (red lines), obtained using (a) the HLLD scheme and (b) the LHLLD scheme.

Table 1. Time- and volume-averaged quantities in the MRI-induced turbulence.

Scheme	P_0	β	w	$\Delta P/P_0$	$\boldsymbol{\lambda}$ for u'	$\boldsymbol{\lambda}$ for B_y
HLLD	3.125×10^0	10^4	7.0×10^{-2}	5.8×10^{-2}	$(1.0, 1.6, 0.64)$	$(0.55, 1.3, 0.49)$
HLLD	3.125×10^2	10^6	1.1×10^{-1}	7.0×10^{-4}	$(1.5, 2.0, 0.66)$	$(0.55, 1.3, 0.48)$
HLLD	3.125×10^3	10^7	1.8×10^{-1}	1.2×10^{-4}	$(1.8, 2.3, 0.68)$	$(0.62, 1.5, 0.50)$
LHLLD	3.125×10^0	10^4	6.3×10^{-2}	5.8×10^{-2}	$(1.0, 1.6, 0.62)$	$(0.54, 1.3, 0.49)$
LHLLD	3.125×10^2	10^6	6.1×10^{-2}	4.4×10^{-4}	$(1.0, 1.4, 0.58)$	$(0.52, 1.2, 0.48)$
LHLLD	3.125×10^3	10^7	6.6×10^{-2}	4.8×10^{-5}	$(1.0, 1.4, 0.57)$	$(0.54, 1.3, 0.49)$

including the Coriolis force within the framework of a local shearing box approximation [15]. We apply the orbital advection method to update the velocity fluctuation $\boldsymbol{u'} = \boldsymbol{u} - \boldsymbol{u}_0$, where $\boldsymbol{u}_0 = -q\Omega x \boldsymbol{e}_y$ corresponds to the background shear flow (with $q = 1.5$ and $\Omega = 1$ representing angular frequency) [16] . Since the velocity fluctuation typically remains subsonic, the solution may be affected by the pressure correction in the present scheme. The initial condition is uniform, $(\rho, \boldsymbol{u'}, \boldsymbol{B}, P) = (1, 0, B_0\boldsymbol{e}_z, P_0)$, with an incompressible uniform random noise imposed to u' to initiate the instability. The ambient magnetic field B_0 is fixed to 0.025, and the gas pressure P_0 is set to $3.125, 312.5, 3125$, yielding initial plasma beta values of $10^4, 10^6, 10^7$. The computational domain covers the range from $-1 < x < 1, -2 < y < 2$, and $-0.5 < z < 0.5$, and it is resolved by $64 \times 256 \times 32$ cells. The wavelength of the fastest growing mode of the MRI, $\lambda_{\rm FGM} = 2\pi B_0/\sqrt{\rho}\Omega = 0.16$, is resolved by approximately five grid points. The boundary condition is periodic in the y and z directions, and shearing periodic in the x direction.

Figure 3 shows the time profiles of the volume-averaged stress $w = \rho u'v' - B_x B_y/4\pi$ obtained using the HLLD and LHLLD schemes. Notably, these results exhibit discrepancies that become more pronounced as the gas pressure (plasma beta) increases. The time- and volume-averaged quantities over the time interval $t = 25 - 50$ are summarized in Table 1. For various pressure values, the stress values remain approximately constant around 0.065 with the LHLLD scheme, while they increase from 0.07 to 0.18 with the HLLD scheme. Considering that the flow remains subsonic within the Mach number range of $0.005 - 0.2$, we can conclude that the pressure-independent solution provided by the LHLLD scheme is indicative of incompressible turbulence.

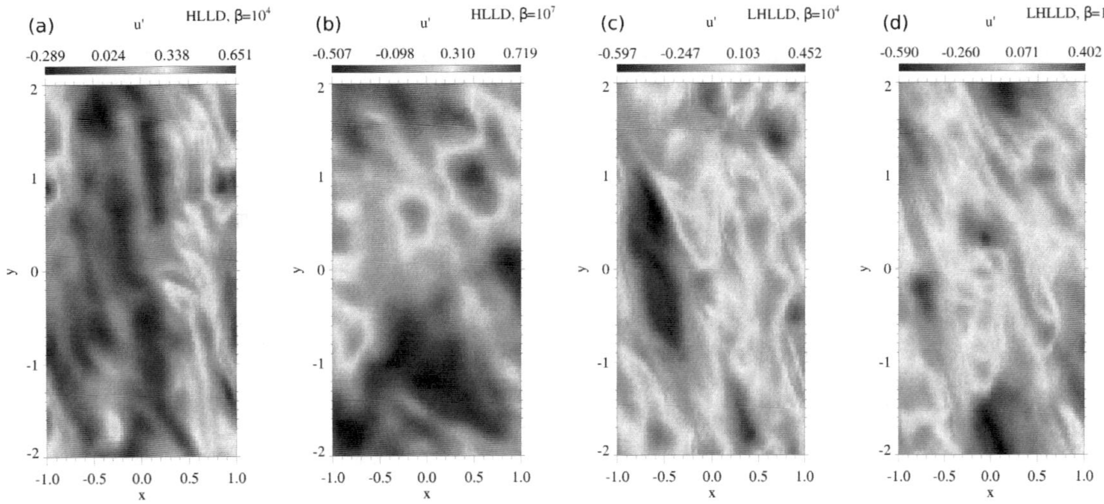

Figure 4. Snapshots of $u'(x, y, z = 0, t = 50)$ in the MRI-induced turbulence with plasma beta values of $10^4, 10^7$, obtained using (a,b) the HLLD scheme and (c,d) the LHLLD scheme.

This conclusion is supported by examining the time- and volume-averaged pressure perturbation $\Delta P = P - P_0$, presented in the fifth column of Table 1. The LHLLD scheme satisfies the asymptotic behavior in the incompressible limit $\Delta P/P_0 = O(M^2) = O(P_0^{-1})$.

The discrepancy in the solutions between the HLLD and LHLLD schemes can be attributed to the dependence of the saturation level of the MRI-induced turbulence on the magnetic Prandtl number [17]. Specifically, ideal MHD simulations of the MRI-induced turbulence are sensitive to the numerical magnetic Prandtl number, which is determined by the scheme design and is not necessarily specified [14]. The numerical magnetic Prandtl number is defined as the ratio of the numerical viscosity ν_n to resistivity η_n, both of which arise from the numerical diffusion terms involved in the momentum and induction equations. For subsonic flows, the numerical viscosity is evaluated from the pressure in the Riemann fan (equation (21)), resulting in $\nu_{n,\mathrm{HLLD}} = c_f \Delta x$ and $\nu_{n,\mathrm{LHLLD}} = c_u \Delta x$. As shown in Figure 4, the velocity profile reveals that small-scale structures become less pronounced in the HLLD scheme as the gas pressure increases, while the structure in the LHLLD scheme remains similar regardless of the pressure. Meanwhile, the numerical resistivity is identical in both schemes, and is evaluated from an explicit expression of the magnetic tension term, yielding $\eta_{n,\mathrm{HLLD}} = \eta_{n,\mathrm{LHLLD}} = \max(|\boldsymbol{u}|, c_{ax})\,\Delta x$ [10]. Consequently, it is anticipated that the numerical magnetic Prandtl number in the HLLD scheme increases with increasing pressure, while it remains approximately constant in the LHLLD scheme.

As a proxy for the numerical diffusion coefficients, we measure characteristic wavelengths defined as follows:

$$(\lambda_x, \lambda_y, \lambda_z) = \frac{2\pi}{(k_x, k_y, k_z)}, \quad (k_x, k_y, k_z) = \sqrt{\frac{1}{\overline{f^2}}\left(\overline{\left(\frac{\partial f}{\partial x}\right)^2}, \overline{\left(\frac{\partial f}{\partial y}\right)^2}, \overline{\left(\frac{\partial f}{\partial z}\right)^2}\right)}, \tag{24}$$

where $\overline{f}$ represents the volume average of the variable f. The time-averaged characteristic wavelengths of u' and B_y are presented in the sixth and seventh columns of Table 1. The wavelength of u' in the HLLD scheme tends to increase as the gas pressure increases, while other quantities remain approximately constant. This suggests that the numerical viscosity and the numerical magnetic Prandtl number in the HLLD scheme are likely to increase with increasing pressure, causing the increase in the saturation level [14]; in contrast, these values remain constant in the LHLLD scheme, resulting in the pressure-independent solution.

4. Conclusion

We presented novel numerical schemes designed for MHD flows in a wide range of Mach numbers. The schemes, the multistate low-dissipation advection upstream splitting method (MLAU) [10] and the low-dissipation HLLD approximate Riemann solver (LHLLD) [11], enhance stability against numerical shock instability and improve the accuracy of low-speed flows in multidimensions by incorporating shock detection and pressure correction. Stringent benchmark tests confirm the performance of the scheme, outperforming conventional shock-capturing schemes. The scheme provides accurate solutions for nearly incompressible flows that remain super-Alfvénic, by virtue of the pressure correction that scales down appropriately to $|\boldsymbol{u}| = c_a$. We refer to this capability as *quasi all-speed* for super-Alfvénic flows. Importantly, this feature ensures that the numerical magnetic Prandtl number remains approximately constant, a critical condition for the numerical study of MHD turbulent phenomena. With this novel ability, the scheme becomes a promising tool to tackle MHD systems, including both high and low Mach number flows.

The source code written in C++ programming language is available on GitHub[1].

Acknowledgments

We would like to thank T. Miyoshi, K. Kitamura, S. Hirose, and T. Sano for their valuable contributions and insightful discussions. This work is supported in part by JSPS KAKENHI Grant Number JP20K04056.

References

[1] Brio M and Wu C C 1988 *Journal of Computational Physics* **75** 400–422
[2] Balsara D S 1998 *Astrophysical Journal Supplement Series* **116** 119–131
[3] Miyoshi T and Kusano K 2005 *Journal of Computational Physics* **208** 315–344
[4] Quirk J J 1994 *International Journal for Numerical Methods in Fluids* **18** 555–574
[5] Liou M S and Steffen C J 1993 *Journal of Computational Physics* **107** 23–39
[6] Liou M S 1996 *Journal of Computational Physics* **129** 364–382
[7] Liou M S 2006 *Journal of Computational Physics* **214** 137–170
[8] Shima E and Kitamura K 2011 *AIAA Journal* **49** 1693–1709
[9] Kitamura K and Shima E 2013 *Journal of Computational Physics* **245** 62–83
[10] Minoshima T, Kitamura K and Miyoshi T 2020 *Astrophysical Journal Supplement Series* **248** 12 (*Preprint* 2004.08012)
[11] Minoshima T and Miyoshi T 2021 *Journal of Computational Physics* **446** 110639 (*Preprint* 2108.04991)
[12] Liou M S 2000 *Journal of Computational Physics* **160** 623–648
[13] Minoshima T, Miyoshi T and Matsumoto Y 2019 *Astrophysical Journal Supplement Series* **242** 14 (*Preprint* 1903.04759)
[14] Minoshima T, Hirose S and Sano T 2015 *Astrophysical Journal* **808** 54–70 (*Preprint* 1506.03524)
[15] Hawley J F, Gammie C F and Balbus S A 1995 *Astrophysical Journal* **440** 742
[16] Stone J M and Gardiner T A 2010 *Astrophysical Journal Supplement Series* **189** 142–155 (*Preprint* 1006.0139)
[17] Lesur G and Longaretti P Y 2007 *Monthly Notices of the Royal Astronomical Society* **378** 1471–1480 (*Preprint* 0704.2943)

[1] https://github.com/minoshim/qasMHD

Journal of Physics: Conference Series **2742** (2024) 012027 doi:10.1088/1742-6596/2742/1/012027

Tackling exascale systems for astrophysics

Lou Roussel-Hard[1,2], Édouard Audit[1], Luc Dessart[2], Thomas Padioleau[1] and Yushan Wang[1]

[1]Université Paris-Saclay, UVSQ, CNRS, CEA, Maison de la Simulation, 91191, Gif-sur-Yvette, France

[2] Institut d'Astrophysique de Paris, CNRS-Sorbonne université, 98 bis boulevard Arago, 75014 Paris, France

E-mail: lou.roussel-hard@cea.fr, edouard.audit@cea.fr, dessart@iap.fr, thomas.padioleau@cea.fr, yushan.wang@cea.fr

Abstract.
Exascale systems are now appearing in the US, Japan, in Europe with two exascale procurements in the coming years and very likely in China. These systems offer great opportunities for the numerical astrophysics community and also raise serious technical challenges. Many legacy codes need to be fully rewritten to benefit from exascale architectures. Performance portability is a key issue as well as the management of the huge amount of data generated by exascale simulations. In this paper we discuss the path we have chosen to exascale and the software engineering solution chosen to address performance portability in the long run and data management. A new radiation hydrodynamics code based on the Kokkos library is presented as well as test runs and preliminary performance analysis. The first target of this new code will be to study supernova ejecta. They are multidimensional fluids, structured on both small and large scales following the non-linear development of complex fluid instabilities during and after the explosion. They contain a wealth of information about stellar composition and the evolution of elements in the interstellar medium. Their study requires significant computing resources that can be met through the use of exascale machines and suitable software.

1. Introduction

The life of a massive star is marked by stages of nuclear fusion in its core, from hydrogen to iron. When a star reaches its final stage, its life ends in a cataclysmic collapse followed by a violent explosion, known as a core-collapse supernova [1], leaving a neutron star at its core. The ejected envelope, originally stratified in composition, undergoes multidimensional fluid instabilities. These instabilities are driven by factors such as the original asymmetry of the explosion engine, the development of Rayleigh-Taylor instabilities around density inversions, or the ^{56}Ni bubble effect [2]. To accurately describe the behavior of this multidimensional fluid, it is necessary to employ large-scale simulations [3]. While the explosion of the star occurs on a large spatial scale, the instabilities act on small scales. Thus we need to follow short time scales of the order of a second, while the evolution of the ejecta must be simulated over the course of days. This results in extremely computationally intensive three dimensional simulations. To efficiently produce numerical results, it is essential to have suitable parallel algorithms along with adequate hardware.

Luckily, we have entered the exascale era with a significant leap in the advances of hardware and heterogeneous architectures. Computational resources are no longer limited in the use of

Central Processing Units (CPUs). Various kinds of Graphics Processing Units (GPUs) have significantly improved the capacity of the modern supercomputers. For example, in the Top500 list [4], more than 60% of the available FLOP/s (Floating-Point Operations Per Second) are delivered by GPUs [5]. We can confirm that GPUs have become a crucial driver of computational performance in high-performance computing (HPC). As a result, when developing a code, we must aim to take advantage of GPUs, and to ensure portability across different hardware architectures.

The code we developed, HERACLES++, is designed to simulate radiative hydrodynamics. It is written in C++ and is an update of the Fortran90 code HERACLES [6]. HERACLES++ is intended to be used in the astrophysics community, with a particular focus on analyzing supernova ejecta properties. It offers both Cartesian and spherical geometries, ranging from one to three spatial dimensions. To achieve good performance and portability, we rely on the Kokkos library [7] which offers an abstraction layer that simplifies development tasks and has a great portability on heterogeneous architectures making it a favorable choice for our project.

The paper is organized as follows: in Section 2 we present an overview of the exascale context. Next, in Section 3, we recall the Euler equations model with gravity and the physics involved in this model. The numerical scheme and its implementation using MPI and Kokkos are explained in Section 4 and Section 5 respectively. Then Section 6 shows several numerical tests and the performance of HERACLES++ is presented in Section 7. Conclusion and perspectives are discussed in Section 8.

2. Exascale context

In the past decade, HPC has experienced a major shift both in terms of hardware and of usage. The emergence of GPU and data-science has deep consequences on the way HPC is used in science. The exascale era will offer the ability to solve large and complex problems, opening up new possibilities for research by combining huge computing power and data analytics. However, to fully exploit these possibilities it becomes essential for scientists to develop software and numerical approaches that can fully harness the benefits of exascale hardware and data science. This includes optimizing applications and developing new programming models to improve performance, long-lasting usability and data management.

The TOP500 ranking [4] is a widely recognized and updated every six months list of the world's top supercomputers. It presents benchmark results such as LINPACK or HPCG. Recently, updated in June 2023, the Frontier system at Oak Ridge National Laboratory (ORNL) in the US, appears to be the first true exascale machine, with a peak performance of 1.102 Exaflop/s. The Jupiter system just announced for 2024 will be the first European supercomputer to surpass the threshold of one trillion calculations per second. Both these systems, and 7 out of the top 10 machines of the Top500, rely on GPU for most of their computing power.

GPU based supercomputers are equipped with chips from two main vendors, namely AMD and NVIDIA, with Intel soon to join. Each of them support different programming languages: CUDA for NVIDIA, HIP/ROCm for AMD and OneAPI for Intel. They are often supplemented with additional models, typically offering a higher-level abstraction, such as OpenMP, OpenACC, Kokkos, or SYCL. To answer this challenge of developing software that can perform efficiently on heterogeneous architectures, we chose to work with the Kokkos library [7]. The Kokkos EcoSystem offers a performance portability solution by providing a unified C++ programming model for different architectures. It is presented as a suite of libraries for fundamental computational algorithms along with tools for profiling and debugging. The library is developed and maintained by a multi-institutional team of HPC experts, scientists, and engineers.

The implementation of Kokkos relies on advanced C++ features such as template programming and function objects. It interfaces with existing shared-memory programming

Journal of Physics: Conference Series **2742** (2024) 012027 doi:10.1088/1742-6596/2742/1/012027

models like OpenMP, CUDA, and HIP, offering a level of abstraction that enhances user-friendliness. It utilizes abstract concepts, such as execution spaces, which determine where a computation is executed, and execution policies, which specify how the computation is executed. Similarly, for memory management, it addresses memory spaces, where data resides, and memory layout, which defines how multidimensional data is organized in memory. These concepts help manage how parallel operations are scheduled, executed, and synchronized providing essential flexibility and control for HPC.

3. Physical model

The basic equations solved by the code are the Euler equations expressing the conservation of mass, momentum, and energy. Additional components such as gravity or heating can be added as source terms. Radiative transfer shall be considered next as it is a key element for supernova ejecta. These equations can be reformulated using the generalized vector of conserved quantities $\mathbf{U} = (\rho, \rho\mathbf{u}, E)$ as:

$$\begin{aligned} \partial_t \mathbf{U} + \nabla \cdot \mathbf{F}(\mathbf{U}) &= \mathbf{S}(\mathbf{U}, \phi), \\ E &= \frac{1}{2}\rho\mathbf{u}^2 + \rho e \end{aligned} \tag{1}$$

where ρ is the density, $\mathbf{u}$ the velocity, E the total energy per volume unit, $\mathbf{F}$ is the flux of conserved quantity $\mathbf{U}$ such as $\mathbf{F}(\mathbf{U}) = (\rho\mathbf{u}, \rho\mathbf{u} \otimes \mathbf{u} + P\mathbf{I}, \mathbf{u}[E + P])$ with $P = P_{eos}(\rho, e, T)$ the pressure, e the internal specific energy and $\mathbf{S}$ a source term.

Concerning the equation of state (EOS), the code is designed to be able to handle any EOS either analytical or tabulated. In the present paper, we are considering two specific equations of state: the perfect gas equation and the perfect gas equation supplemented with radiation which represents a simplified implementation of radiation in a purely hydrodynamic code.

$$P_{gas} = \rho e(\gamma - 1) \qquad \text{and} \qquad P_{gas+rad} = P_{gas} + \frac{1}{3}a_r T^4 \tag{2}$$

where γ is the adiabatic index, T represents the gas temperature and a_r the radiative constant. The equation $P_{gas+rad}$ applies to gases when the mean free path of photons is significantly smaller than the size of the system. The radiation is then at local thermal equilibrium with the gas and the temperatures of both the gas and radiation are equal. This flexibility in the choice of the equation of state allows for the adaptation of modeling to the specific characteristics of the system under study, whether it involves standard conditions of a perfect gas or situations involving radiative processes.

The code supports the treatment of passive scalars, which is valuable for understanding mixing in heterogeneous media like supernovae ejecta. The passive scalars, f_{x}, can behave like tracers, facilitating the observation and analysis of fluid behavior and evolution within the medium without altering the fundamental properties of the fluid itself. They are also necessary when including processes like chemistry or heating from the radioactive decay of ^{56}Ni.

Their evolution is given by:

$$\partial_t(\rho f_{\mathsf{x}}) + \nabla \cdot (\rho\mathbf{u} f_{\mathsf{x}}) = 0 \qquad \text{with} \qquad \sum_i f_{\mathsf{x}i} = 1 \tag{3}$$

with x representing the different species followed by the passive scalar

The code is also adapted for various gravitational source terms such as a point mass or a constant gravitational field. Self-gravitating fluids will only be considered at a later stage since they are not really necessary for supernova ejecta.

Journal of Physics: Conference Series **2742** (2024) 012027 doi:10.1088/1742-6596/2742/1/012027

4. Numerical scheme

In this section, we will introduce the notation for space-time discretization. For simplicity, we will concentrate on a one-dimensional problem. The positions of the mesh nodes are defined by $x_{i+1/2}$ and the center of the cell $(x_{i-1/2}, x_{i+1/2})$ by x_i. The size of the cell is defined as δx_i. The time interval between t^n and t^{n+1} is denoted as δt. The notation U_i^n represents the average quantity at time t^n for the cell x_i.

The MUSCL-Hancock (Monotone Upstream-centered Schemes for Conservation Laws) scheme [8] is a commonly used numerical method in fluid mechanics and numerical simulations to solve partial differential equations. This method is a practical choice for achieving second-order accuracy in numerical simulations. Its ability to accurately represent gradients and discontinuities, along with its adaptability and conservation properties, make it a reliable tool for various applications, such as fluid dynamics and compressible flows. The main idea behind this scheme is to achieve a better approximation of solution gradients to prevent undesirable oscillations near discontinuities. This method consists of two steps before advancing to the next time step. The first step is a reconstruction of quantities on each side of the interface with a slope limiter of our choice, as referenced in [9]. The quantities will then evolve over a half time step. After these steps, we have the quantity $U_{i\pm1/2}^{n+1/2}$.

Finally, the Godunov scheme [10] will reconstruct the values at the next time step:

$$U_i^{n+1} = U_i^n + \frac{\delta t_n}{\delta x_i}\left(F_{i-1/2}^{n+1/2} - F_{i+1/2}^{n+1/2}\right) + \delta t\, S_i^n \tag{4}$$

with F a Riemann solver describing the flow at the cell interface.

The code is built in a modular way such that various slope limiters and Riemann solvers can be used. In the test cases presented in this paper, the HLL solver and either a minmod or a van Leer slope limiter were used [8].

The code is designed for a logically Cartesian grid. By default the grid is regular, however it is possible for the user to implement any grid spacing that would be needed for a particular problem, such as a logarithmic grid or other specialized grid. This applies to Cartesian, cylindrical or sphercical coordinate systems in either one, two or three spatial dimensions. The same flexibility is allowed for boundary conditions. Free-flow, reflective and periodic boundary conditions are already implemented and user defined boundary conditions can easily be implemented.

The numerical scheme in three dimensions is similar to the one-dimensional scheme in all three directions, eventually taking into account geometrical terms. The global scheme fully couples all directions. See [6] for more details.

5. Implementation

We have chosen to design our code using MPI to deal with distributed memory parallelism and Kokkos for shared memory computations. The global computational domain is divided equally into subdomains and each subdomain is assigned to one MPI (Message Passing Interface) process [11]. Within the subdomain, we use the Kokkos library [7] to exploit shared memory parallelism.

Our code implementation has a modular structure where each module represents a given step for the hydrodynamic scheme. For example, we have a module for the face reconstruction step, a module for the Riemann solver, etc. To be more specific, a module is composed by one or several Kokkos kernels. A kernel is a C++ functor, which is executed following the pattern `Kokkos::parallel_for` or `Kokkos::parallel_reduce`. For example, Equation 4 is implemented as :

```
Kokkos::parallel_for(
    "Parallel_for_example",
```

```
Kokkos::MDRangePolicy<Kokkos::Rank<3>>({0, 0, 0}, {nx, ny, nz}),
KOKKOS_LAMBDA(int i, int j, int k){
// F_L = compute_FL(i,j,k);
// F_R = compute_FR(i,j,k);
// dv = compute_dV(i,j,k);
U_new(i, j, k) = U(i, j, k) + dt/dv * (F_L - F_R) + dt * S(i, j, k);});
```

In this code snippet, `U_new`, `U` and `S` are all `Kokkos::View` which is the core data type used in the Kokkos library. The `MDRangePolicy` defines how to iterate over a multi-dimensional container.

Another aspect of our code worth mentioning is the use of the PDI [12] library that aims to decouple high-performance simulation codes from Input/Output (I/O) concerns. With PDI, we describe the I/O operations in a dedicated YAML file instead of interleaving them with the simulation code. We don't need to recompile the code when changing the I/O parameters. For the configuration of a simulation, we load a file with all the necessary parameters for the numerical scheme and the domain decomposition. Then, for the initialization of the physical values, we can either use PDI to read them from an HDF5 file or initialise them inside HERACLES++. The output values are written with PDI in HDF5 and XMF files. PDI is also designed to easily couple simulation codes with (in-situ) data-analytics tools which is a great asset for future usage of the code.

6. Hydrodynamics tests

The code has undergone extensive testing in one, two and three dimensions, covering a variety of hydrodynamical tests. We provide an overview of the tests conducted. For simplicity and speed, we assume an ideal gas equation of state. The two and three dimensional cases have been run using GPUs. The numerical scheme employed is the one previously described, which is the MUSCL scheme with the HLL Riemann solver using second-order discretizations. The slope limiter is applied to the primitive variables using a van Leer slope limiter for the one and two dimensional tests and a minmod slope limiter for the three dimensional cases. The tests are conducted using a regular Cartesian grid.

6.1. Sod shock tube

The classical shock-tube test is designed to demonstrate the code's shock-capturing capabilities. We consider the initial condition proposed in [8] for a perfect gas with the adiabatic index $\gamma = 1.4$. The initial condition consists of two states in a domain x from 0 to 1:

$$(\rho, u, p)(x, t=0) = \begin{cases} (1.0, 0, 1.0) & \text{if } x < 0.5 \\ (0.125, 0, 0.1) & \text{if } x > 0.5 \end{cases} \quad (5)$$

where ρ is the density, u the velocity and p the pressure. The boundary conditions are transmissive. This test is not particularly challenging for modern HPC code but it is interesting as it exhibits all fundamental hydrodynamical waves and is well-documented. The exact solution consists of a propagating rarefaction wave on the left side, a rightward-propagating contact discontinuity and a rightward-propagating shock.

Figure 1 shows the analytic solution and the results from our code. The code correctly captured all of the different waves. The contact wave is the most diffuse, captured within approximately 4 grid points, while the shock wave spans around 2 grid points. The rarefaction wave is well described by our scheme.

6.2. Kelvin-Helmholtz instability

The Kelvin-Helmholtz instability is a multi-dimensional instability arising from a velocity shear within a fluid. We consider a two dimensional perfect fluid in a rectangular domain, same as

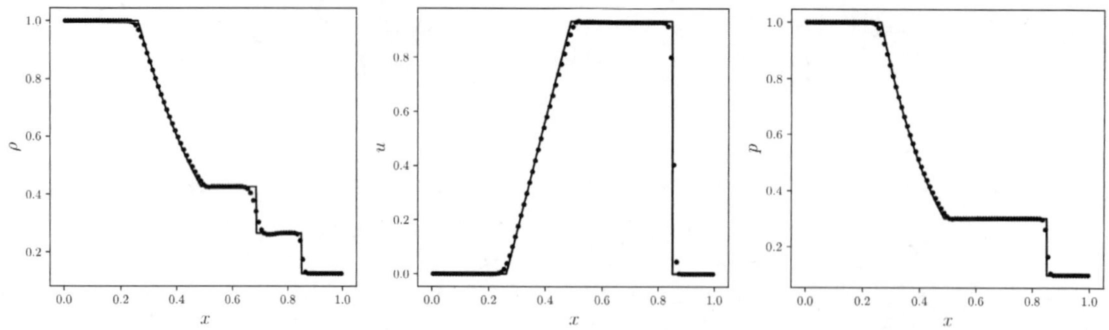

Figure 1: Sod shock tube profiles at $t = 0.2$ of the density (left), velocity (middle), and pressure (right). The solid line is the analytic solution while the dots show the result obtained with a 100 grid points resolution.

[13], $x \in [0, 4]$ and $z \in [0, 2]$. The initial conditions include a density jump:

$$\begin{aligned} \rho &= 1 + \frac{\delta\rho}{2\rho_0}\left[\tanh\left(\frac{z - z_1}{a}\right) - \tanh\left(\frac{z - z_2}{a}\right)\right] \\ v_x &= u_0 \times \left[\tanh\left(\frac{z - z_1}{a}\right) - \tanh\left(\frac{z - z_2}{a}\right)\right] \\ v_y &= \mathrm{A}\sin(\pi x) \times \left[\exp\left(-\frac{(z - z_1)^2}{\sigma^2}\right) + \exp\left(-\frac{(z - z_2)^2}{\sigma^2}\right)\right] \\ P &= 10 \end{aligned} \tag{6}$$

with $a = 0.05$ $\sigma = 0.2$, $z_1 = 0.5$ and $z_2 = 1.5$. The initial amplitude of the perturbation in the vertical velocity is $A = 0.01$ and the initial density perturbation is $\delta\rho/\rho_0 = 1.0$. The initial lateral velocity constant u_0 is 1.0 and the adiabatic index is $\gamma = 5/3$ for an adiabatic gas. We ignore the gravity for this test. The boundary conditions are periodic in the x direction and transmissive in the z direction. The instability results in the creation of small symmetrical vortices, which cascade into turbulence. This is a critical test, the Kelvin-Helmholtz instability is a fundamental and crucial instability in fluids with shear, especially in astrophysics, as for the Rayleigh-Taylor instability in the next section.

Figure 2 shows the colormaps for the density at the different times. The initial density perturbations evolve to form characteristic vortices. At a qualitative level, these results compare satisfactorily with those shown in [5].

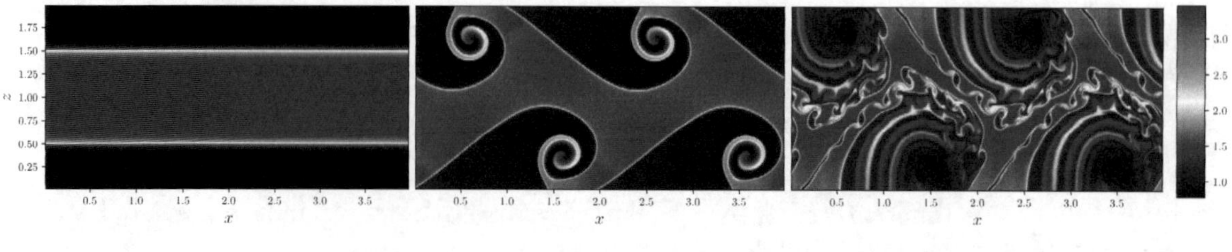

Figure 2: Colormaps density of the Kelvin-Helmholtz instability over the grid at $t = 0$ (left), $t = 3.4$ (middle) and $t = 6$ (right). The (x, y) domains contains $(2048, 1024)$ zones.

6.3. Rayleigh-Taylor instability

Our first three dimensional test concerns the nonlinear development of the Rayleigh-Taylor instability in three dimensions. The initial setup is taken from [13]. The domain is a box with dimensions of $L = 10$, where $(x, y) \in [-L/2, L/2]$ and $z \in [-L, L]$. The gravitational acceleration is constant $g_z = -2$ and the adiabatic index is set to $\gamma = 5/3$. The initial density is stratified such as:

$$\rho(z) = \rho_0 \Big(1 - \frac{\gamma - 1}{\gamma} \frac{\rho_0\, g_z\, z}{\rho_0}\Big)^{\frac{1}{\gamma-1}} \quad \text{with} \quad \rho_0 = \begin{cases} \rho_h = 3 & \text{if } z \geq 0, \\ \rho_l = 1 & \text{otherwise.} \end{cases} \tag{7}$$

The top and bottom density are set so that the Atwood number $A = (\rho_h - \rho_l)/(\rho_h + \rho_l)$ is $1/2$. The initial pressure is set to maintain hydrostatic equilibrium:

$$P = P_0 \Big(\frac{\rho}{\rho_0}\Big)^{\gamma} \quad \text{with} \quad P_0 = 2\pi(\rho_h + \rho_l) g\, L \tag{8}$$

We track the evolution of the fluid using a passive scalar field f_{x}. It is initialized to be 1 for positive z and 0 otherwise. For the boundary conditions, we assume periodicity in the x and y directions and reflectivity in the z direction. At time $t = 0$, the interface between the two fluids is perturbed:

$$\begin{aligned} h(x, y) = \frac{1}{H} \sum_{8 \leq k_x^2 + k_y^2 \leq 16} \Big[& a_{\mathbf{k}} \cos(k_x X) \cos(k_y Y) + b_{\mathbf{k}} \cos(k_x X) \sin(k_y Y) \\ & + c_{\mathbf{k}} \sin(k_x X) \cos(k_y Y) + d_{\mathbf{k}} \sin(k_x X) \sin(k_y Y) \Big] \end{aligned} \tag{9}$$

where $X = 2\pi x/L$, $Y = 2\pi y/L$ and the wave numbers k_x and k_y have values of $[0, 1, 2, 3, 4]$. The coefficients $a_{\mathbf{k}}$, $b_{\mathbf{k}}$, $c_{\mathbf{k}}$ and $d_{\mathbf{k}}$ are sampled from a uniform distribution with values ranging from -1 to 1, excluding the boundaries. The normalized coefficient H is adjusted after obtaining the coefficients so that $h_{rms} = 3 \times 10^{-4} L$. The initial perturbation generates small buoyant bubbles that interact and grow into larger bubbles as the system evolves, as shown in Figure 3.

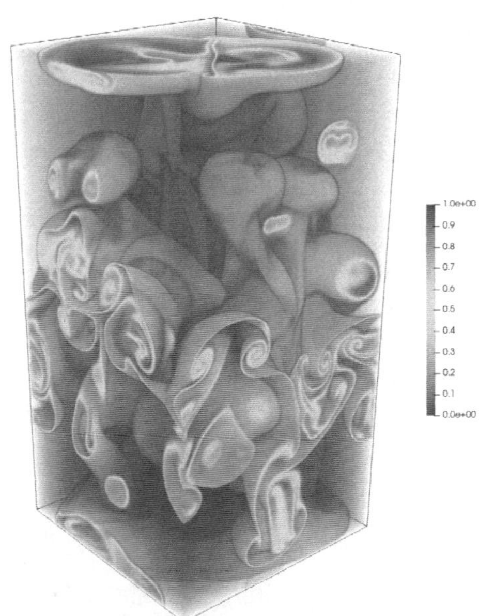

Figure 3: 3-D volume rendering for the passive tracer concentration of the 3-D Rayleigh-Taylor test at time: $A\, g_z\, t^2/L \sim 3$. Red values indicate higher concentration and blue lower. The number of zones in (x, y, z) are $(256, 256, 512)$.

6.4. *Off-center Sedov blast wave*

Our final test is the classical Sedov self-similar blast wave [14] in spherical coordinates and in three dimensions, but here with an off-center placement of the initial blast. This test allows us to test our geometrical source terms and is a relevant representation of the initial phase of a supernova blast. The initial conditions are similar to [15], with some adjustments. The domain is a portion of a sphere around the equator: $r \in [0.5, 1.5]$, $(\theta, \phi) \in [\pi/4, 3\pi/4]^2$. Initially, we consider gas at rest with $\rho_0 = 1$, $P_0 = 10^{-2}$ and $\gamma = 5/3$. The blast is off-center and occurs at the coordinates $r_e = 1$, $(\theta_e, \phi_e) = \pi/2$. To initiate this explosion, we introduce an amount of total energy $E_1 = 1$ in a single cell . We perform the test on a regular grid, and the boundary conditions are transmissive in all directions.

From [14] , the analytical position of the shock is given by:

$$r_{sh}(t) = \xi_0 \left(\frac{E_1 \, t^2}{\rho_0} \right)^{1/5} \tag{10}$$

with $\xi_0 \approx 1.15$ for an adiabatic equation of state with $\gamma = 5/3$.

Figure 4 shows the result and the comparison with the analytical solution. As explained in [13], small-scale axis-related artifacts appear in the density of off-axis blasts. However, the evolution of the blast remains spherical.

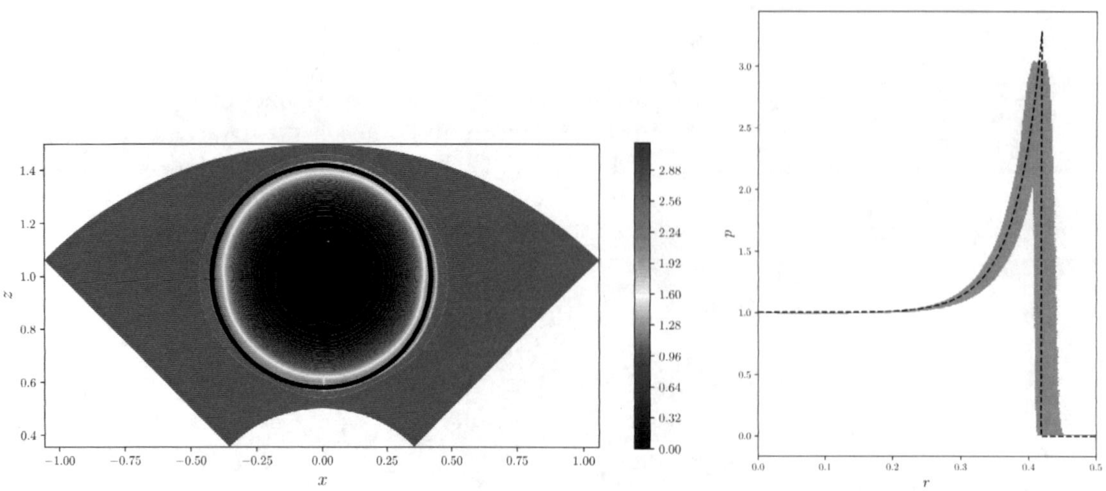

Figure 4: 3-D spherical Sedov blast wave at $t = 0.08$. Left: 2-D density, slice at $\phi = \pi/2$ in the Cartesian referential. The analytic shock radius (r_{sh}) is overplotted in black. Right: the pressure distribution as a function of the distance from the blast center with the analytic solution in black. The (x, y, z) domain contains $(256, 256, 512)$ zones.

7. Performance

The tests are performed on the French supercomputer Adastra [4] hosted at CINES in Montpellier. This cluster is equipped with AMD MI250X GPUs and is designed with 4 GPUs per node. Each GPU has a 128GB memory and a theoretical peak memory bandwith of 3276 GB/s. The MI250X GPU is composed of 2 GCD (Graphic Complex Die) and we assign one MPI process to each GCD. For the scaling tests, we use the 5.5.1 version of rocm with the HIPCC compiler driver. The Kokkos version is 4.1.00 and we enable the Kokkos option `Kokkos_ARCH_VEGA90A`. We use the 3-d Rayleigh-Taylor simulation with size (n_x, n_y, n_z), and the initial conditions presented in the previous section. We run the test for 50 iterations. The performance is evaluated for the temporal loop without the initialization and I/O. It shows the number of cells updated per second:

$$\text{perf} = \frac{n_x\, n_y\, n_z\, n_{itr}}{t_{elapsed}} \tag{11}$$

with n_{itr} the number of iterations and $t_{elapsed}$ the elapsed time.

7.1. Performance for one MPI process

For the first test, we are going to evaluate how the performance evolves with the size of the simulation using a single GCD. For maximum efficiency, the size of the problem must be large enough to expose sufficient parallelism. In our case, we can see in Figure 5 that the performance increases rapidly until the size of the problem reaches 20M cells. For problem of larger size the performance remains roughly constant. We have done several runs for each point and the variation in performance is less than 1%.

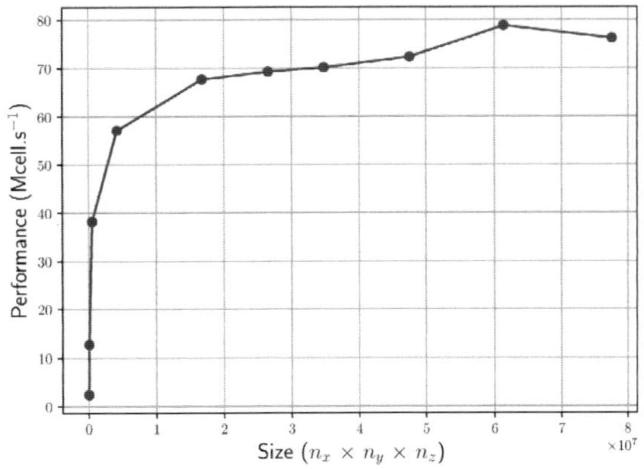

Figure 5: Performance in Mcells per second based on the size of the problem running with one MPI process (one GCD). The size of n_x, n_y and n_z are roughly the same.

7.2. Weak scaling

The weak scaling test shows the ability of the code to improve its performance while the workload increases in proportion to the available computing resources. The problem size per GCD is fixed to $352 \times 352 \times 384$ for one GCD. While doubling the number of GCDs, and the total size of the problem doubles as well. Figure 6 shows approximately an efficiency of 96% up to 512 GPUs.

7.3. Strong scaling

In the strong scaling test, we fix the total problem size to $416 \times 416 \times 448$. Then we increase the number of GPUs. Figure 7 shows that the acceleration is close to ideal for a small number of GPUs. As expected in most strong scaling tests, the speedup slows down with increasing number of GCDs. Indeed, the problem size per GCD becomes too small to saturate the bandwidth, and the MPI exchanges take more time due to the increased number of ghost cells.

8. Conclusion

With the results obtained so far, we believe that the technical choices made for building a sustainable and user friendly version of HERACLES++ were very relevant. Keeping strongly on-board the astrophysicist community while making a radical change to C++ was a challenge. But the use of multi-dimensional Kokkos views and the modularity of the code design

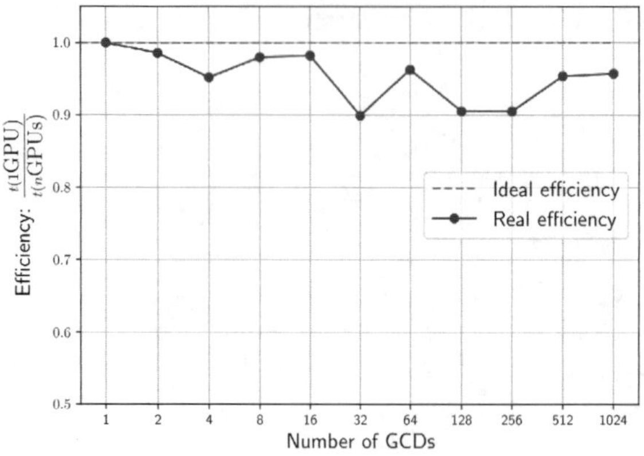

Figure 6: Weak scaling: efficiency of the code, compared to one GCD, when the size of the problem and of the number of MPI process double.

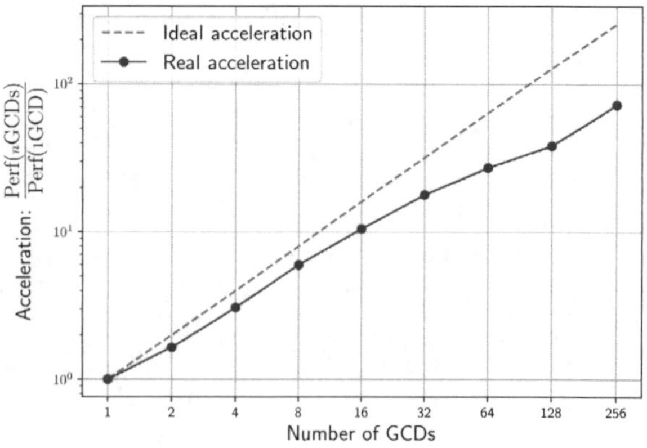

Figure 7: Strong scaling: acceleration of the performance, compared to one GCD, when the size of the problem remains constant and the number of MPI process increases.

make it quite accessible to (astro)physicists with mainly a Fortran background. Performance portability was also a key issue and preliminary results are encouraging. Without any specific work on optimisation, the same source code already has very acceptable performances on various architectures. Several improvements are already underway to further improve both the performance and the memory footprint. The first results concerning supernovae ejecta are in preparation.

References

[1] Baade W and Zwicky F 1934 *Proceedings of the National Academy of Science* **20** 254–259
[2] Woosley S E 1988 **330** 218
[3] Müller B 2020 *Living Reviews in Computational Astrophysics* **6** URL
[4] Top500 list (june 2023) URL `https://www.top500.org/lists/top500/2023/06/`
[5] Herten A 2023 Many cores, many models: GPU programming model vs. vendor compatibility overview (*Preprint* 2309.05445)
[6] González M, Audit E and Huynh P 2007 **464** 429–435
[7] Trott C R *et al.* 2022 *IEEE Transactions on Parallel and Distributed Systems* **33** 805–817 ISSN 1558-2183 conference Name: IEEE Transactions on Parallel and Distributed Systems
[8] Toro E 2009 *Riemann Solvers and Numerical Methods for Fluid Dynamics: A Practical Introduction* (Springer Berlin Heidelberg) ISBN 9783540498346
[9] Berger M and Aftosmis M 2005 *43rd AIAA Aerospace Sciences Meeting and Exhibit - Meeting Papers*
[10] Godunov S K and Bohachevsky I 1959 *Matematičeskij sbornik* **47(89)** 271–306
[11] Kjolstad F B and Snir M 2010 Ghost cell pattern *Proceedings of the 2010 Workshop on Parallel Programming Patterns* ParaPLoP '10 (New York, NY, USA: Association for Computing Machinery) ISBN 9781450301275
[12] PDI URL `https://pdi.dev/master/`
[13] Skinner M A, Dolence J C, Burrows A, Radice D and Vartanyan D 2019 *ApJS* **241** 7
[14] Sedov L I 1946 *Journal of Applied Mathematics and Mechanics* **10** 241–250
[15] Lesur, G R J, Baghdadi, S, Wafflard-Fernandez, G, Mauxion, J, Robert, C M T and Van den Bossche, M 2023 *AA* **677** A9

Journal of Physics: Conference Series **2742** (2024) 012028 doi:10.1088/1742-6596/2742/1/012028

Methods and Challenges of Including Dust Evolution in Magnetohydrodynamic Simulations

Jonathan D. Slavin

Center for Astophysics | Harvard & Smithsonian, 60 Garden Street, MS 04, Cambridge, MA 02138, USA

E-mail: jslavin@cfa.harvard.edu

Abstract. Dust plays many important roles in the interstellar medium, affecting the heating, cooling and chemistry in all the different phases of the medium. Up until recently though, if dust was included in simulations at all, it was typically assumed that it was tightly coupled to the gas. More recent work has allowed for the dust to decouple from the gas, though this brings with it challenges for including all the dust destruction processes, especially grain-grain collisions. We discuss the various methods of simulating dust evolution in the context of magneto-hydrodynamical simulations of the interstellar medium and their strengths and weaknesses. In particular we present a method for improving on the tight coupling approach that shows promise for simulating the evolution of small grains subject to shocks.

1. Introduction

Dust plays a crucial role in determining the thermal state of the interstellar medium (ISM) via a variety of mechanisms:

- photoelectric heating
- depletion of important coolants (e.g. ions of C and O) from the gas phase
- cooling of hot gas by conversion of thermal energy into infrared radiation.

In addition dust in dense ISM is important for star and planet formation because of its role in catalyzing molecule formation. Molecules are important for cooling gas in those dense clouds that are optically thick to UV and optical radiation, which in turns allows the dense clouds to cool further and condense on their way to star and planet formation. By locking up metals, dust is also important for the transport of metals, both within galaxies and from galactic disks to halos. Dust formed in supernovae carry the metals synthesized in the progenitor and in the explosion into the ISM. Similarly, dust formed in the dense winds of AGB stars enriches the ISM with metals. Finally dust absorbs and scatters starlight, especially in the far UV, which is important for the ionization of low first ionization potential elements such as Mg, Fe and Si.

2. Why Simulating Dust is Hard

Given the importance of interstellar dust as outlined above, there has been increasing interest in including it in simulations of the evolution of the ISM of galaxies. However, there are a number of characteristics of dust that make it intrinsically difficult to include in numerical magnetohydrodynamical calculations. These include:

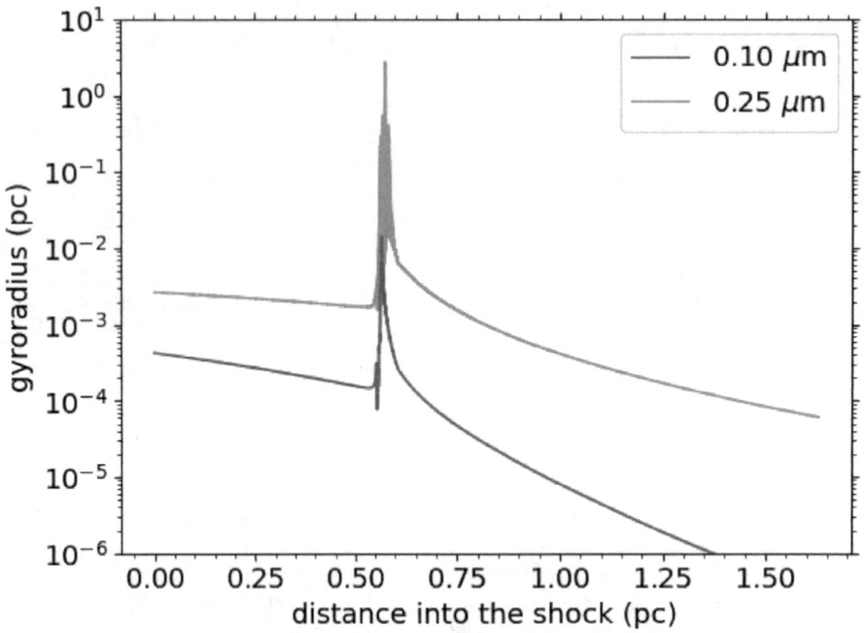

Figure 1. Comparison of the gyroradii for 0.1 μm and 0.25 μm silicate grains as they move through a steady 200 km s^{-1} radiative shock. The big peak in gyroradius occurs at the start of the shock cooling zone where the grains are betatron accelerated as the magnetic field is compressed along with the gas. We see the extremely large range of scales over which grain trajectories need to be followed. The grain charges range from $\sim -350 - +1600$ with the highest charges in the hot shock region (on left) and the largest negative charge in cooling zone at ~ 0.55 pc into the shock.

- dust grains are charged with the charging depending on a variety of factors including the electron temperature, the electron density, the interstellar radiation field and the speed of the dust relative to the surrounding plasma,
- the charge-to-mass ratio is quite low for grains compared with ions, but also spans a wide range, leading to a wide range in gyroradii (see Figure 1),
- for grains on the small end of the grain size distribution (which are the most numerous, though they carry less mass) their gyro-periods are short, demanding short timesteps to calculate their motions,
- grain-grain interactions such as shattering and vaporization can be important for grain evolution and destruction and mainly involve collisions of small grains with large grains
- dust instabilities, so called "resonant drag instabilites" can cause a high degree of dust clumping that doesn't coincide with gas clumping [1]

These factors especially come into play in highly dynamical situations in which shocks and outflows occur.

3. Grain destruction in shocks

A primary motivation for including dust in simulations of the ISM is to study grain destruction, especially as it occurs when shocks are present. The mechanisms of grain destruction are

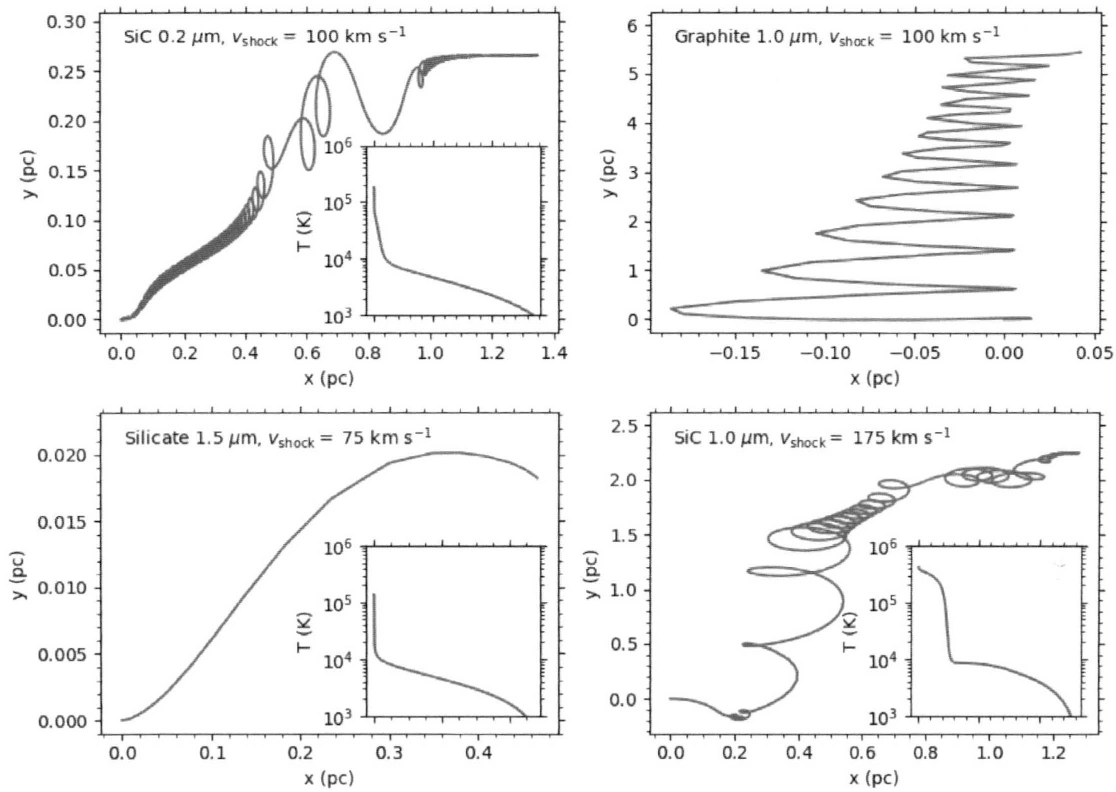

Figure 2. Grain trajectories behind steady radiative shocks for various shock speeds and grain types [from 2]. A range of behaviours is seen including relatively tight coupling (upper left), reflection upstream leading to Fermi acceleration (upper right), and moderate decoupling (lower right) and finally almost full decoupling of a very large grain that passes through the shock only slightly deflected. Note that the x-axis limits for the inset plots are the same as for the corresponding larger plots.

sputtering of atoms off of grains by ion impacts and vaporization of grains in grain-grain collisions. Thermal sputtering occurs behind fast shocks that heat the gas to temperatures above about $\sim 10^5$ K. So-called non-thermal or inertial sputtering occurs when grains are moving at a substantial speed relative to the gas. Vaporization in grain-grain collisions occurs when such collisions are at speeds high enough that one of the grains is completely destroyed [see 3]. When a grain is moving with the gas and enters a fast shock, the gas is quickly slowed, in the shock frame, to $\sim \frac{1}{4}v_{\rm shock}$. The grain is not slowed quickly and thus attains a gyro velocity of $\sim \frac{3}{4}v_{\rm shock}$, the difference between the pre-shock speed in the frame of the shock and the post-shock speed of the grain relative to the gas (for high mach number shocks). In radiative shocks, some distance behind the shock, the gas cools and condenses and the component of the magnetic field perpendicular to the shock direction gets amplified (in addition to the factor of about four amplification at the shock front). This in turn leads to betatron acceleration of the grains and higher gas-grain relative velocities. In some cases grains can be reflected back upstream and undergo Fermi acceleration leading very large grain velocities, possibly helping with the injection problem for cosmic rays. Figure 2 shows some example trajectories calculated by [2] for various grain types and shock speeds. The trajectory in the upper right shows a grain that has been Fermi accelerated.

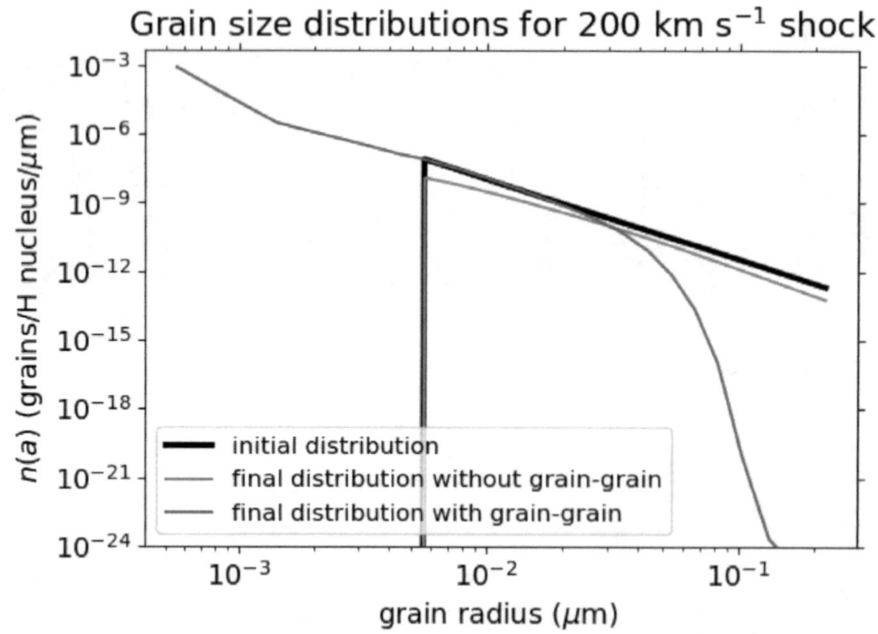

Figure 3. Comparison of the effects of sputtering and grain-grain collisions on the grain size distribution. The size distributions are for grains that have passed through a 200 km s^{-1} shock with only grain-grain collisions or sputtering turned on. Note that the overall level of grain destruction for the two cases is not greatly different, despite the large differences in size distributions.

Shocks not only destroy grains but also change the grain size distribution. Observations of dust in the ISM, especially extinction of starlight, have lead to the inference that grains have a distribution of sizes ranging from $\sim 0.005 - 0.25\,\mu$m in the diffuse ISM, with roughly a power law shape in size, $n(a) \propto a^{-3.5}$, where $n(a)$ is the number of grains per grain size and a is the grain radius [e.g., 4]. This implies that most of the surface area of the grains is in small grains and most of the mass is in large grains. Sputtering reduces the grain size by the same amount in radius, Δa, per unit time regardless of grain size, so the fractional decrease in mass is larger for small grains. Thus sputtering tends to shift the grain size distribution to smaller sizes removing more grains from the small grain end. Grain-grain collisions leads to shattering of grains and vaporization. Shattering doesn't destroy grains, i.e. return grain material back to the gas phase, but does reshape the grain size distribution by shifting mass from large grains to small grains. Small grains are stopped more quickly by gas drag, which plays a part in promoting grain-grain collisions because of the relative velocity of the small and large grains. The differing effects of sputtering and grain-grain collisions are illustrated in Figure 3.

4. Methods of Calculating Grain Evolution in Hydro/MHD Simulations

The interest in understanding grain evolution in the ISM has lead to a variety of methods of including grains in hydrodynamical and MHD simulations. The three main types of calculation are:

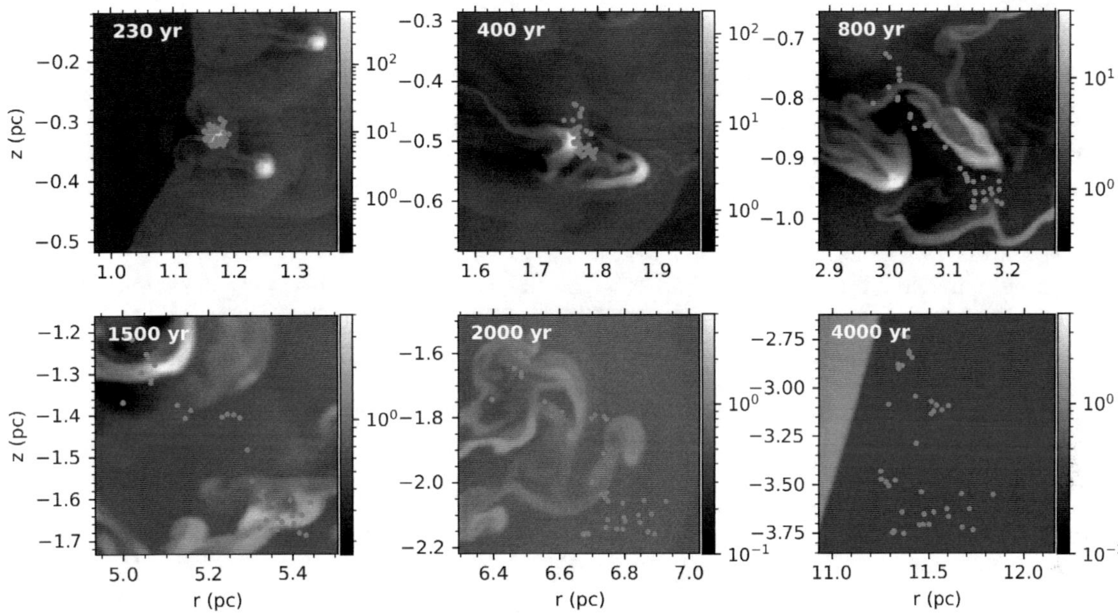

Figure 4. Evolution of dust in an HD simulation of the reverse shock overrunning an ejecta clump in an expanding supernova remnant [from 5]. The green dots are representative dust grains. There is no magnetic field in the simulation. Note that the grains, which start out close together, are widely scattered by the last frame. In this simulation grains are formed in dense ejecta clumps and escape the remnant into the ISM.

- tight coupling assumption: grains are assumed to move with the gas because of their small gyroradii and gas drag [e.g. 6, 7],
- pressureless fluid method: dust is treated as a fluid (or several fluids) that can move relative to the gas and interacts with it via drag and possibly Lorentz forces [e.g. 8, 9],
- free particle/kinetic approach: grains are treated as particles that are allowed to move relative to the gas subject to drag and (sometimes) Lorentz forces [e.g. 5, 10, see Figure 4 for an example].

Note that within these broad categories there are several variations among the approaches used. Each of the approaches listed above has its advantages and disadvantages. Though a thorough exploration of these would require more space than we have here, we briefly describe these pros and cons of each method.

For the tight coupling method the advantages are: 1) it is easy to include a range of grain sizes, for example using bins of grain sizes [3], and grain-grain interactions, 2) gyro speed can be included as an internal variable, 3) there are no stringent time constraints – one only needs to resolve the grain destruction and slowing timescales. The main disadvantage of the method is that large grains, i.e. grains that decouple from the gas on scales of the variations in gas properties, cannot be modeled accurately. Calculations of grain destruction using steady shocks [e.g. 3], give the grains an initial gyro speed equal to $\frac{3}{4}v_{\rm shock}$, however in hydro/MHD calculations one doesn't know *a priori* when the grains will enter a shock. Thus coupled grains, which are essentially tracer particles with extra characteristics, do not naturally experience small scale decoupling at shock fronts and thus will not get the boost to their gyro speeds (carried as an internal variable) that they should, We discuss a method for overcoming this limitation below.

Journal of Physics: Conference Series **2742** (2024) 012028 doi:10.1088/1742-6596/2742/1/012028

The multi-fluid approach treats dust as a separate fluid that couples to the gas. This approach has the advantages that: 1) dust can slip relative to the gas, 2) using multiple grain size bins/fluids allows for inclusion of grain-grain interactions, 3) small scale decoupling occurs naturally. The main disadvantages are: 1) it assumes that all grains of a given size in the same region have the same velocity, 2) if grain gyro radii are large, grain fluids could overlap in space raising the question of how to define the fluid velocity, 3) very large decoupling, such as Fermi acceleration at a shock front does not seem possible to accommodate. More generally dust grains are not coupled tightly to each other; collisions between grains are infrequent so a fluid approximation is not justified.

The free particle/kinetic approach treats grains as free particles subject to interactions with the gas and, potentially, the magnetic field. The advantages are that: 1) large grains can be handled accurately without any assumptions about a uniform grain velocity, 2) particles are correctly accelerated at shock fronts, including any possible Fermi acceleration. The disadvantages mainly concern grain-grain interactions since those require knowing the characteristics of the other grains in the vicinity. Also, the accuracy of such calculations are dependent on the number of grains used and so could become computationally expensive for a large simulation. An additional issue is that in an MHD calculation, one needs to resolve the gyro periods which could be very short for small grains.

5. Improving the Tight Coupling Approximation

As detailed above, the tight coupling approach could, at least for small grains, be a good approximation that would allow inclusion of all the main processes that affect grain evolution, including grain-grain collisions. The lack of small-scale decoupling at the shock front, however is a stumbling block. In general the gyro speed of grains evolves via electro-magnetic acceleration and gas drag:

$$\frac{dv_\perp}{dt} = \frac{1}{2}\frac{v_\perp}{B}\frac{dB}{dt} - \frac{F_d}{m_{gr}} \tag{1}$$

where $v_\perp$ is the gyro speed (grain speed perpendicular to the magnetic field direction), B is the total magnetic field strength, F_d is the drag force ($\propto v_\perp^2$), and m_{gr} is the grain mass. An implicit assumption is that except for gyrating around the magnetic field, the grain does not move relative to the gas. Said another way, the guiding center of the grain motion moves with the gas. One thing to note is that the betatron acceleration from the first term on the right is only effective once the grain has a significant gyro velocity. We note that we are only concerned here with the velocity perpendicular to the magnetic field since, motions along the field cannot increase the gyro velocity. Thus this method is incapable of following grain motion along the magnetic field that is decoupled from the gas motion. The drag force then should only use the gyro velocity of the grains.

It is clear that grains should experience an effective boost to their gyro velocity any time that the gas speed perpendicular to the magnetic field changes quickly over distances smaller the grain's gyro radius. Using this insight and via experimentation we have found that an effective method is to 1) add $\Delta v_{\rm gyro} = \Delta v_{\rm gas}$ to the $v_{\rm gyro}$ iff $\Delta v_{\rm gas} > 0.1 c_{\rm sound}$, 2) when this boost is applied, betatron and drag terms are turned off. ($c_{\rm sound}$ is the local sound speed.) Note that $\Delta v_{\rm gas}$ here is only the component of the gas speed perpendicular to the magnetic field. Thus during time steps that the criterion is met the grains increase their gyro speeds by an amount $\Delta v_{\rm gas}$, but this only occurs near shocks. If shocks were infinitely sharp in the simulation, one might want the factor of 0.1 above to be closer to 0.75, however with the shock spread over a few zones, the factor of 0.1 has worked well so far.

Our development of this approach was done using the MHD code FLASH, version 4.6.2 with some additional physics that we have added as enhancements to the code. We add optically thin radiative cooling and heating via a table look-up. With regards to dust processes, we

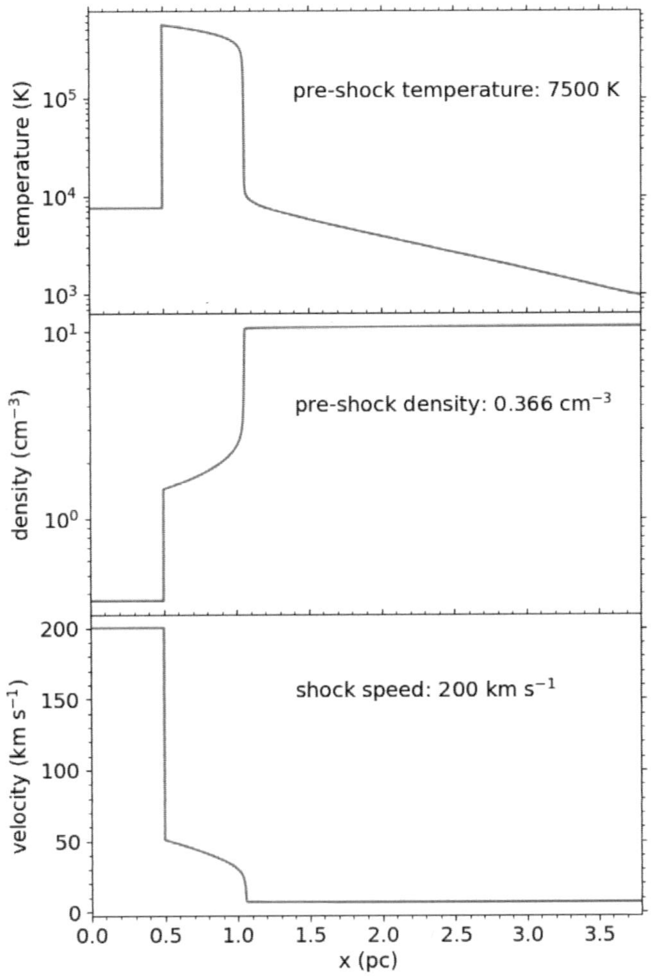

Figure 5. Shock profile used for testing grain evolution code. The calculation takes place in the shock frame with the upstream gas flowing in from the left and the shock at 0.5 pc. The gas flows off the grid on the right with outflow boundary conditions.

have implemented, for tightly coupled grains, time advancement of position and velocity using mapping of gas properties from the grid to the grain positions, which is already included in the code, but with the addition of a calculation of the grain charge using a table look-up scheme and the calculation of the gyro velocity in the way discussed above. We generated the grain charge table using code we have written which incorporates code generously provided to us by J. Weingartner [see 11], extended to include relative gas-grain motions. Gas (collisional) drag is calculated and leads to slowing of the gyro speed. Thermo-kinetic sputtering of the grain is also included. We have implemented both the tight coupling scheme and a kinetic scheme in which we calculate the grain trajectories via a Boris pusher method. The shock profile is set up initially via a steady shock calculation. Our results presented here are for a 200 km s^{-1} shock, which we found to be stable and not subject to the cooling overstability that affects somewhat slower shocks [12]. The shock profile is shown in Figure 5. The grains are spread over several zones upstream of the shock and moving with the gas. The calculation is in 1D, though for the

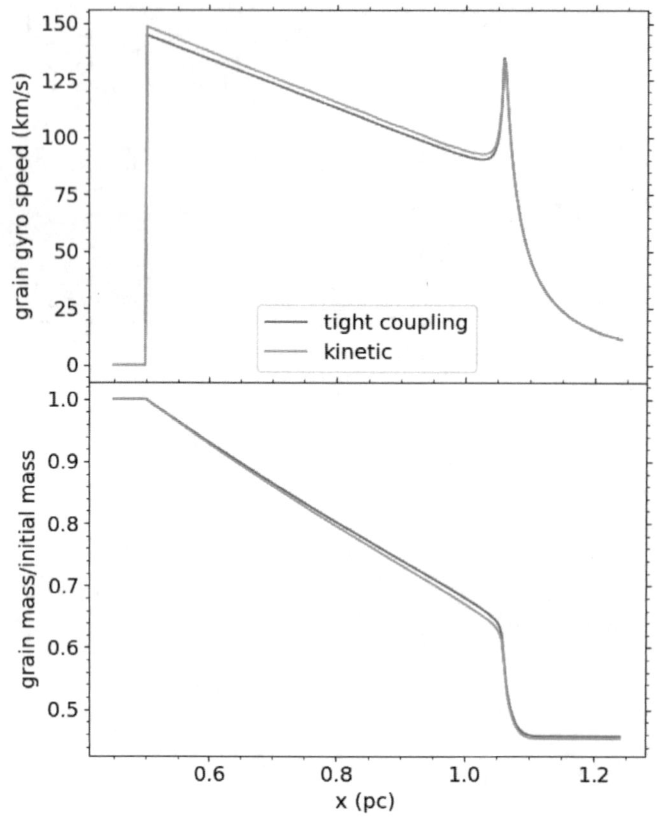

Figure 6. Comparison of the grain gyro speed (top) and grain mass (bottom) for our tight coupling and kinetic grain calculations. As can be seen there is excellent agreement, with the gyro velocity boost method performing very well in giving the tight coupling grains the appropriate gyro velocity after entering the shock.

free particle scheme, the grains are allowed to move in 3D. Since the magnetic field is assumed to be in the z direction and the gas motion is in the x direction, the grain motion is only in x and y.

Figure 6 shows a comparison of our results for the grain gyro speed and grain mass for coupled and kinetic assumptions for 0.1 μm silicate grains. As can be seen, the agreement is excellent for both the gyro speed and the grain mass. Very little tuning was done to achieve these results, which leads us to believe that they are robust. Further testing will be necessary to confirm that, especially in 2D and 3D and for oblique shocks.

6. Summary

Including dust grains in magneto-hydrodynamical calculations of the evolution of the ISM is complicated by the large range of size scales involved and because of their self-interaction via grain-grain collisions. Various methods for including grains have their strengths and weaknesses. Some methods are fairly accurate for small grains but not for large grains and vice versa. A kinetic approach, which could be accurate for large grains is expensive for small grains because of their small gyro radii.

A tight coupling approach is appropriate for small grains and can allow for relatively easy inclusion of grain-grain collisions, however, for use in an MHD code, a mechanism is needed to allow for the boost to the gyro speed imparted by small scale decoupling of the grains from the gas at shock fronts. We have found a method that allows for the correct behavior at a shock front without explicit shock tracking and depends only on the change in gas speed and local sound speed. For a radiative shock that is stable, we have shown that the grain gyro speed and mass match closely those found for a kinetic grain treatment.

The ultimate goal of this work is to find a method that is accurate and not too expensive for calculating the evolution of grains over the whole observed size range and including all the physical processes expected to be important in such environments. One possible avenue is to include small grains with a tight coupling approximation and large grains with a kinetic treatment in the same simulation. Then the erosion of the large grains via collisions with the small grains could be calculated. Such calculations could improve our understanding of grain destruction in clouds that are overrun by supernova remnant shocks, which is a key component of dust evolution in the ISM.

Acknowledgments

This research was supported by NASA Astrophysics Theory grant 80NSSC22K0810. The author thanks the organizers of ASTRONUM 2023 for the invitation to present this work. The author also thanks John Raymond and Eli Dwek for helpful conversations.

References

[1] Squire J and Hopkins P F 2018 *ApJL* **856** L15

[2] Slavin J D, Jones A P and Tielens A G G M 2004 *ApJ* **614** 796–806

[3] Jones A P, Tielens A G G M and Hollenbach D J 1996 *ApJ* **469** 740

[4] Mathis J S, Rumpl W and Nordsieck K H 1977 *ApJ* **217** 425–433

[5] Slavin J D, Dwek E, Mac Low M M and Hill A S 2020 *ApJ* **902** 135

[6] Silvia D W, Smith B D and Shull J M 2010 *ApJ* **715** 1575–1590

[7] Martínez-González S, Wünsch R, Silich S, Tenorio-Tagle G, Palouš J and Ferrara A 2019 *ApJ* **887** 198

[8] Monceau-Baroux R and Keppens R 2017 *A&A* **600** A134

[9] Kirchschlager F, Schmidt F D, Barlow M J, De Looze I and Sartorio N S 2023 *MNRAS* **520** 5042–5064

[10] Moseley E R, Teyssier R and Draine B T 2023 *MNRAS* **518** 2825–2844

[11] Weingartner J C, Draine B T and Barr D K 2006 *ApJ* **645** 1188–1197

[12] Langer S H, Chanmugam G and Shaviv G 1981 *ApJL* **245** L23–L26

AUTHOR INDEX

Institute of Physics
Dirac House, Temple Back
Bristol BS1 6BE UK

ISSN: 1742-6588
ISBN 978-1-7138-9730-9

CURRAN ASSOCIATES INC.
proceedings.com